CALCULUS FOR ENGINEERS

Preliminary Edition
Volume 1

Robin Carr & Bill Goh
Drexel University

SAUNDERS COLLEGE PUBLISHING
Harcourt Brace College Publishers

Fort Worth Philadelphia San Diego New York Orlando San Antonio
Austin Toronto Montreal London Sydney Tokyo

QA
303
C37
1995
V.1

Copyright ©1995 by Harcourt Brace & Company

All rights reserved. No part of this publication may be reproduced or transmitted in any form or by any means, electronic or mechanical, including photocopy, recording, or any information storage and retrieval system, without permission in writing from the publisher.

Requests for permission to make copies of any part of the work should be mailed to: Permissions Department, Harcourt Brace & Company, 8th Floor, Orlando, Florida 32887.

Printed in the United States of America.

The E4 progect is supported in part by The Science and Engineering Education Directorate of the National Science Foundation, Grant Number: USE-8854555.

Any opinions, findings, and conclusions or recommendations expressed in this material are those of the authors and do not necessarily reflect the views of the National Science Foundation.

Carr/Goh: CALCULUS FOR ENGINEERS, Preliminary Edition, Volume 1

ISBN 0-03-011644-9

456 017 987654321

Cover credit: © Gregory Ochocki/Photo Researchers, Inc.

Preface

Imagine a prospective <u>music</u> student who has applied to a first class music academy eager to play the trumpet. The academy's brochure describes the following curriculum. In the freshman year, students study the physics and mathematics of brass instruments to fully appreciate the subtle harmonics and overtones of trumpets and horns. In the sophomore year, students learn about materials and the construction of brass instruments.

In the junior year students learn to read music and <u>finally</u> in the senior year, students are offered their first chance to actually practice on a musical instrument.

As strange as this scenario seems, it illustrates the kind of educational experience that has been offered to undergraduate <u>engineering</u> students in most western universities since Sputnik jolted the western psyche in 1957. Partly due to the shock caused by the launching of Sputnik, and partly due to funding constraints, the focus of engineering curricula shifted away from laboratories, design and the fundamental craft and art of engineering and moved towards a host of textbook-based "scientific" courses. The swing has been so extreme, that today, even though <u>design</u> is the central activity of engineers, an engineering student's only exposure to design usually occurs in a design course offered in the senior year. Is it any wonder that by the end of the sophomore year, two out of three engineering undergraduates have dropped out - most having never had any genuine engineering experience?

Our book, Calculus for Engineers, is an attempt to facilitate a new kind of educational experience for engineering students - one in which engineering lies at the intellectual centerpiece of the curriculum. The musical analogy of the new experience is that the student will be given a shiny brass trumpet to practice on from the very first day! Calculus for Engineers (CAFÉ) is not just a traditional calculus sequence with some physical examples thrown in. We believe that a calculus sequence designed for engineers must highlight and embrace the central philosophy, art, craft, paradigms and fundamentals of engineering. This can be expressed more succinctly with the following simple statement:

> **Calculus for Engineers is by definition a calculus experience designed around the engineering student.**

We now list a number of examples that show how this philosophy is reflected in the design of CAFÉ - Calculus for Engineers.

- **Integrated Design**
  Above all else, engineering means the <u>design</u> of products, processes or services which better mankind. Thus we have attempted to integrate design throughout the examples, exposition and exercises! For example, the very first section of CAFÉ

introduces the mathematical concepts of lines and slopes since these will later be needed to define the derivative as the slope of the tangent line. To integrate design, students use these simple mathematical concepts to design a cafeteria tray dispenser. (As each dinner tray is removed, the stack of trays magically rises back to the same height.) What design makes this possible?

- **Engineering Analysis**

    Given a design for a circuit, mechanical linkage, robot arm or sensor, students learn to model its behavior using the appropriate mathematical paradigm.

- **Optimization**

    Few of the designs conceived by an engineering team actually get built. Only those designs which are optimal in some sense ever see the light of day. The optimal design may be the least costly, the most efficient or the most profitable. The identification of optimal designs is a perfect context in which to teach the standard calculus material known as 'maximum and minimum problems'. In CAFÉ, students discover the optimal design for a lion's cage, greenhouse, window frames of various geometries, hot tubs and kites.

- **Control**

    The control of processes and production is critical to many engineering endeavors. Every calculus book has a section on related rates. However, in our section on related rates, students discover how to control the orientation of a solar panel so that it tracks the sun for optimal efficiency. The rotation of a crucible containing molten metal is controlled so that it pours at a constant rate.

- **Sensors and Measurement, Product Specification**

    Sensor technology has been declared by the US government to be an area of critical importance to the nation's economic future. The average new car contains several hundred sensors including 'metasensors' which sense what the other sensors are doing. Thus, we consider it important to discuss the design of a number of sensors and gauges such as the tire gauge explored in Volume I. A traditional calculus sequence introduces intervals before discussing the domain and range of functions. In CAFÉ, intervals are first used in a discussion of product specification and tolerances. In what interval might the resistance of a resistor nominally rated at 100 Ohms lie if its fourth band is silver? In what interval may the resistance lie if it is to deliver a power specified in a given range?

- **Manufacturing Processes and Visualization**

    Many manufacturing processes are real world applications of fundamental mathematical principles. For example, students discover that an *xy*-graph is the graph of a function if and only if the graph can be stacked upon itself without intersecting. Cast in this form, the more usual vertical line test becomes familiar to any engineer who was worried about the packaging of inexpensive products like paper cups which are always sold stacked one inside another. Thus even the most basic mathematical notion, that of a function, can be enriched by embracing engineering paradigms. Students also see how the mathematical concept of level curves (contour lines), which is used to visualize surfaces in three dimensional space, is actually applied in the technology of stereolithography in which a solid object is 'printed' one layer at a time.

    Many other examples of engineering design and paradigms can be seen in every section of CAFÉ.

Although CAFÉ is clearly designed for engineering students, the level of mathematical rigor has not been sacrificed. Complete proofs are offered for all the theorems and the treatment of such topics as the "Mean Value Theorems" is more substantial than that offered in most calculus books.

This work was done as part of Drexel University's "Enhanced Educational Experience for Engineers" project which has been supported by the National Science Foundation. We were guided by reports like the "National Action Agenda", by Drexel colleagues from many different disciplines and by numerous conversations with engineers all over the country. Our greatest gratitude goes to Professor Donald Thomas (Drexel MEM) who has given so much help and advice to us over the last five years.

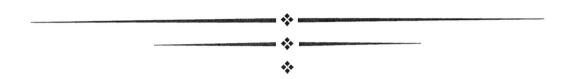

Acknowledgments

We would like to thank a few special people who have made this book possible. First and foremost, our greatest appreciation goes to two of our E^4 students, Catherine L. Miller and Wynn Sanders, who more than anyone else have helped to make *Calculus for Engineers* a reality. Each has worked tirelessly, weekday and weekend, day and night, to refine the text, to create and enhance the graphics, and to help in the design and creation of fresh and exciting problems. They have given us many ideas and taught us new computer skills and techniques that make *Calculus for Engineers* better and more visually appealing. Additionally, Timothy Ruiz has made significant artistic contributions to our project. Tim helped design the chapter covers and produced many beautiful and accurate graphics. Finally, Sam Albert has been instrumental in the preparation of the answers and solutions and in checking their accuracy. Thanks Cathy, Wynn, Sam, and Tim for all your help.

We wish to thank Dr. Robert Quinn and Provost Dennis Brown who have created the opportunity for us to develop this book and to help in the creation of an engineering curriculum designed around the engineering student.

We consulted with many of our science and engineering colleagues, but a great debt is owed to Don Thomas (MEM). Additional thanks go to Jennifer Atchison, Wayne Hill and Stuart Harper for facilitating our work (and for encouragement when it counted most) and as always, a heartfelt thanks goes to Pat Christie.

Finally, we wish to thank Jay Ricci and Alexa Barnes of Saunders College Publishing for their enthusiastic support and shared vision of a calculus experience designed for the engineering student.

CAFE Preliminary Edition

Volume One

Table of Contents

Chapter One - Elementary Engineering Functions   1

§ 1.1	Linear Functions	3
§ 1.2	Intervals and Tolerance Analysis	17
§ 1.3	Functions, Parabolas and Tangent Lines	29
§ 1.4	The Algebra of Functions	45
§ 1.5	Trigonometric Functions	61
§ 1.6	Exponential and Logarithms	77

Chapter Two - Limits and Derivatives   87

§ 2.1	Tangent Lines and Derivatives	89
§ 2.2	Limits	105
§ 2.3	Continuity of Functions	123
§ 2.4	Differentiation Rules	135
§ 2.5	Derivatives of Trigonometric Functions	149
§ 2.6	Chain Rule and Implicit Differentiation	161

Chapter Three - Applications of the Derivative   177

§ 3.1	The Natural Exponential and Logarithm	179
§ 3.2	Linear Approximations and Newton's Method	199
§ 3.3	Related Rates	213
§ 3.4	Optimization	227

Chapter Four - Introduction to Integration   243

§ 4.1	The Area Problem	245
§ 4.2	Area and Definite Integrals	259
§ 4.3	Fundamental Theorems	277
§ 4.4	Substitution	293
§ 4.5	Mean Value Theorem (Part I)	305
§ 4.6	Mean Value Theorem (Part II)	323
§ 4.7	Integration by Parts	341
§ 4.8	Approximate Integration	353

Chapter Five - Geometric Applications of Integration   373

§ 5.1	Volume (Part I)	375
§ 5.2	Volume (Part I)	389
§ 5.3	Arclength	405
§ 5.4	Area of Surfaces of Revolution	423

Answers to Odd Numbered Problems   439

Chapter One

Elementary Engineering Functions

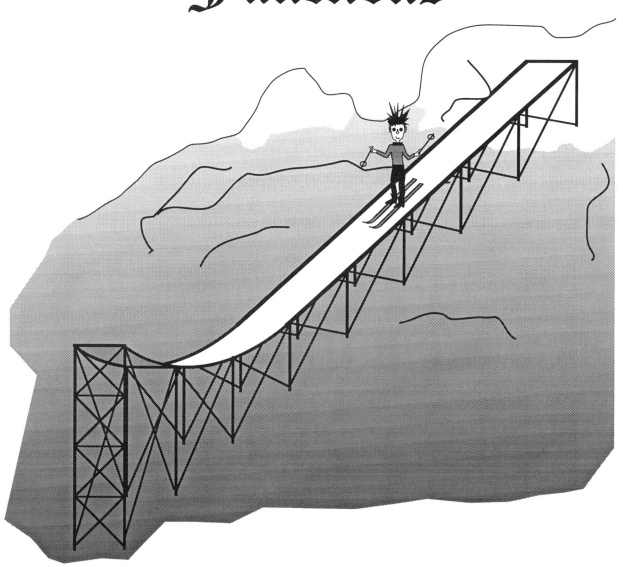

§1.1 Linear Functions

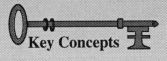

Key Concepts

- Linear Relations
- Parallel Lines
- Slope of a Line
- Perpendicular Lines

Linear functions play an especially important role in calculus and in engineering. In this section we introduce the concept of linear functions and explore their many applications to engineering design, physical phenomena, calibration, unit conversion and sensors.

Definition

A <u>linear</u> function is a function of the form $y = mx + b$. The parameter m is called the <u>slope</u> of the line and b is the <u>vertical intercept</u>.

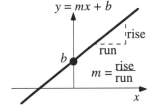

The following example seems a dynamite way to begin our exploration of calculus.

In centuries past, American engineers carved the great railroads into a reluctant earth with considerable sweat and a little help from powerful explosives like dynamite and TNT (trinitrotoluene). Often, the charge was controlled by lighting a **fuse**. A fuse is a length of readily combustible material that is lighted at one end causing the explosive to detonate when the flame has traveled to the other end.

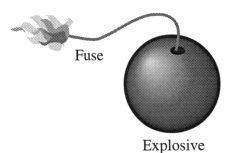

Example 1 - Explosive Fuse

A fuse has length 10 meters and burns at a rate of two centimeters per second when lit. What length of fuse (in cm) remains after t seconds and when will the explosive detonate?

Solution

In t seconds the length of fuse burnt is $2t$ centimeters. Since we started with an initial length of $10\ m \times \dfrac{100\ cm}{1\ m} = 1000\ cm$, the length of fuse remaining is

$$\boxed{L(t) = 1000 - 2t}.$$

The vertical intercept (1000 cm) corresponds to the initial length of fuse while the slope (–2 cm/sec) is the rate at which the length of fuse decreases. The explosive will detonate when the length of the fuse is zero. This is called the horizontal intercept and is seen from $L(t) = 1000 - 2t = 0$ to be 500 seconds.

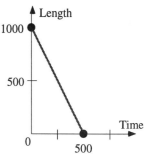

Parallel and Perpendicular Lines

Two lines in the plane are called <u>parallel</u> if they do not intersect.
Two lines intersecting at right angles are called <u>perpendicular</u>. (See illustration).
Parallel lines have the same slope and the product of slopes of two perpendicular lines is –1.
For example, the utility poles shown in the left illustration are all perpendicular to the level ground and parallel to each other. After an earthquake, the ground has tilted to a slope m. The poles remain perpendicular to the ground and each have slope $-1/m$.

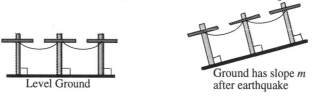

Example 2 - Pitch of Threaded-Screw

Each time the jack undergoes a complete revolution, the threaded screw advances 2 millimeters. This distance between the threads of the screw is called the <u>pitch</u>!
If the tire jack has an initial height of 200 millimeters, find its height after x turns.

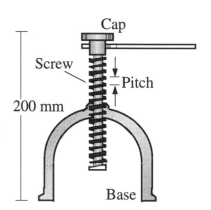

§1.1 Linear Functions   CAFÉ   Page 5

Solution

The height will be a linear function of the number of turns x so we may write:
$$h(x) = mx + b$$

The intercept b is the initial height so $b = 200$ mm. The slope is just the pitch of the screw, so $m = 2$ mm. Thus:
$$\boxed{h(x) = 2x + 200}$$

Don't Forget

We will be using the concept of <u>pitch</u> many times. The pitch of a singly-threaded screw is the distance between the threads or the distance the screw advances in one complete revolution.

Example 3

A bicyclist races with a constant velocity along the x–axis starting at the point x_0 and finishing the race at the point x_f. Find an expression for his location at time t and the time at which he finishes the race.

Solution

Let the bicyclist's location (coordinates) at time t be denoted by $x(t)$. Now $x(t)$ will be a linear function of t, so we may write
$x(t) = mt + b$.
The intercept b is the initial location x_0. The slope is his velocity. Thus $x(t) = x_0 + vt$. He finishes at a time t satisfying: $x_f = x_0 + vt$.
Thus

$$\boxed{t = \frac{x_f - x_0}{v}}$$

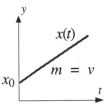

Hooke's Law

An important example of a physical phenomenon obeying a linear rule is the response of a spring to various applied weights or forces. In the diagram, the spring is shown first at its equilibrium length with no weights attached. When one weight is attached the spring stretches a length Δx. When a second identical weight is attached, the spring stretches a total length of $2\Delta x$. The extension of the spring is proportional to the applied weight or force. Thus we can write:

Hooke's Law
$$F = k\Delta x$$

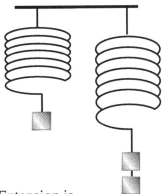

Extension is proportional to weight added.

To extend a spring a total distance Δx beyond its equilibrium position, we must apply a force $F = k\Delta x$. The proportionality constant k is called the spring constant.

Example 4

The equilibrium length of a spring is 6 inches. When a weight of two pounds is attached to the spring, its length increases to 7 inches. Find the spring constant and express the length of the spring as a linear function of the applied force F.

Solution

From Hooke's Law, the spring constant is $k = \dfrac{F}{\Delta x} = \dfrac{2 \text{ lb}}{1 \text{ in}} = \boxed{2 \text{ lb/in}}$.

Thus when a force F is applied, the extension of the spring will be $\Delta x = \dfrac{F}{k} = \dfrac{F}{2}$. Since the equilibrium length of the spring is 6, the length L when a force of F pounds is attached will be $L = \dfrac{F}{2} + 6$. The slope is the reciprocal of the spring constant ($1/k$) and the vertical intercept is the equilibrium length of the spring.

We already know enough about lines and slope to design some useful products.

Example 5 - Design of a Cafeteria Tray Dispenser

A simple engineering application of Hooke's Law is seen in the design of a cafeteria tray dispenser. The tray dispenser holds many trays as indicated in the figure. As one tray is removed from the top, another almost magically takes its place. This feature means that a tray is always in easy grasp at the top of the dispenser. How does it work?

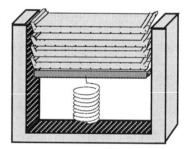

Solution

The trays rest on a platform supported by a system of springs. (Only one spring is shown representing the effective spring constant.) The tray is designed so that its equilibrium position is at the top of the dispenser when no trays are loaded. Let each tray have a thickness Δx and a weight Δw. When each tray is removed from the dispenser, the force on the spring(s) due to the weight of the trays is reduced by Δw. The spring expands a distance Δx causing the next tray to rise into the place of the tray that has just been removed. Thus the tray dispenser will work provided the effective spring constant is selected to be $k = \dfrac{\Delta w}{\Delta x}$.

Pressure

Anyone who has accidentally been stepped on by a high-heel shoe on a crowded bus understands the difference between force and pressure. Pressure is defined as force per unit area.

$$P = \frac{F}{A}$$

Example 6

Consider a four pound cube all of whose edges are two inches in length resting on a table. What pressure does the cube exert on the table's surface?

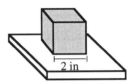

Solution

The force F is just the weight of the cube so $F = 4$.
This force is distributed over the bottom of the cube which has area $A = 4$ in^2.

Thus the pressure on the table surface is $P = \dfrac{F}{A} = \dfrac{4 \text{ lb}}{4 \text{ in}^2} = \boxed{1 \dfrac{\text{lb}}{\text{in}^2}}$.

What would the pressure on the table be if n of these cubes were stacked one atop the other? Explain why a nail may cause a jogger wearing sneakers serious injury yet a person wearing pajamas can sleep comfortably on a bed of nails.

Example 7 - Design of a Tire Gage

How does a tire gage work? The illustration reveals the secrets of how at least one major manufacturer of tire gages has designed this useful instrument.

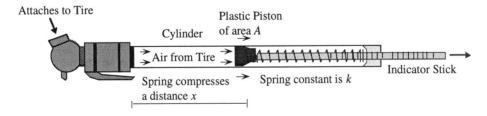

A spring is nested inside a metal cylinder. When a tire's pressure is measured, a leak-proof plastic piston is thrust against the spring compressing it a distance x proportional to the pressure inside the tire. This causes an indicator stick to be pushed out the end and the tire pressure can be read. What should the spring constant k be and how is the indicator stick calibrated?

Solution

Let P denote the air pressure inside the tire and let P_0 denote atmospheric pressure. If A denotes the cross-sectional area of the cylinder, then the effective force transmitted to the spring due to the tire's air pressure acting on the valve is $F = (P-P_0)A$. This force causes the spring to compress a distance x given by Hooke's Law.

$$F = (P-P_0)A = kx \quad \text{or} \quad P(x) = \frac{k}{A}x + P_0$$

Thus the pressure inside the tire is a <u>linear function</u> of the distance x by which the spring is compressed. The intercept corresponds to atmospheric pressure and the slope is the ratio $\frac{k}{A}$.

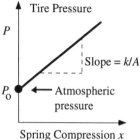

Example 8 - Choice of Spring Given Required Performance

Find the spring constant k of a spring to be used in a tire gage so that an increase in tire pressure of 10 lb/in² extends the indicator a distance of one inch. (Assume that atmospheric pressure is $P_0 = 14.7$ lb/in² and that the inside radius of the cylinder is 1/5 inch.)

Solution

By Hooke's law, the spring constant k satisfies:

$$F = (P-P_0)A = k\Delta x \quad \text{or} \quad (10)\pi\left(\frac{1}{5}\right)^2 = k \cdot 1$$

The spring should be selected to have a spring constant $k = \boxed{\frac{2\pi}{5} \text{ lb/in}}$.

Example 9 - Conversion between Temperature Scales

Temperature is commonly measured in either the Fahrenheit or Celsius scale. Both these temperature scales use the boiling and freezing point of water (at atmospheric pressure) as reference points.
The Celsius scale defines the freezing point as 0°C and the boiling point as 100°C.
The Fahrenheit scale defines the freezing point as 32°F and the boiling point as 212°F.

a) Find a linear equation to convert a temperature reading from Celsius to Fahrenheit.
b) At what temperature do the two scales agree?

§1.1 Linear Functions   CAFÉ   Page 9

Solution

The Fahrenheit reading F is a linear function of the Celsius reading C and so can be written in the slope-intercept form:

$$F = mC + b$$

From the given data we know two points on this line corresponding to the readings at the freezing and boiling points of water.

$$P_1 = (0, 32) \quad \text{and} \quad P_2 = (100, 212)$$

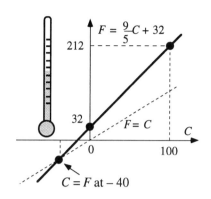

Point P_1 identifies the vertical intercept to be $b = 32$.

The slope is: $\quad m = \dfrac{\text{rise}}{\text{run}} = \dfrac{F_2 - F_1}{C_2 - C_1} = \dfrac{212 - 32}{100 - 0} = \dfrac{9}{5}$

Using the slope-intercept formula, we find $\boxed{F = \dfrac{9}{5} C + 32}$

To find when the two scales are equal we substitute $F = C$ into the above equation and discover that $C = F = -40$. Unfortunately, mercury (used in many thermometers) freezes at $-39°F$, causing damage to the thermometer just before this interesting point can be seen!

Pressure Variation with Depth

As any scuba diver knows, the pressure under water increases linearly with depth. (The increasing pressure can cause considerable pain on the eardrums.) In addition to discomforting divers, the increasing pressure has caused the demise of many dams and reservoirs! Given the engineering importance of this effect, let us find an expression for the pressure P as a function of depth x.

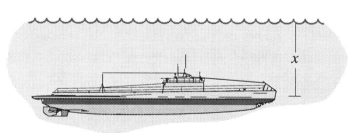
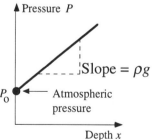

At the surface, the pressure is of course one atmosphere and the depth is zero so $P(0) = P_o = 10^5 \text{ N/m}^2$.

Consider a column of water extending from the surface down to a depth of x meters and having a cross-sectional area of 1 m². The increase in pressure at depth x must support the weight of the above x cubic meters of water. Since the density of water is $\rho = 10^3 \text{ kg/m}^3$, the weight of x cubic meters of water is $W = \rho g x$ where $g = 9.8$ m/s² is the acceleration of gravity. Thus:

Pressure at Depth x
$$P(x) = \rho g x + P_0$$

Archimedes Principle

A simple consequence of this linear relationship between pressure and depth is <u>Archimedes Principle</u>.

> Any body partially or completely submerged in a fluid is supported by a <u>buoyant force</u> equal to the weight of the fluid displaced by the body.

When a wooden cube is placed in water, part of it will be above and part below the water line. The portion of the cube under the water is called the <u>draft</u> (d). There is a buoyant force (B) acting on the block of wood due to the difference in pressure between the top and bottom faces of the cube (pressures acting on the sides cancel out). In mathematical terms:

$$B = PL^2 - P_0L^2 = (P - P_0)L^2$$

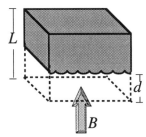

where P_0 is atmospheric pressure and P is the pressure at the bottom face. Recalling the relationship $P(x) = \rho g x + P_0$ where x is the depth d we obtain:

$$(P - P_0)L^2 = \rho g d L^2$$

Now $V = dL^2$ is the submerged volume of the cube and $\rho d L^2$ is the mass of the water displaced by the cube. Thus $\rho g d L^2$ is the weight of the fluid displaced by the submerged body. Thus the buoyant force is equal to the weight of the displaced water!

Bouyant Force
$$B = \rho g d L^2$$

§1.1 Linear Functions   CAFÉ   Page 11

WARMUP EXERCISES

1) Find the equation of the line through each of the following pairs of points. Express the line in the slope-intercept form $y = mx + b$.

   a) $(0,0)$ and $(1,2)$ b) $(1,1)$ and $(-1,1)$

   c) (a,b) and (c,d), $a \neq c$ d) (a,a) and (b,b), $a \neq b$

2) Find the equation of the line of slope $m = 2$ passing through the indicated point. State the vertical and horizontal intercepts for each line.

   a) $(0,0)$ b) $(1,0)$ c) $(0,b)$ d) $(c,0)$

3) Find the equation of the line <u>parallel</u> to the line $y = 3x + 7$ and passing through the indicated point.

   a) $(0,7)$ b) $(0,b)$ c) $(c,0)$

4) Find the equation of the line perpendicular to the line $y = 3x + 7$ and passing through the indicated point.

   a) $(0,7)$ b) $(0,b)$ c) $(c,0)$

INTERMEDIATE EXERCISES

5) A locomotive pulls three freight cars along a straight track at a velocity of 20 meters per second. At time 0, cars A, B and C are located at a distance of 10, 20 and 30 meters from the origin, respectively. Find the position of each freight car at time t and plot all three positions on the same graph. Why are the three graphs all parallel? The graph of a person walking along the track in the opposite direction is perpendicular to the three graphs for A, B and C. What is the speed of the person?

6) A cowboy chases after a feisty cow along a straight path. His horse runs at a speed of 30 m/sec and the cow at 20 m/sec.

   a) If the cow begins with a 30 meter head start, express the distance between the cow and cowboy as a function of the time t.
   b) Plot the location of the cowboy, the location of the cow and the distance between them on the same graph.
   c) At what time does the cowboy catch up to the evasive bovine?

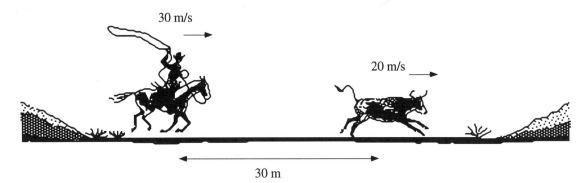

7) For each complete revolution, the threaded screw of the clamp advances 3 millimeters. Recall that this distance between the threads of the screw is called the <u>pitch</u>! If the initial separation between the faces of the vise is 5 cm, how many turns will be required to hold a 2 cm block firmly in place?

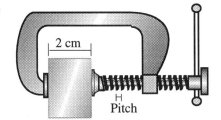

8) The threaded screw of the wrench has a pitch of 1.2 millimeters. If the initial separation between the faces of the wrench is 1.2 cm, how many turns will be required to firmly hold a hexagonal steel nut having a diameter of 0.6 cm?

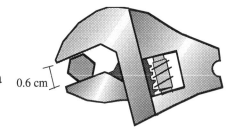

9) **Spike and His Shadow**
   a) Spike, who is six feet tall from tip of hair to toe, stands a distance x away from a twenty-four foot high street light. Express the length L of his shadow as a function of x.
   b) Express the length of Spike's shadow if he was outside in the daytime and the sun was at an angle of θ degrees with the horizon.

   (Hint: $\tan \theta = \dfrac{\text{opposite}}{\text{adjacent}}$)

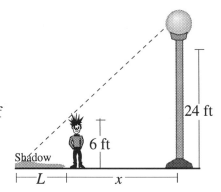

§1.1 Linear Functions   CAFÉ   Page 13

10) A cylindrical rain gage consists of a graduated cylinder and an enlarged opening to collect the rain. Assume the radius of the opening is twice the radius of the cylinder. Express the height of water collected as a function of the amount of rain that has fallen (in inches). What is the advantage to enlarging the opening?

11) Pure spring water is collected in plastic tanks and then shipped to customers. If the water flows into a cylindrical empty tank at the rate of 0.1 m³/min, express the height x of water in the tank as a function of time t (in minutes). Assume the height of the tanks is H meters and the radius is R meters. At what time t will a tank overflow if left unattended? Plot the height x of water both before and after it begins to overflow.

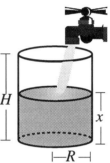

12) **Sonar**
A research vessel is mapping the topography of the ocean floor using sonar. The sound waves are directed vertically downwards and the time t for the waves to return is measured. The speed of sound in sea water is 1533 m/s. Express the depth d (in meters) of the ocean floor in terms of the return time t.

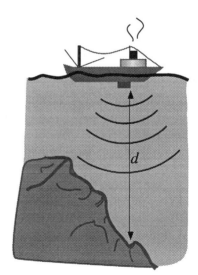

Hooke's Law

13) The equilibrium length of a spring is 10 inches. A three pound weight extends the spring an additional 2 inches. What is the spring constant and what is the equation for the length of the spring as a function of the applied weight F?

14) A spring is 10 inches long when a one pound weight is attached and is 12 inches long when a two pound weight is attached. Find the spring constant and the equilibrium length of the spring. Express the length as a function of the applied weight.

15) A spring is 8 inches long and has a spring constant $k = \frac{1}{2} \frac{\text{pound}}{\text{inch}}$.
   a) Express the length as a function of the applied weight.
   b) At what applied weight will the length of the spring be 10 inches?

16) When a current of I amperes flows through a resistor rated at R Ohms, the total voltage drop V (in volts) across the resistor satisfies Ohm's Law:
$$V = IR$$

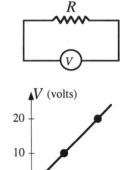

Thus the voltage is a linear function of the current with the resistance equal to the slope.

a) What is the voltage drop across a 10 Ohm resistor when a current of 2 amperes is flowing through it?

b) What resistance R corresponds to the graph in the illustration?

Engineering Design Problems

17) A paper towel dispenser is to be manufactured so that the spring exerts a force of 5 pounds on the smaller rod when no force is applied and a force of ten pounds when compressed for insertion.
Find the spring constant if the smaller rod must be compressed one inch for insertion.

18) Cafeteria Tray Dispenser
A stack of 100 cafeteria trays weighs 25 pounds and has a total height of 20 inches. Explain how to design a cafeteria tray dispenser for these particular trays and find the required spring constant if it is to function properly.

19) Tire Gauge
Assume that atmospheric pressure is $P_0 = 14.7 \text{ lb/in}^2$ and that the inside radius of the cylinder is 1/4 inch. Select a spring constant for the tire gauge so that an increase in tire pressure of 5 lb/in^2 extends the indicator a distance of one inch.

§1.1 Linear Functions   CAFÉ   Page 15

20) **Ammonia Based Temperature Scale**
Both the Fahrenheit and Celsius scales are based upon the melting and boiling points of water. Coincidentally, the pressure at sea level on a planet in the Andromeda Galaxy is the same as Earth's, but there, liquid ammonia (NH_3) is the most abundant fluid on the planet instead of water. The inhabitants have devised their own temperature scale based on the melting ($-77.7°C$) and boiling points ($-33.4°C$) of ammonia.

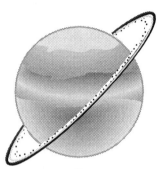

In their <u>A</u>mmonia scale, the melting point of ammonia is defined as $0°A$ and the boiling point is $100°A$. What would these ammonia breathing creatures measure for the melting and boiling points of water?

21) At what depth is the pressure on a scuba diver equal to two atmospheres?

22) A spring scale reads the weight of a copper cube of side 0.2 m as it slowly slips into water. Determine the reading of scale (in kg) as a function of the draft x in (in m) in the domain $0 \leq x \leq 0.2$. Assume the density of water is $1000 \frac{kg}{m^3}$ and the density of copper is $8.0 \times 10^3 \frac{kg}{m^3}$.
Remember, the force supplied by the scale is in newtons

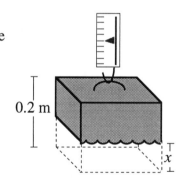

23) **Eureka !**
A piece of copper–gold alloy weighs x newtons in air. When the alloy is suspended from a spring balance and submerged in water, the balance reads $\frac{9x}{10}$ newtons. Express the weight of gold in the alloy as a function of x. The density of water is $1000 \frac{kg}{m^3}$, the density of gold is $19.3 \times 10^3 \frac{kg}{m^3}$, and the density of copper is $8.9 \times 10^3 \frac{kg}{m^3}$.

24) **Absolute Zero**
The table gives the pressure P of a confined volume of an ideal gas at two temperatures. The pressure is a linear function of temperature.

Pressure (N/m²)	Temperature (°C)
1.5×10^5	0
2.051×10^5	100

a) Express the pressure P as a linear function of T.

b) At what temperature is the pressure zero? (This temperature is called absolute zero.)

25) Gaps are left between railroad tracks because the steel expands in length when heated. Common steel railroad tracks have a length of 30 m when their temperature is 0 °C. A safety standard requires that the gap between the tracks be zero when the temperature is 200 °C. Assume the length L of the track expands linearly with temperature and that the linear expansion coefficient (α) is $\frac{11 \times 10^{-6}}{°C}$.

Determine the length of the gap G (in meters) as a function of the temperature T. (Hint: The length of the track at T satisfies $L = L_0(1 + \alpha(T - T_0))$, where $L_0 = 30$ m, and $T_0 = 0\,°C$.)

§1.2 Intervals and Tolerance Analysis

Key Concepts

❖ Interval ❖ Absolute Value

Intervals and Product Specification

The physical dimensions and performance characteristics of many products are often specified to lie in precise intervals. Even baseball, the national pastime, is not immune to the mathematics of intervals. Consulting the "Official Baseball Rules" we find the following specifications.

"The ball should be a sphere formed by yarn wound around a small sphere of cork, rubber, or similar material covered with two stripes of white horsehide or cowhide, tightly stitched together. It shall weigh not less than 5 nor more than $5\frac{1}{4}$ ounces avoirdupois and measure no less than 9 nor more than $9\frac{1}{4}$ inches in circumference."

Thus the weight w of the baseball is specified to lie in the interval $5 \leq w \leq 5\frac{1}{4}$. Another notation for the allowed interval is $[5, 5\frac{1}{4}]$.

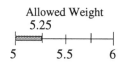

Further, the circumference C of the baseball is specified to lie in the interval $9 \leq C \leq 9\frac{1}{4}$. This interval can also be written as $[9, 9\frac{1}{4}]$.

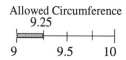

What exactly is an interval?

❖

Definition of Interval

An <u>interval</u> is a set of real numbers containing <u>no gaps</u>.
The graph of an interval looks like an unbroken piece of the real number line.

An interval is said to be closed if it contains its endpoints. The set of all points x satisfying $a \leq x \leq b$ is the typical closed interval. We denote this closed interval using the square bracket notation $[a,b]$.

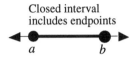
Closed interval includes endpoints

An interval is said to be open if it does not contain its endpoints. The set of all points x satisfying $a < x < b$ is the typical open interval. We denote this open interval using the round bracket notation (a,b).

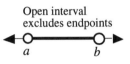
Open interval excludes endpoints

Sometimes we need to consider unbounded intervals, i.e. those that continue indefinitely to the left or to the right. Such intervals are denoted using the infinity symbol. For example, the set of positive real numbers is denoted by $(0,\infty)$ or by the inequality $x > 0$.

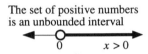
The set of positive numbers is an unbounded interval

Gap Test

A subset of the real line is an interval if and only if all the numbers between any two points in an interval are also in the interval. (This is just another way to say that an interval contains no gaps.)

Engineering Examples of Intervals

I) **Safety Intervals**
A manufacturer of an aerosol can specifies its use only for temperatures in the interval from the freezing point of water (32°F) to 120°F. Safe use is only guaranteed in the interval [32,120].

II) **Performance Intervals**
A research laboratory thermometer might be guaranteed to be accurate within 0.1°F only over the limited interval from the freezing to the boiling point of water. Accuracy is only guaranteed over the interval [32, 212].

III) **Tolerance Interval**
The dimensions of a machined part are advertised as 10.0 cm ± 0.1 cm. The manufacturer guarantees the parts will all lie in the interval [9.9, 10.1]. If you want more accurate parts, say lying in the interval [9.999, 10.001] you may have to pay a lot more for these parts even though they contain no more raw material.

IV) **Percentage**
The percentage of gold in a suspect coin must be in the interval [0,100]. The square brackets denote that the endpoints are included so this is a closed interval. Indeed the coin could be counterfeit containing no gold whatsoever (0%) or conceivably, every atom in the coin could be a gold atom (100% - not very likely however.)

Gold Coin

§1.2 Intervals & Tolerance Analysis   CAFÉ

The Absolute Value Function

We would like to introduce a new function that has many applications to the description of intervals, tolerances and performance specifications like those we have just considered. It is called the absolute value function.

> **Definition: The Absolute Value Function**
>
> The absolute value of a real number x is its distance from the origin and is denoted either as abs(x) or as $|x|$.
>
> If x is positive, its distance from the origin is equal to itself.
> If x is negative, its distance from the origin is equal to $-x$.

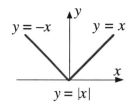

The <u>absolute value</u> of a number is its distance from the origin.

Thus the absolute value function can be written in the following piecewise or split form:

$$|x| = +x \text{ if } x \geq 0 \quad \text{and} \quad |x| = -x \text{ if } x \leq 0$$

For example, $|-3| = +3$ $|+3| = +3$ $|0| = 0$

Things we can describe with the absolute value function.

1) The <u>distance</u> between two points a and b on the real axis is Distance(ab) = $|a-b|$.
    For example: Distance($-3, 2$) = $|(-3)-2|$ = $|-5|$ = 5.

    Using the absolute value function, we need not worry whether a lies to the left or to the right of b.

2) The <u>maximum</u> of two numbers a and b can be written as Max(a,b) = $\frac{a+b}{2} + \frac{|a-b|}{2}$.
    For example: Max($-3, 2$) = $\frac{(-3)+2}{2} + \frac{|(-3)-2|}{2}$ = $\left(-\frac{1}{2}\right) + \frac{5}{2}$ = 2.

An interval can be described using endpoint notation as in [a, b] or using absolute value notation as follows.

Absolute Value Notation for Intervals

Every closed interval [a, b] can be expressed in absolute value notation as $|x - c| \leq r$ where c is the center of the interval and r is the 'radius' of the interval.
The <u>center</u> of the interval is the average of the endpoints
$c = \frac{a+b}{2}$.

The radius is the distance of either endpoint from the center and is given by $r = \dfrac{b-a}{2}$. Our original notation $[a, b]$ emphasizes the endpoints of the interval. The absolute value notation emphasizes the center and radius of the interval. Similarly, an open interval (a,b) can be expressed in absolute value notation as $|x-c| < r$.

Example 1 - Baseball Specifications in Absolute Value Notation

Express the official weight and circumference specifications for a baseball in the absolute value notation $|x-c| \leq r$. Recall that these state: 'It (the ball) shall weigh not less than 5 nor more than $5\frac{1}{4}$ ounces avoirdupois and measure no less than 9 nor more than $9\frac{1}{4}$ inches in circumference.'

Solution

The circumference C of the ball must lie in the closed interval $\left[9, 9\frac{1}{4}\right]$. The center of this interval is $\dfrac{9 + 9\frac{1}{4}}{2} = 9\frac{1}{8}$ and the radius of the interval is $\dfrac{9\frac{1}{4} - 9}{2} = \dfrac{1}{8}$. Thus, the official circumference specifications in absolute value notation are $\boxed{\left|C - 9\frac{1}{8}\right| \leq \frac{1}{8}}$ inches.

The weight w of the ball must lie in the closed interval $\left[5, 5\frac{1}{4}\right]$. The center of this interval is $c = 5\frac{1}{8}$ and the radius of the interval is $r = \frac{1}{8}$. Thus, the official weight specifications in absolute value notation are $\boxed{\left|w - 5\frac{1}{8}\right| \leq \frac{1}{8}}$ ounces.

The absolute value function has many interesting and useful properties. Since the absolute value function has been defined using a "split formula", the proofs of these properties usually involve breaking the proof into a number of cases.

Properties of the Absolute Value Function

Let a and b be real numbers. Then the absolute value function satisfies each of the following properties.

Property 1:	$|ab| = |a||b|$
Property 2:	$|a+b| \leq |a| + |b|$ (Triangle Inequality)
Property 3:	$|-a| = |a|$
Property 4:	$|a| = |b|$ if and only if $a = \pm b$

Hints for proving some of these properties are given in the exercises.

Percentage Notation for Specifying Intervals ($x \pm p\%$)

Another common notation for specifying performance and tolerance intervals uses percentages. For example, a manufacturer sells resistors rated as 1000 Ohms plus or minus 10 percent. Since 10 % of 1000 is one hundred, the actual resistance is only guaranteed to lie in the interval $[900, 1100]$.

Example 2 - Resistor Tolerances

Resistor manufacturers specify resistance using color coded bars. The resistance tolerance is indicated by the color of the fourth band. The maximum percentage error of the resistor is given by the following color convention.

Gold - 5%   Silver - 10%   No Band - 20%

The fourth band in an assorted package of 2000 Ohm resistors is sometimes gold, sometimes silver and occasionally not present. In each case, express the possible values for the resistance using interval notation and absolute value notation.

Solution

First note that 5% of 2000 is 100, 10% is 200 and 20% is 400. The possible values for each resistor are summarized in the following table.

Color	Percentage Tolerance	Interval	Absolute Value Notation
Gold	5 %	$[1900, 2100]$	$|R - 2000| \leq 100$
Silver	10%	$[1800, 2200]$	$|R - 2000| \leq 200$
No Band	20%	$[1600, 2400]$	$|R - 2000| \leq 400$

Control of Function Output

Very often, an engineer must control the value of a function. For example, an appliance manufacturer may require a heating element to deliver a certain power P which is a function of the element's resistance R ($P = \frac{V^2}{R}$). A baseball manufacturer must control the ball's circumference to meet the official specifications. The absolute value function and the language of intervals are important tools for assuring the proper specifications are met. The next few examples illustrate how the output of functions can be controlled within given tolerances.

Example 3 - Baseball

To what accuracy must a baseball manufacturer control the radius of the ball to meet the official circumference specifications? (Give the answer to three decimal places.)

Solution

The official circumference specifications are $\left|C - 9\tfrac{1}{8}\right| \leq \tfrac{1}{8}$. If the radius of the ball is r, its circumference is $C = 2\pi r$. Thus the radius of an official ball must satisfy the inequality $\left|2\pi r - 9\tfrac{1}{8}\right| \leq \tfrac{1}{8}$, or expressing this as an interval:

$$-\tfrac{1}{8} \leq 2\pi r - 9\tfrac{1}{8} \leq \tfrac{1}{8}$$

Solving the inequality for r we find:

$$\frac{9}{2\pi} \leq r \leq \frac{9\tfrac{1}{4}}{2\pi}$$

To three decimals we see that the radius (in inches) must lie in the interval $\boxed{[1.432, 1.472]}$.

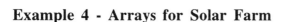

Example 4 - Arrays for Solar Farm

A subcontractor is bidding to build the square solar arrays for a large 'Solar Farm' in Hawaii. The specifications require that each square solar panel have an area A satisfying $|A - 4.0| \leq 0.1$ square meters. To what accuracy must the length x of the sides of the square panel be controlled to achieve this accuracy? (State answer to three decimal places.)

Solution

The target value for the area is $|A - 4.0| \leq 0.1$ square meters.
Expressing the allowed area in interval form gives $3.9 \leq A \leq 4.1$
Since the area of a square of sides x is $A = x^2$, the inequality can be rewritten:

$$3.9 \leq x^2 \leq 4.1$$

Since x must be positive, we find that: $\sqrt{3.9} \leq x \leq \sqrt{4.1}$ or $1.975 \leq x \leq 2.025$

This can be expressed in absolute value notation as $\boxed{|x - 2| \leq 0.025}$.
The length of the sides of the square array must not deviate from 2 meters by more than 2.5 centimeters.

§1.2 Intervals & Tolerance Analysis   CAFÉ   Page 23

Example 5 - Unit Conversion

An American manufacturer of laboratory heaters has subcontracted to a European firm the work on a small microprocessor that controls the temperature F (in degrees Fahrenheit). The microprocessor has several settings, one of which keeps the temperature of the samples within the interval $|F - 167| \leq 18$. The European engineers need to know the equivalent specification in degrees Celsius.

Solution

The conversion from the Celsius scale is given by $F = \frac{9}{5}C + 32$. Thus the temperature specification $|F - 167| \leq 18$ is equivalent to $\left|\left(\frac{9}{5}C + 32\right) - 167\right| \leq 18$.

Therefore, $\qquad -18 \leq \left(\frac{9}{5}C - 135\right) \leq 18.$

Solving for C we find the allowed interval:

$$\boxed{65 \leq C \leq 85} \quad \text{or} \quad \boxed{|C - 75| \leq 10}$$

Example 6 - Design of Heating Element

A heating element of resistance R is to deliver 120 ± 10 watts of power when operated at precisely 120 Volts. What is the allowed range of values for its resistance?

Solution

The power dissipated by the resistor is given by $P = \dfrac{V^2}{R} = \dfrac{120^2}{R}$ and is required to satisfy the specification $|P - 120| \leq 10$. Thus: $\left|\dfrac{120^2}{R} - 120\right| \leq 10$

Therefore $\left|\dfrac{120}{R} - 1\right| \leq \dfrac{1}{12}$ or $\dfrac{11}{12} \leq \dfrac{120}{R} \leq \dfrac{13}{12}$.

Don't Forget

If a and b are positive numbers satisfying $a \leq b$ then $\dfrac{1}{a} \geq \dfrac{1}{b}$. Notice how the inequality reversed direction!

Solving for the resistance and reversing the inequalities we find:

$$\boxed{\dfrac{1440}{11} \geq R \geq \dfrac{1440}{13}} \quad \text{or} \quad 110.8 \leq R \leq 130.9$$

A 120 Ohm resistor with a gold fourth band (5%) will have values ranging from 114 to 126 Ohms and would be a good choice.

WARMUP EXERCISES

1) Dinosaur Dimensions

Paleontologists claim that the infamous velociraptor of the mega-movie <u>Jurassic Park</u> was not as large as portrayed. Amazingly, another member of the raptor family, Utahraptor, was recently discovered with dimensions matching those of the movie dinosaur. Utahraptor had deadly fore and aft claws and was able to kill even the largest dinosaurs. However, the most dangerous meat eater of its time was Tyrannosaurus Rex. It was capable of killing any plant eater with a single bite. Express all the following estimates using the absolute value notation: $|x - c| \leq r$.

a) The velociraptor's true length is estimated to be only 6 to 9 feet, its weight from 60 to 140 pounds and its top speed from 35 to 40 mph. (Wolf-sized)

b) The larger Utahraptor's weight is estimated to have been 1,000 to 1,200 pounds and its length from 15 to 20 feet. (Polar bear-sized)

c) The deadly T. Rex is estimated to have had a length from 44 to 50 feet and a weight from 3 to 5 <u>tons</u>. (The mother of all carnivorous dinosaurs!)

In exercises **2-8**, express the given inequality by using the absolute value notion.

2) $-3 \leq x \leq 3$ 3) $0 \leq x \leq 3$ 4) $-3 \leq x \leq 0$ 5) $2 \leq x \leq 4$

6) $-1 < x < 3$ 7) $7 < x < 11$ 8) $x < -2$ or $x > 2$

In exercises 9-12 express each interval using absolute value notation.

9) $[0,100]$ 10) $[32,212]$ 11) $(-1,1)$ 12) $(-20,-12)$

In the following exercises, the specifications for the length of a manufactured part are given. Express the length of the part in three other forms: as a percentage, using absolute value notation and in interval notation.

13) $L = 2\ cm \pm 1cm$.   14) $L = 20\ cm \pm 1cm$

15) $L = 200\ cm \pm 1cm$ 16) $L = 200\ cm \pm 10cm$

In each of the following exercises an interval is presented which could arise in the everyday workings of an engineer. Express each interval using interval notation $[a,b]$, percentage notation $(x \pm p\%)$ and absolute value notation $|x - c| \leq r$.

17) A resistor R is rated at 100 ohms plus or minus 10%.

§1.2 Intervals & Tolerance Analysis   CAFÉ

18) In a commercial dyeing process, best color results are obtained if the fabric is immersed in the dye vat from 15 to 20 minutes.

19) The speed limit on a highway is 55 miles/hour. A certain highway officer stops motorists if they speed 5 miles/hour above the speed limit or if they are 15 miles/hour below the speed limit.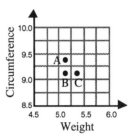

20) The preparation of many foods requires precise timing. One commercial manufacturer of spaghetti recommends adding the spaghetti to boiling water and to continue boiling for 13 to 15 minutes. They also make a thinner brand of spaghetti for which the recommended cooking time is only 9 to 11 minutes.

21) Which of the three baseballs is honest according to the specifications of the Official Baseball Rules. Recall that these state:

   'It (the ball) shall weigh not less than 5 nor more than $5\frac{1}{4}$ ounces avoirdupois and measure no less than 9 nor more than $9\frac{1}{4}$ inches in circumference.'

22) Express the official baseball specifications using percentages. ($x \pm p\%$)

23) **Using Intervals For Prospecting**
A geological engineer is prospecting for the mineral beryl, whose varieties include emerald, aquamarine, morganite, golden beryl and several others. It is known that the specific gravity of the beryl family ranges from 2.63 to 2.80. The specific gravity of a number of sample crystals have been measured. Which could belong to the beryl family?

Sample Crystal	Specific Gravity
Crystal A	2.8 ± 0.2
Crystal B	3.1 ± 0.2
Crystal C	3.5 ± 0.5
Crystal D	2.4 ± 0.1

The specific gravity of a sample is the ratio of its weight to the weight of an equal volume of water. Thus the specific gravity of water is 1.

Half-Open Intervals
Intervals are sometimes open at one end and closed at the other. For example, the set $0 \leq x < 1$ includes the left endpoint 0 but does not include the endpoint 1 on the right. We denote this interval with the notation [0,1). The following grading scheme contains many such intervals.

24) A certain professor uses the following grading scheme.
Students receiving 90 or more points will be given an A.
Those less than 90 but equal to or greater than 80 will receive a B.
Those less than 80 but equal to or greater than 70 will receive a C.
Those less than 70 but equal to or greater than 60 will receive a D.
An F is given to any score less than 60.
Express each grade interval using half-open intervals.

Plot each function in the following exercises on the domain $-5 \leq x \leq 5$.

25) $y = \dfrac{x + |x|}{2}$ 26) $y = \dfrac{x - |x|}{2}$ 27) $y = \dfrac{3x - |x|}{2}$ 28) $y = |x-1| + |x|$

Each equation in the following exercises establishes a relation between the variables x and y. Create a graph showing the points satisfying each relation.

29) $y = |-x|$ 30) $y = -|x|$ 31) $x = |y|$
32) $x = -|y|$ 33) $|x| = |y|$ 34) $|x| \cdot |y| = 0$

35) Show that the absolute value function can also be given by the formula $|x| = \sqrt{x^2}$. Note that the square root symbol refers to the positive square root only.

Prove each of the following properties of the absolute value function.

36) Prove for all real numbers a that $|-a| = |a|$.
Divide the proof into two cases depending on the sign of a.

37) Prove for all real numbers a and b that $|a| = |b|$ if and only if $a = \pm b$.

38) Prove for all real numbers a and b that $|a+b| \leq |a| + |b|$. (Triangle Inequality)

INTERMEDIATE EXERCISES

A piecewise linear function is a function whose graph consists of a number of line segments. Such functions often arise in control theory and signal processing. Often, we can write these functions using the absolute value function. For example, the function illustrated to the right can be written as $y = \dfrac{x + |x|}{2}$ instead of using the more cumbersome split or piecewise definition

$$y = \begin{cases} 0 & x < 0 \\ x & x \geq 0 \end{cases}.$$

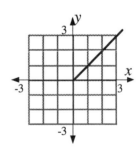

Express each of the following functions in two ways.
First use a split or piecewise definition and then use the absolute value function as above.

39) 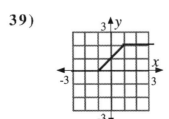   40) 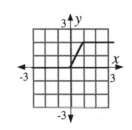   41)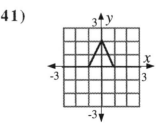

§1.2 Intervals & Tolerance Analysis CAFÉ

In the following exercises find all real values of x satisfying the given inequalities. Express the answer using both interval notation and using the absolute value function, where possible.

42) $x^2 < 9$

43) $x(x-6) < 0$

44) $x(x+6) \geq 0$

45) $\dfrac{3}{x-1} \leq x+1$

46) $x^2 > x$

47) $x^3 > x$

48) Which of the following identities are true for all real numbers x and y? Provide a counterexample, if the identity is not true.

   a) $|x + y| = |x| + |y|$
   b) $|x| = x$
   c) $|2x| = 2|x|$
   d) $|xy| = y|x|$

49) Express the minimum of two real numbers a and b in terms of the absolute value function. Compare the answer with that given for the maximum in the text.

50) A 'ramp function' is a function which rises linearly from a constant value to another constant value over a fixed interval. It may represent the approach from a highway onto a bridge, a ramp for disabled people to enter public buildings or a ladder or stairwell from one level to another in a building. Let m be the slope of the middle line segment. Verify that the illustrated ramp function can be written in terms of the absolute value function by the equation:

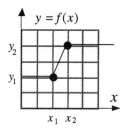

$$f(x) = \dfrac{m}{2}(|x-x_1| - |x-x_2|) + \dfrac{y_1+y_2}{2}$$

51) The specifications for an incubator require that it maintain the temperature C (in Celsius) within the interval $|C - 30| \leq 5$. Express this specification in degrees Fahrenheit.

52) A photocell is used to determine the velocity of a particle by measuring the time t for the particle to travel from point A to point B, a distance of precisely 2 meters away. How accurately must the time interval be measured to determine a velocity of 100 meters per second within 1 m/sec?

53) **Temperature Control**
The temperature of a certain computer lab must be maintained in the range of 65-70 degrees Fahrenheit. Find the equivalent specification in degrees Celsius.

54) Window Design
The shape of a church window consisting of four identical squares is given as shown. It is required that the total area A of the window should satisfy $|A - 5| \leq 0.1$ square meters. To what accuracy must be the total length L of the framing material be controlled to meet this requirement?

55) A spherical water tank is designed to hold $100 \text{ m}^3 \pm 1\text{m}^3$ volume of water. What is the allowed range of values for its radius?

56) The optimal design of a plastic waste basket in the form of a cylinder without the top lid requires that the height is equal to the radius. Assume now that the total surface area (the area does not include the lid) of the plastic that makes the basket is $3 \text{ m}^2 \pm 0.1\text{m}^2$. Determine the allowed range of the radius for its construction.

§1.3 Functions, Parabolas & Tangent Lines

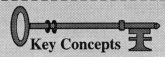
Key Concepts

- ❖ Notion of a Function
- ❖ Vertical Line Test
- ❖ Graph of a Function
- ❖ Quadratic Functions

What is a function?

Functions are simply the most important concept in mathematics. We have already met a number of examples of functions such as linear functions and the absolute value function.

Definition

A <u>function</u> is a mathematical rule or process that assigns a uniquely determined output to every value x in a set of inputs. The set D of all inputs is called the <u>domain</u> of the function. The set of all outputs is called the <u>range</u> of the function.

This notion of a function as a process is represented in the adjacent diagram. The domain is symbolized as a container holding the allowable inputs and the function as a process or machine that determines the unique output for each input value.

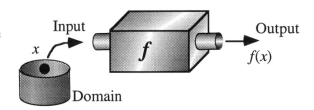

❖

Example: Hula-Hoops

Consider a machine that manufactures Hula-Hoops. Each input to the machine is a plastic tube or segment of length x that is then curved and bent into a Hula-Hoop. This same physical process can represent a number of different functions.

For example:

1) **The Identity Function**
   The circumference C of the circular Hula-Hoop is the same as the length of the original segment x. Thus $C(x) = x$ so the output is the same as the input!

2) **Linear Function**
   The radius R of the Hula-Hoop is the function $R(x) = \dfrac{x}{2\pi}$

3) **Quadratic Function**
   The area A of the Hula-Hoop is $A(x) = \pi R^2 = \dfrac{x^2}{4\pi}$.

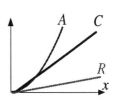

To completely determine a function we must also specify the allowable inputs or the domain. Since the length of a strip cannot be negative and since you can't make a Hula-Hoop from nothing, the most natural domain in the above process is the set of positive real numbers.

$$D = \{x \in \mathbf{R} \mid x > 0\}$$

Thus the circumference, the radius and the area are each functions that map the set of positive numbers into the reals.

Graphical, Numerical and Symbolic Representations of Functions

Functions can be represented in a number of ways.

Tabular Representation of Functions

One of the simplest representations of a function is to create a table or list with the input values in the left column and the output values in the right column. This is particularly useful for finite sets. Let us consider the function which assigns to each month the number of days in that month (excluding leap years.) The domain consists of the finite set of twelve months. The range is the set of three points $\{28, 30, 31\}$.

INPUT	OUTPUT
January	31
February	28
March	31
April	30
May	31
June	30
July	31
August	31
September	30
October	31
November	30
December	31

Since we have excluded leap years, each month has a precise number of days. Thus the rule which assigns to each month the number of days it contains in a normal year is a function.

However, if we were also to enter a row for February in leap years, the table would contain two values for February, 28 and 29. In this case, the table would not correspond to a function.

> **Tabular or Numerical Test for Functions**
>
> A complete table of input-output pairs corresponds to a function provided the same input does not appear in two rows with different outputs.

Graphical Representation of Functions using Cartesian Coordinates

Most of the functions we have been plotting so far have graphs that look like lines or line segments. How do we plot a function that does not consist of line segments? The <u>graph</u> of a function $y = f(x)$ consists of all points (x,y) in the Cartesian plane such that x is in the domain of f and $y = f(x)$ is the corresponding output.

For example, consider the function $y = x^2$ which gives the area of a square having edges of length x.

Each of the points $(0,0), (1,1), (2,4)$ and $(3,9)$ lies on the graph of the function. Adding more points, we see that the graph of the function $y = x^2$ looks like the curve drawn to the right.

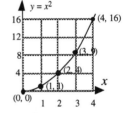

Similarly, the graph of the <u>relation</u> $x = y^2$ consists of all points (x,y) in the plane satisfying this equation. However, for each value of x, there are two corresponding values for y. For example both the points $(4,2)$ and $(4,-2)$ occur on the graph of $x = y^2$. Thus this relation does not correspond to a function.

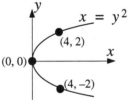

Vertical Line Test

How can one tell just by looking whether a graph corresponds to a function? The graph will be a function provided there are no ordered pairs having the same first coordinate but a different second coordinate. This would mean that the graph would have two points with the same x value, say (x,y_1) and (x,y_2). Provided we plot the x-axis horizontally and the y-axis vertically, these two points will be easy to spot since one must be directly above the other. If we can draw a vertical line through two points of the graph, then the graph does not correspond to a function.

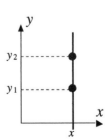

> **Vertical Line Test**
>
> A graph corresponds to a function if and only if each vertical line intersects the graph at most once.

Example 1

Which of the following two graphs corresponds to a function?

Solution

The graph of $y = |x|$ is a function since each vertical line intersects the graph in at most one point. However the graph of $x = |y|$ is seen not to correspond to a function since each vertical line to the right side of the y-axis intersects the graph in two points. It fails the vertical line test.

Stackable Objects and Functions

Engineers who design packaging for products have another perspective on the vertical line test. For inexpensive products like paper cups, a large fraction of their cost is incurred in shipping and handling. Thus reducing their packaging volume by <u>stacking</u> can lead to significant savings. The cross-section of the paper cup shown to the right passes the vertical line test and so its graph could be that of a function. Notice also that the cups easily stack.

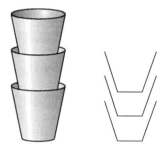

Paper cup passes vertical line test. The cups stack easily as shown

However, the cross-section of the flask fails the vertical line test and the flasks cannot be stacked as shown without breaking!

Since stacking requires placing objects <u>vertically</u> above each other without one copy touching that above it, we see that objects which can stack are precisely those whose cross-sections pass the vertical line test. A graph corresponds to a function precisely when copies of the graph can be vertically stacked as with the paper cups.

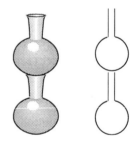

Flask fails vertical line test.

Example 2 - Pneumatic Press

In a simple manufacturing process, discs and plates are made from plastic or metal slugs. Each slug has height x and a radius of 10 mm. The slug is squeezed under a pneumatic press until it forms a disc of thickness 1 mm. Slugs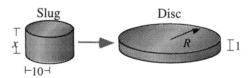
exceeding 25 mm in height cannot fit into the pneumatic press. What is the radius of each disc produced in terms of the height x of the input slug? Describe the domain and range.

Solution

We will assume that the volume of the metal or plastic has not changed under the manufacturing process. Such materials are said to be incompressible.
Thus $V_{slug} = V_{disc}$ (Incompressible assumption)

Using this formula we find:

$$V_{slug} = \pi \, 10^2 \, x = 100\pi x \text{ mm}^3$$
and
$$V_{disc} = \pi R^2 \cdot 1 = \pi R^2 \text{ mm}^3$$

Don't Forget

The volume of a cylinder is $V = \pi r^2 h$.

Solving for the radius R of the disc we find: $\boxed{R(x) = 10\sqrt{x}}$

To produce a disc of height 1 mm by this process, the slug must be at least this high and since the height of the slug cannot exceed 25 mm the domain for this process is $1 \leq x \leq 25$. The range is $[10, 50]$.

Variables and Constants

Variables not surprisingly are quantities that can vary over some set of values. Consider a function like $y = \pi x^2$ which could be interpreted as giving the area of a circle of radius x. In this case, x is considered the **independent variable** and y is considered the **dependent variable**. That is the area of the circle depends on its radius. Both x and y can vary over the set of positive real numbers. However the symbol π is not a variable. It represents the value of pi which is about 3.1415926535. Such symbols are called **constants**. Other common constant are numbers like 2 or zero and physical constants such as c which usually denotes the speed of light in a vacuum.

Quadratic Functions

Engineers use many functions as part of their everyday work both in design and analysis. We have already encountered some of the many applications both of linear functions ($y = mx + b$) and the absolute value function $y = |x|$. The next class of functions we shall explore are the quadratic functions.

Definition
A quadratic function is a function of the form $f(x) = ax^2 + bx + c$ where a, b and c are real constants and $a \neq 0$. (We exclude the case $a = 0$ since the function would then just be a line.)

The parabola opens upwards if $a > 0$

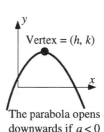
The parabola opens downwards if $a < 0$

The graph of a quadratic function is a parabola. If $a > 0$, the parabola opens upwards. If $a < 0$, the parabola opens downwards. The point (h, k) at the peak of the parabola is called the vertex. It is often useful to write the equation of the parabola in terms of the vertex coordinates, also known as completing the squares.

Vertex Formula

$$f(x) = ax^2 + bx + c = a(x-h)^2 + k$$

The following sample problem illustrates an engineering application of <u>completing the squares</u> to the design of enclosures and cages.

Example 3 - Optimal Design

Zoo officials have reluctantly decided that they can afford at most 100 meters of fencing to build a rectangular enclosure for a panda exhibit. What choice of length and width for the rectangular enclosure will leave the most room?

Solution

Let the length of the enclosure be x and let the width be w. Since the total perimeter is 100 meters we have

$$100 = 2x + 2w \quad \text{or} \quad w = 50 - x$$

Thus the total area of the rectangular enclosure is

$$A(x) = xw = x(50-x).$$

The area is an example of a quadratic polynomial. To find the maximum area we regroup the terms of A by completing the squares.

$$A(x) = 625 - (25-x)^2$$

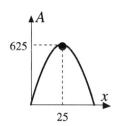

Since the area is in the form 625 minus some nonnegative quantity, we see that the maximum area is 625 when the length is $x = 25$. The width $w = 50 - x$ is also 25 so that the optimal enclosure is a 25×25 square.

The method used in the previous example to find the maximum of a quadratic function is called 'completing the square' and can be applied to any quadratic function.

Vertex Formula - Completing the Square

Consider any quadratic function $f(x) = ax^2 + bx + c = a(x-h)^2 + k$. The vertex (h,k) of the parabola has coordinates $h = -\dfrac{b}{2a}$ and $k = c - \dfrac{b^2}{4a}$.

> If $a > 0$, k is the minimum value of the quadratic.
> If $a < 0$, k is the maximum value of the quadratic.

Proof

Recall that the form $f(x) = a(x-h)^2 + k$ is called the <u>vertex formula</u> for the parabola. We will derive the vertex formula by rearranging the quadratic into a form where its maximum or minimum can be seen by inspection.

$$f(x) = ax^2 + bx + c = a\left(x^2 + \frac{b}{a}x + \frac{b^2}{4a^2}\right) + \left(c - \frac{b^2}{4a}\right) = a\left(x + \frac{b}{2a}\right)^2 + \left(c - \frac{b^2}{4a}\right)$$

Now the variable x only occurs in the square term $\left(x + \frac{b}{2a}\right)^2$.

The minimum value of the square $\left(x + \frac{b}{2a}\right)^2$ is 0 when $x = -\frac{b}{2a}$.

Thus if $a > 0$, the quadratic assumes a minimum value of $c - \frac{b^2}{4a}$ when $x = -\frac{b}{2a}$.

If $a < 0$, the quadratic assumes a maximum value of $c - \frac{b^2}{4a}$ when $x = -\frac{b}{2a}$.

End of Proof

Example 4

Complete the square to find the minimum value of the quadratic function
$f(x) = x^2 - 4x + 5$.

Solution

Rearranging the quadratic we find:
$$f(x) = x^2 - 4x + 5 = (x-2)^2 + 1$$
The variable x occurs only in the square term $(x-2)^2$.
The minimum value of the square term is 0 when $x = 2$.
Thus the minimum of $f(x)$ is 1 and this also occurs when
$\boxed{x = 2}$.

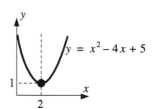

The Quadratic Formula

In many applications it is important to know when a function is zero. For complicated functions the problem of locating the zeroes can be very challenging. Fortunately for quadratic functions the zeroes can be found from a simple formula.

Formula for the roots of a quadratic polynomial

Consider the quadratic function $f(x) = ax^2 + bx + c$.
The quantity $D = b^2 - 4ac$ is called the <u>discriminant</u> and $f(x)$ has either 2, 1 or 0 roots depending on the sign of D.

If $D > 0$ then the function has exactly <u>two</u> real roots given by the formula:

$$x = \frac{-b \pm \sqrt{b^2 - 4ac}}{2a}$$

If $D = 0$ the function has exactly <u>one</u> real root given by:

$$x = \frac{-b}{2a}.$$

If $D < 0$ the function has <u>no</u> real roots since we cannot take the square root of a negative number.

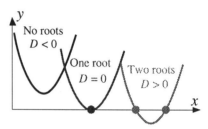

In the exercises, you will derive this formula by completing the square.

Example 5 - Finding the roots of a quadratic

Find <u>all</u> real roots for each of the quadratic functions below.

a) $f(x) = x^2 + 1$ **b)** $f(x) = x^2 + 2x + 1$ **c)** $f(x) = x^2 - x - 1$

Solution

a) In the case of $f(x) = x^2 + 1$, $a = c = 1$ and $b = 0$ so the discriminant is $D = b^2 - 4ac = -4$. There are no real roots.

b) In the case of $f(x) = x^2 + 2x + 1$, the discriminant is $D = b^2 - 4ac = 0$ so there is exactly one real root given by $x = \dfrac{-b}{2a} = \boxed{-1}$.

c) In the case of $f(x) = x^2 - x - 1$, the discriminant is $D = b^2 - 4ac = +5$ so there are two real roots given by $x = \dfrac{-b \pm \sqrt{b^2 - 4ac}}{2a} = \boxed{\dfrac{1 \pm \sqrt{5}}{2}}$.

Example 6

Find where the graph of the parabola $y = x^2$ intersects the line $y = 3x - 2$.

Solution

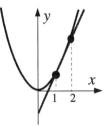

Since the two y values must be equal at any intersection point we have $x^2 = 3x - 2$. Bringing all the terms over to the left we find:

$$x^2 - 3x + 2 = (x-1)(x-2) = 0.$$

Thus the line intersects the parabola when $\boxed{x = 1}$ and when $\boxed{x = 2}$.

As the above example illustrates, most lines through a point on a parabola also intersect the parabola at a second point. However, there are two very special lines which do not. Clearly, a <u>vertical</u> line through any point cannot intersect the parabola at any other point because a parabola is a function and so it satisfies the <u>vertical line test</u>. The second line is called the <u>tangent</u> line and is defined below.

Definition: Tangent Lines of a Quadratic

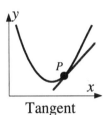
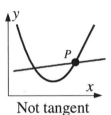

The <u>tangent line</u> at a point P on a parabola $y = ax^2 + bx + c$ is the unique non-vertical line through P which intersects the graph of the parabola at no point other than P.

Tangent   Not tangent

To find the tangent line at a point P on a parabola, we consider any line of slope m through P and set up a quadratic equation for its intersection point(s) with the parabola. The value of m that results in a single root of the quadratic equation is obtained by setting the discriminant D equal to zero as in the next example.

Example 7

Find the tangent line to the parabola $y = x^2 + 1$ at the point $P = (1,2)$.

Solution

First consider any line through the point $P = (1,2)$ with slope m. Using the point-slope formula, the equation for this line is:

$$\frac{y-2}{x-1} = m \quad \text{or} \quad y = mx + (2-m)$$

The intersection point(s) of the line with the parabola can be found by solving the equation:

$$x^2 + 1 = mx + (2-m) \quad \text{or} \quad x^2 - mx + (m-1) = 0$$

Since the tangent line must intersect the parabola only at the point P, we are looking for a value of m which will make the discriminant D of the above quadratic equation equal to zero. Thus the slope m of the tangent line must satisfy:

$$D = b^2 - 4ac = m^2 - 4(m-1) = (m-2)^2 = 0.$$

Therefore the tangent line has slope $m = 2$ and so is given by the equation $\boxed{y = 2x}$.

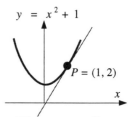

The tangent line intersects the curve at just one point.

Polynomials

Linear and quadratic functions are special cases of polynomial functions.

Definition

A <u>polynomial</u> of degree n is a function of the form $p(x) = a_0 + a_1 x + a_2 x^2 + \cdots + a_n x^n$ where $a_0, a_1, \cdots, a_n$ are real constants and the leading coefficient a_n is not zero. The real constants are called the <u>coefficients</u> of the polynomial.

Thus a constant function is a polynomial of degree zero, a non-constant linear function is a polynomial of degree one and a quadratic function is a polynomial of degree two.

A <u>root</u> or <u>zero</u> of a polynomial is any real number where the value of the polynomial is zero. Graphically this means that the graph of the polynomial touches the x-axis at each root. A <u>monomial</u> is a special case of polynomials where all coefficients except for the leading coefficient are equal to zero. Some examples of monomial functions are:

$$x, \; x^2, \; x^3, \; x^{1000}$$

Even & Odd Functions

Definition
A function $f(x)$ is said to be <u>even</u> if it satisfies the identity $f(-x) = f(x)$ for all x.

A function $f(x)$ is said to be <u>odd</u> if it satisfies the identity $f(-x) = -f(x)$ for all x.

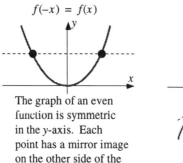

The graph of an even function is symmetric in the y-axis. Each point has a mirror image on the other side of the

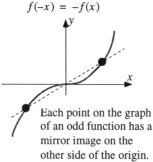

Each point on the graph of an odd function has a mirror image on the other side of the origin.

The graphs of polynomial functions provide the motivation for the names of even and odd functions. Since $(-x)^{2n} = x^{2n}$, every even integer power of x is an even function and similarly, every odd power of x corresponds to an odd function. A polynomial is even if and only if it contains no odd powered monomials. Thus the generic even polynomial is a combination of even powers of x.

$$p(x) = a_0 + a_2 x^2 + a_4 x^4 + \cdots + a_{2n} x^{2n}$$

In contrast, the graph of an odd function is not symmetric with respect to any axis. Instead, the graphs of odd functions are said to be symmetric with respect to the origin. That is, the graph appears the same if it is rotated about the origin by 180°. Another way to express the symmetry of odd functions is to note that the origin always lies on the midpoint of the line segment connecting the points $(x, f(x))$ and $(-x, f(-x))$.

Question: What is the symbolic form of the generic odd polynomial?

WARMUP EXERCISES

1) Which sets of ordered pairs could have been generated by a function?

    a) $\{(0,0),(1,1),(2,2),(3,3)\}$ b) $\{(-2,2),(-1,1),(0,0),(1,1),(2,2)\}$

    c) $\{(1,0),(2,0),(3,0),(4,0)\}$ d) $\{(2,-2),(1,-1),(0,0),(1,1),(2,2)\}$

2) Which of the following letters could correspond to the graph of a real function of x?

V   W   X   Y   Z

3) Which of the following graphs could correspond to functions of x?

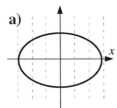 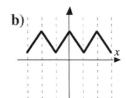 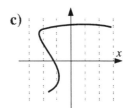 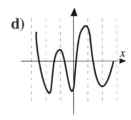

a)   b)   c)   d)

Create a graph of the given equation and use the vertical line test to indicate whether or not the graph could be that of a function of x.

4) $y = |x|$ 5) $x = |y|$ 6) $y = 0$ 7) $x^2 + y^2 = 1$

8) $x = 1$ 9) $xy = 1$ 10) $4y = x^2$ 11) $4x = y^2$

12) The graph of an even function is symmetric about the y-axis.
Why do we not define a class of <u>functions</u> which are symmetric about the x-axis?

13) The following <u>equation</u> is often seen on student exams.
Is it true that $\sqrt{a^2 + b^2} = a + b$ for all real numbers a and b?

14) A metal strip of length x is made into a hoop by bending it into the shape of a circle. The manufacturers create a 1 cm overlap at the two ends of the strip to create a good join. Express the radius and area of the metal hoop as a function of x.

15) A folding machine inputs paper sheets of length x and folds them in half in a process that eventually manufactures envelopes. The machine can fold sheets up to 14 inches in length. Describe the effect of folding on sheet length in terms of a function $f: D \to R$. Specify a reasonable domain and range.

§1.3 Functions... CAFÉ Page 41

16) Various shaped drinking glasses are shown in cross-section.

a) Which of the glasses passes the vertical line test for being a function?
b) Which passes the vertical line test after it has been turned upside down?
c) Which can be stacked one above the other as in a paper cup dispenser?
d) Can you prove that a graph corresponds to a function if and only if the corresponding object is vertically stackable?

17) Renowned abstract artist Nibor Rrac captures the essence of functionality in his painting "Entangled Functions". Which of the five bold brush strokes could correspond to the graph of a real function of x? (Apply the vertical line test.)

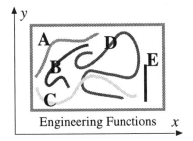

INTERMEDIATE EXERCISES

Sketch each quadratic function, then express it in the vertex form: $y = a(x-h)^2 + k$
Find its maximum or minimum.

18) $y = x^2 - x$ **19)** $y = -x^2 + x$

20) $f(x) = x^2 + 3x + 2$ **21)** $f(x) = x^2 - 1$

Find the zeroes of each of the following quadratic functions, if any exists.

22) $y = x^2 - x$ **23)** $f(x) = 10(x-2)(x+3)$

24) $f(x) = x^2 + 3x + 2$ **25)** $f(x) = 2x^2 - 3x + 1$

26) $f(x) = x^2 + 2x + 2$ **27)** $f(x) = x^2 - 2x + 2$

28) Find the points of intersection of the parabola $y = x^2 - 1$ and the lines:

     a) $y = x + 1$ b) $y = x - 1$ c) $y = 1 - x$

29) Find the points of intersection of the parabolas $y = x^2 - 1$ and $y = 1 - x^2$.

30) Find where the graph of the quadratic function $y = x^2 + 1$ intersects the following functions. Sketch each curve and indicate the intersection point(s) on your sketch.

    a) $\quad y = 2x \quad$ **b)** $\quad y = 4x \quad$ **c)** $\quad y = x^2 + 2x + 3$

31) Find the slope of the tangent line to the parabola $y = x^2 - 1$ at the point $P = (1,0)$.

32) Find the equation of the tangent line to the parabola $y = x^2 - 1$ at the point $P = (-1,0)$.

33) Show that the slope m of the tangent line to the parabola $y = x^2$ at the point $P = (a, a^2)$ is $m = 2a$.

34) Show that the tangent line at the vertex of the parabola $y = ax^2 + bx + c$ is horizontal.

35) Consider the quadratic function $f(x) = ax^2 + bx + c$ with $a \neq 0$.
The quantity $D = b^2 - 4ac$ is called the <u>discriminant</u>. Show that if $D > 0$ then the function has exactly <u>two</u> real roots given by:

$$x = \frac{-b \pm \sqrt{b^2 - 4ac}}{2a}$$

36) For which values of the parameter c does the quadratic function $y = x^2 + c$ have:

    a) No real roots   **b)** Exactly one root   **c)** Two Roots

37) For which values of the parameter b does the quadratic $y = x^2 + bx + 1$ have:

    a) No real roots   **b)** Exactly one root   **c)** Two Roots

38) For which values of the parameters a, b and c is the quadratic function $f(x) = a + bx + cx^2$ odd? For which values is it even?

39) Wire of various diameters can be formed by drawing thicker wire through a die, which is a metal block containing small conical holes. Metal that can be drawn in this way is said to be ductile.

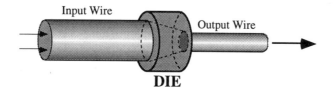

DIE

Let the diameter of the input wire be D and the diameter of the output wire be d. For a given die these will be constants. Assuming a length x of thicker wire is input to the die, what will the length y of the resulting wire be? The total volume of wire remains constant during this process.

§1.3 Functions...   CAFÉ   Page 43

Optimization Problems: (Use the method of completing the square.)

40) During the molting season, geese are temporarily flightless and vulnerable to predators. To protect the geese, one thousand meters of fencing have been purchased to construct a rectangular enclosure along the river's edge. The portion of the enclosure along the river will not need to be fenced since it offers a natural protection from most predators. What should the dimensions (length and width) of the enclosure be to create the greatest area for the flightless geese?

41) Two hundred meters of fencing have been purchased to construct a rectangular barrier around a swimming pool having dimensions of 20 by 30 meters. A patio is to be constructed between the fence and the pool. What should the length x and width y of the barrier be to maximize the area of the patio around the pool? Repeat the problem if the dimensions of the pool are 10 by 20 meters. (Does the size of the pool matter?)

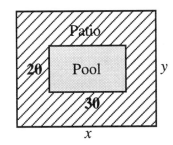

42) A fish is caught by a young eagle at the surface of a pristine mountain lake causing a circular ripple that travels outward at a speed of 2.5 m/sec.
Express the area enclosed by the ripple as a function of time t (in seconds).

43) A camping tent has the shape of a triangular prism. (All five faces of the tent, including the floor, are constructed from canvas.)

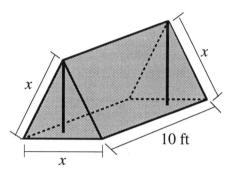

a) Determine the area of canvas required for its construction as a function of the width x.

b) Determine the area of canvas required if the two supporting poles are of height 5 ft.

Don't Forget

Recall the area of a triangle of base b and height h is $A = \frac{1}{2}bh$.

44) The glass roof and walls of a greenhouse have the dimensions shown. If the total volume of the greenhouse is $V = 600$ m^3, determine the area of glass required for its construction as a function of x (in meters).

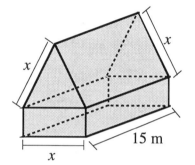

45) Number of Divisors

The number $8 = 2^3$ has four positive divisors: $1, 2, 2^2$ and 2^3.

a) Find a formula for the number of divisors of 2^n.
b) Find a formula for the number of divisors of 6^n.
c) Find a formula for the number of divisors of 30^n.
d) Find a formula for the number of divisors of 12^n.

46) Crystal Growth

The commercial growth of crystals is usually begun with a seed crystal. Even if the seed crystal is imperfectly formed, the resulting crystals can be almost perfect. In the illustration, a seed crystal is unevenly shaped, being twice as long as it is wide. Let us suppose that this first layer of the crystal has 2 molecules and that the next layer has 10 molecules. Find the number of molecules in the nth layer.

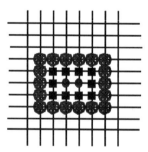

47) When an object is dropped from rest, the distance D it falls in time t is given by the equation: $D(t) = \frac{1}{2}gt^2$ where g is the acceleration of gravity ($g = 32$ feet/sec^2). Thus $D(t) = 16t^2$. How long does it take for an object dropped from rest to fall 1 foot, one hundred feet and 1000 feet?

§ 1.4 The Algebra of Functions

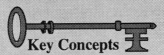
Key Concepts

* Arithmetics of Functions
* The Unit Step Function
* Composition of Functions
* Piecewise Functions

Building New Functions from Old

Just as concrete and steel beams can be combined into many different engineering structures, functions can be combined to yield yet more functions. Functions can be added, subtracted, multiplied and divided in much the same way that ordinary numbers are combined. First however, note that the ordinary addition of numbers is a <u>binary</u> operation.
That is two numbers x and y are inputted and then added to create a single output number $(x+y)$ as shown in the above illustration.

Addition is a binary operation with two input ports

Definition

Let f and g be functions and let x be an input point in the domains of <u>both</u> functions. The sum of f and g is another function, denoted by $f+g$, which given an input x yields as output the sum of the outputs of f and g at x. This may be expressed in symbolic form by the equation:

$$(f+g)(x) = f(x) + g(x)$$

Example: If $f(x) = 2x$ and $g(x) = x^2$ then $(f+g)(x) = 2x + x^2$

We can view the addition of functions schematically as shown to the right. A single input x is given to each function f and g. The outputs $f(x)$ and $g(x)$ are then fed into the two input ports of the addition machine resulting in the sum $f(x) + g(x)$.

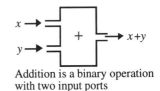

$$(f+g)(x) = f(x) + g(x)$$

The same kind of definition can be made for any other binary arithmetic operation such as subtraction, multiplication and division (provided one takes care not to divide by zero.) Each of these binary operations on functions can be represented schematically as in the above case of addition. (Just change the plus sign to a subtraction, multiplication or division sign.)

The Arithmetics of Functions

Let f and g be functions and let x be an input point in the domains of both functions. Each arithmetic operation $\{+,-,\times,/\}$ can be extended pointwise to an operation on functions as follows:

Sum: $\quad (f+g)(x) = f(x) + g(x)$ $\qquad$ Difference: $\quad (f-g)(x) = f(x) - g(x)$

Product: $\quad (fg)(x) = f(x)g(x)$ $\qquad$ Quotient: $\quad (f/g)(x) = \dfrac{f(x)}{g(x)}$

Since division by zero is undefined, the domain of the quotient f/g must exclude any point where $g(x) = 0$.

Don't Forget

Example 1

Consider the two linear functions $f(x) = 2x + 1$ and $g(x) = 3x + 2$. Find the sum, difference, product and quotient of these functions. Which of these new functions are not linear? What is the domain of each function?

Solution

The Sum:
Using the pointwise definition, we see that the sum of the two linear functions is:

$$(f+g)(x) = f(x) + g(x) = (2x+1) + (3x+2) = 5x + 3$$

The sum is defined for all real inputs x. Notice how the slope of the sum is equal to the sum of the slopes.

The Differences:
Next we find the difference, but which difference? Should we subtract f from g or g from f? As posed, the question is ambiguous so let us find both possible differences.

$$(f-g)(x) = f(x) - g(x) = (2x+1) - (3x+2) = -x - 1$$

$$(g-f)(x) = g(x) - f(x) = (3x+2) - (2x+1) = x + 1$$

§1.4 The Algebra of Functions

Both differences are again linear functions and their domain is the set of real numbers.

The Product:
The product of the two linear functions is:
$$(fg)(x) = f(x)g(x) = (2x+1)(3x+2)$$

The product is not a linear function, but is a quadratic polynomial.

The Quotients:
With the quotient we again encounter an ambiguity in the question. Should we divide f by g or g by f? Let us find both possible quotients.

First, $(f/g)(x) = \dfrac{f(x)}{g(x)} = \dfrac{2x+1}{3x+2}$. The quotient is not a linear function. Further, since division by zero is undefined, the quotient function cannot be evaluated whenever the denominator is zero. Since the root of $g(x)$ is $x = -\dfrac{2}{3}$, the domain of the quotient function f/g must exclude this point.

The other possible quotient is $(g/f)(x) = \dfrac{g(x)}{f(x)} = \dfrac{3x+2}{2x+1}$. In this case the denominator equals zero when $x = -\dfrac{1}{2}$ so the domain of the quotient function g/f must exclude this point.

Example 2 - Domains

Let $f(x) = \sqrt{1-x^2}$ and $g(x) = \sqrt{x-1}$.
Find the domain for each of the new functions $f+g$, $f-g$, $g-f$, fg, g/f and f/g.

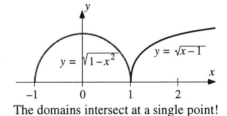
The domains intersect at a single point!

Solution

Since the real square root function is undefined for negative inputs, the domain of $f(x) = \sqrt{1-x^2}$ is the closed interval $[-1,1]$ and the domain of $g(x) = \sqrt{x-1}$ is the interval $x \geq 1$.

The intersection of these two domains consists of the single point $x = 1$.
Since any point in the domain of the new functions must belong simultaneously to the domain of both f and g, the only possible point in the domain of these functions is $x = 1$.
The domain of each of the new functions $f+g$, $f-g$, $g-f$ and fg consists of the single point $x = 1$. However, this point is excluded from the domains of both g/f and f/g since $f(1) = 0$ and $g(1) = 0$. (Division by zero is undefined.)
Thus there is not even a single point where the quotient functions can be defined.
Usually things are not as extreme as in this example.

In addition to the graphical and symbolic approaches to understanding how functions are combined pointwise, the next sample problem explores this new idea numerically.

Example 3

The following table gives values of two functions f and g at certain points in their domain. Find the values of the functions $f+g$, fg and f/g at these domain points.

Input (x)	$f(x)$	$g(x)$
0	0	undefined
1	1	1
2	4	0
3	9	3

Solution

Input (x)	f	g	$f+g$	fg	f/g
0	0	undefined	undefined	undefined	undefined
1	1	1	2	1	1
2	4	0	4	0	undefined
3	9	3	12	27	3

Note that the combined functions are only defined where both f and g are defined.

Composition of Functions

The arithmetic operations are not the only way that one can combine functions. The following sample problem illustrates the composition of functions.

Example 4 - Composition of Functions

A seashore resort is worried that the relentless erosion of its beaches will devalue beachfront property, threaten the natural wildlife and decrease summer tourism. An engineer specializing in beachfront hydrology has been hired as a consultant to help solve the problem. The consultant charges $100 per hour plus $200 expenses for each day worked.

Weekday	Hours Worked
Monday	8
Tuesday	4
Wednesday	10
Thursday	8
Friday	12

What is the consulting bill for each day?

§1.4 The Algebra of Functions

Solution

The consulting charge C is a linear function of the time T worked each day:

$$C(T) = 200 + 100\,T$$

The time T worked each day can be viewed as a function having a domain equal to the set of five weekdays. To find the <u>Bill</u> for each day, first apply the function T to find the number of hours worked and then apply the function C to find the cost to the community. When two functions are applied one after the other like this it is called <u>composition</u>.

Bill(*Monday*)	= $C(T(Monday))$	= $C(8) = 200 + 100 \times 8$	= \$1,000
Bill(*Tuesday*)	= $C(T(Tuesday))$	= $C(4) = 200 + 100 \times 4$	= \$600
Bill(*Wednesday*)	= $C(T(Wednesday))$	= $C(10) = 200 + 100 \times 10$	= \$1,200
Bill(*Thursday*)	= $C(T(Thursday))$	= $C(8) = 200 + 100 \times 8$	= \$1,000
Bill(*Friday*)	= $C(T(Friday))$	= $C(12) = 200 + 100 \times 12$	= \$1,400

Notice that the resulting function <u>Bill</u> is not obtained by combining algebraically the outputs of the other two functions as we have done until now in this section. Instead the output of one function (T = hours worked) is used as the input to the second function (C = cost). Let us generalize the situation with the following definition.

Definition of Composition: (Input the Output)

The <u>composition</u> of two functions f and g, denoted by $g \circ f$, is that function whose output is obtained by inputting the output of f into g. Symbolically this is expressed by the formula:

$$(g \circ f)(x) = g(f(x))$$

The formula only makes sense if the first output $f(x)$ belongs to the domain of g. Thus the domain D of the composition $g \circ f$ consists of those inputs in the domain of f which produce outputs in the domain of g. The composition of two functions is illustrated in the following figure.

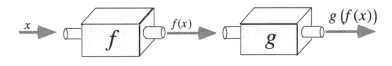

Example 5 - Composing functions in different orders.

Consider two functions $f(x) = \sqrt{x}$ and $g(x) = x^2$.
Find symbolic expressions for $g \circ f$ and $f \circ g$ and give the domain for each composition.

Solution

Although the domain of $g(x) = x^2$ is the entire set of real numbers, the domain of $f(x) = \sqrt{x}$ is restricted to the set of nonnegative real numbers since we cannot take the square root of a negative number.

1) The domain of $(g \circ f)(x) = (\sqrt{x})^2 = x$ is the set of nonnegative real numbers, since x must first be input into the square root function.

2) However, the domain of $(f \circ g)(x) = \sqrt{x^2} = |x|$ is the entire set of real numbers.

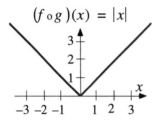
$(f \circ g)(x) = |x|$

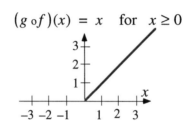
$(g \circ f)(x) = x$ for $x \geq 0$

We will now introduce a new function that has many applications in signal processing, microprocessor design and the control of machines. As we apply this new function to engineering problems you will notice how its use involves composition and all of the algebraic operations on functions that we have just met.

The Unit Step Function

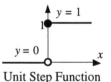
Unit Step Function

Many machines operate in and on/off mode. Examples include light switches and inexpensive air conditioners. A useful function for modeling the performance of such on/off machines is the <u>unit step function</u>.

This function is defined by the split formula: $u(x) = \begin{cases} 0 & \text{if } x < 0 \\ 1 & \text{if } x \geq 0 \end{cases}$

Examples: $u(-1) = 0 \quad u(0) = 1 \quad u(2) = 1$

Comment: Although we have defined the value of the unit step function at zero to be one, in applications it is sufficient to leave the function undefined at this point.

Example 6 - State of a Machine

An air conditioner is set to turn on only if the temperature is greater than or equal to 80°F. Let 0 indicate that the machine is off and let 1 indicate that the machine is on. Express the state of the machine as a function of the temperature T.

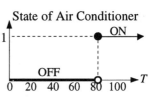

State of Air Conditioner

Solution

The unit step function jumps at the origin from 0 to 1. We can move the jump over to the point $T = 80$ where the air conditioner turns on, by <u>shifting</u> the argument by 80.

$$\text{State}(T) = u(T-80)$$

Air Conditioner

To verify that this works, just consider the two cases where T is above and below 80 degrees. Thus the state of the machine is the <u>composition</u> of the linear function $T - 80$ with the unit step function.

Piecewise Functions & The Unit Step Function.

Working with functions having a split or piecewise definition can become quite cumbersome. The unit step function is the key to obtaining an explicit symbolic expression for such functions that is valid on all intervals. There is no need to bother with case by case analysis of the function as shown in the examples below!

Example: Function defined in two pieces.

Suppose that a function $f(x)$ is defined piecewise over two intervals so that:

$$f(x) = \begin{cases} f_1(x) & \text{if } x < a \\ f_2(x) & \text{if } x \geq a \end{cases}$$

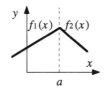

This function is the same as: $\boxed{f(x) = f_1(x) + [f_2(x) - f_1(x)]\, u(x-a)}$

To verify that this works, just consider the two cases where x is greater than or less than the point a which separates the intervals. Notice how this expression involves combining functions in many ways. We see in this one expression addition, subtraction, multiplication and composition of functions!

Example: Function defined in three pieces.

Suppose that a function $f(x)$ is defined piecewise over three intervals so that:

$$f(x) = \begin{cases} f_1(x) & \text{if } x < a \\ f_2(x) & \text{if } a \leq x < b \\ f_3(x) & \text{if } b \leq x \end{cases}$$

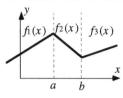

Using the step function we can write this as:

$$\boxed{f(x) = f_1(x) + [f_2(x) - f_1(x)]\, u(x-a) + [f_3(x) - f_2(x)]\, u(x-b)}$$

A similar expression holds for a function defined piecewise over any number of intervals.

Example 7 - Spike's Summer Job

Spike's summer job pays $20 per hour if he works less than 40 hours each week, and time and a half for all time equal to or beyond 40 hours. Let $Pay(t)$ denote his total earnings during a week in which he has worked t hours. Express $Pay(t)$ in terms of the unit step function.

Solution

We first express his pay using a split or piecewise formula:

$$Pay(t) = \begin{cases} 20t & \text{if } t < 40 \\ 30t - 400 & \text{if } t \geq 40 \end{cases}$$

Using the equation $f(x) = f_1(x) + [f_2(x) - f_1(x)]\, u(x-a)$, we can write the pay in terms of the unit step function u.

$$Pay(t) = 20t + u(t-40)\,((30t - 400) - 20t) = 20t + u(t-40)(10t - 400)$$

Example 8

An energy-saving air conditioner has three settings. Off, medium and high.
The machine is off if the temperature if less than 75°F (State = 1)
The machine is on medium if the temperature lies in the interval [75, 85). (State = 2)
The machine is on high if the temperature is greater than or equal to 85°F. (State = 3)
Express the state of the machine in terms of the temperature T.

Solution

The state of the machine is given by a formula split over three intervals. Thus it can be written in the form:

$$f(x) = f_1(x) + [f_2(x) - f_1(x)]\, u(x-a) + [f_3(x) - f_2(x)]\, u(x-b)$$

where $a = 75$, $b = 85$, $f_1 = 1$, $f_2 = 2$ and $f_3 = 3$.

$$\text{State}(T) = 1 + u(T-75) + u(T-85)$$

Air Conditioner

Example 9 - Highway Construction

Highway engineers often have to join different tracks of roadways having different slopes. Often the connecting portion has a parabolic shape. The parabola is chosen so that the slope of the road remains continuous avoiding jars and shocks.

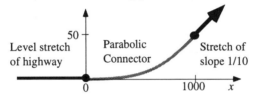

The resulting height of the highway surface is given by a function defined separately over three intervals.

$$f(x) = \begin{cases} 0 & \text{if } x < 0 \\ \dfrac{x^2}{20{,}000} & \text{if } 0 \leq x < 1000 \\ \dfrac{x}{10} - 50 & \text{if } 1000 \leq x \end{cases}$$

Write an expression for the height of the highway's surface using the unit step function.

Solution

We can write this as: $f(x) = f_1(x) + [f_2(x) - f_1(x)]\, u(x-a) + [f_3(x) - f_2(x)]\, u(x-b)$
where $a = 0$ and $b = 1000$. Replacing f_1, f_2 and f_3 by their values above:

$$f(x) = \frac{x^2}{20{,}000}\, u(x) + \left[\frac{x}{10} - 50 - \frac{x^2}{20{,}000}\right] u(x-1000)$$

WARMUP EXERCISES

Find the domain for each of the indicated functions.

1) $f(x) = \sqrt{x} + \dfrac{1}{x}$
2) $f(x) = \sqrt{4-x^2}$
3) $f(x) = \sqrt{x + \dfrac{1}{x}}$
4) $f(x) = \sqrt{x - \sqrt{x}}$
5) $f(x) = \sqrt{x^2-4}$
6) $f(x) = \dfrac{\sqrt{1-x^2}}{\sqrt{x^2-1}}$
7) $f(x) = \dfrac{1}{x-4}$
8) $f(x) = \dfrac{1}{x^2-4}$
9) $f(x) = \sqrt{1-x^2}\,\sqrt{x^2-1}$

10) Consider the following table of values for the functions f and g.

Input (x)	f(x)	g(x)
0	0	1
1	1	0
2	0	1
3	1	0

   a) Find the values of the functions $f+g$, fg and f/g at the input points {0, 1, 2, 3}.

   b) Compare the domains of the functions $\dfrac{fg}{g}$ and f.

11) Two functions f and g are the same if they have the <u>same domain</u> D and $f(x) = g(x)$ for all x in D.

   a) Show that the two functions $f+g$ and $g+f$ are the same.

   b) Let the domain of f be the entire set of real numbers R and let the domain of g be the set of nonnegative numbers $x \geq 0$. Are the functions f and $(f+g)-g$ the same?

12) Let $f(x)$ and $g(x)$ be functions having domain R, the set of real numbers. A <u>zero</u> or <u>root</u> of the function f is a point x such that $f(x) = 0$. Each of the following statements describes a property of the zeroes of one of the functions $f+g$, $f-g$, fg or $\dfrac{f}{g}$. Match each function with the correct description of its roots.

   i. The graphs of f and g intersect.
   ii. The graph of f touches the x-axis and g is not zero.
   iii. The graph of either f or g touches the x-axis.
   iv. The graph of f intersects the graph of $-g$.

§1.4 The Algebra of Functions   CAFÉ

13) Consider two linear functions $f(x) = m_1 x + b_1$ and $g(x) = m_2 x + b_2$.
    a) Find symbolic expressions for each of the functions $f+g, f-g, fg$ and f/g.
    b) Which of the four new functions are linear?
    c) Compare the slope of each new linear function to the slopes of f and g.
    d) What is the domain of each of the new functions?

14) Consider two linear functions $f(x) = m_1 x + b_1$ and $g(x) = m_2 x + b_2$.

    a) Find symbolic expressions for each of the functions $g \circ f$ and $f \circ g$.

    b) Under what condition will the two compositions be equal?

15) Using only the four binary operation symbols and two schematics for the functions f and g, create schematic diagrams showing how an input variable x is processed by each of the function combinations fg, $f-g$ and $\dfrac{f+g}{f-g}$.

16)   a) Link together any of the functions below so that an input of x produces an output of $|x|$.
    b) Link together the functions below so that an input of a and b produces an output of $\sqrt{a^2 + b^2}$.

    Input x → f → Output $\sqrt{x}$

    Input x → g → Output x^2

    Input x, Input y → h → Output $x + y$

17) Consider three functions $f(x) = 2x$, $g(x) = \dfrac{1}{x}$ and $h(x) = x^2$. It is possible to compose these functions in a total of $3! = 6$ different orders. Obtain symbolic expressions for all six possibilities.

18) Consider the four functions $f(x) = -x$, $g(x) = x^2$, $h(x) = 4-x$, $p(x) = \dfrac{1}{x}$, and $q(x) = \sqrt{x}$. Express each function below as a composition of two or more of these functions.

    a) $\dfrac{1}{x-4}$ b) $\dfrac{1}{x^2-4}$ c) $\sqrt{x^2-4}$ d) $\sqrt{4-x^2}$

INTERMEDIATE EXERCISES

19) A company pays new employees $15 per hour if the time worked is less than 35 hours a week, and time and a half for all time equal to or beyond 35 hours. Express the pay as a function of the hours worked using the unit step function $u(x)$.

20) In each part of this problem, express the air conditioner's setting as a function of the temperature T using the unit step function $u(x)$.

Air Conditioner

   a) An energy-saving air conditioner has three settings. Off, medium and high.
   The machine is off if the temperature T if less than 70°F.   (State = 1)
   The machine is on medium if the temperature lies in the interval [70, 80) (State = 2)
   The machine's on high if the temperature's greater than or equal to 80°F (State = 3)
   Hint: $f(x) = f_1(x) + [f_2(x) - f_1(x)]\, u(x-a) + [f_3(x) - f_2(x)]\, u(x-b)$

   b) An energy-saving air conditioner has four settings. Off, low, medium and high.
   The machine is off if the temperature T if less than 70°F (State = 1)
   The machine is on low if the temperature lies in the interval [70, 75). (State = 2)
   The machine is on medium if the temperature lies in the interval [75, 80). (State = 3)
   The machine's on high if the temperature's greater than or equal to 80°F. (State = 4)

Each of the following functions is written in terms of the unit step function.
Give an equivalent split or piecewise definition of the function.

21) $f(x) = x\, u(x-1) + x^2\, u(x-2)$

22) $f(x) = x\, u(x-1) + x\, u(x-2)$

23) $f(x) = f_1(x) + f_2(x)\, u(x-a)$

24) $f(x) = f_1(x) + f_2(x)\, u(x-a) + f_3(x)\, u(x-b)$, $a < b$?

25) On a hot day, the dimensions of a rectangular metal plate are increasing due to thermal expansion. Let T denote the temperature in Celsius. The length of the plate is $L(T) = 10 + 0.00002T$ feet and the width of the plate is $W(T) = 20 + 0.00004T$ feet. Find an expression for the area of the plate as a function of temperature.

§1.4 The Algebra of Functions — CAFÉ

26) Salt water containing 2 kg of salt per cubic foot is being added to a tank originally containing 100 cubic feet of fresh water at the rate of 1 cubic foot per minute. When will the concentration in the tank reach 1 kg of salt per cubic foot? Note the volume of water in the tank at time t is $V(t) = 100 + t$ and the amount of salt in the tank at time t is $S(t) = 2t$. The concentration is $C = S/V$.

27) A cylindrical tank having dimensions shown originally contains 2 m³ of brine solution at a salt concentration of 2 kg/m³. A valve is opened and fresh water is added to the tank at a rate of 1/4 m³/min.

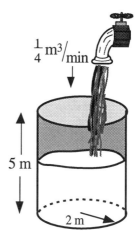

    a) Express the concentration of the salt in kg/m³ as a function of the time t (in minutes) up to the time of overflow.

    b) Repeat the problem if instead of fresh water being added, a more dilute brine of concentration 1 kg/m³ was added.

28) A standard baseball diamond is a square with all sides equaling 90 feet. A batter hits a home run and races around the diamond with a speed $v = 18$ ft/sec.

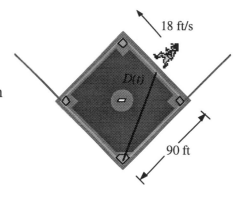

    a) Express the distance $D(t)$ between the batter and the home plate as a function of time t in seconds using a split or piecewise definition for your answer. Since there are four edges to the baseball diamond, you will have to express the distance from home plate over four different intervals.

    b) Express the distance by a single expression using the unit step function. Hint: The following formula may prove useful in part **(b)**:

$$f(x) = f_1(x) + [f_2(x) - f_1(x)]\,u(x-a) + [f_3(x) - f_2(x)]\,u(x-b) + [f_4(x) - f_3(x)]\,u(x-c)$$

ADVANCED EXERCISES

The remaining exercises illustrate how the need to combine functions arises in many engineering applications.

29) The walls of a cylindrical lead container for storing radioactive waste are 1/2 foot thick on the sides, top and bottom. Thus the outer diameter D is 1 foot more than the inner diameter d. The container stores 20π cubic feet of waste. (Volume of interior chamber.)

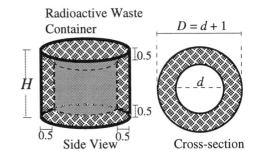

a) Express the height H of the container as a function of the outside diameter D.
b) Determine the external volume of the container as a function of D.

30) A grain silo is constructed in the shape of a circular cylinder of radius r with a hemispherical top. The bottom is just an earth surface. When full, the silo holds 250 cubic meters of cattle feed.

a) What is the height h of the cylindrical portion of the silo in terms of r?
b) Determine the area of sheet metal required for its construction as a function of r.
c) A second grain silo of similar shape is to be constructed from 1000 square meters of sheet metal. Determine the volume V of the silo as a function of the radius r. (Assume no material is wasted.)
The surface area of a sphere with radius r is $A = 4\pi r^2$. The surface area of a cylinder is $A = 2\pi rh$.

31) A rectangle is inscribed in a semicircle as shown. The radius of the semicircle is R and the height of the rectangle is x. Express the area A of the rectangle as a function of x.

32) A fruit can has the shape of a cylinder. Assume the total surface area S of the tin can is 2 square feet. Express the volume V of the can as a function of its radius R.

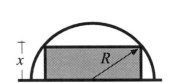

33) A large stone is released from rest over the edge of a cliff crashing on the valley floor below with a big bang. The total time t for the stone to crash and the resulting sound waves to return to the top of the cliff is recorded. Express the height H of the cliff as a function of t.
Assume the velocity of sound is 343 m/s.
(Hint: use the fact that $t = 0$ when $H = 0$ to determine the correct formula for t.)

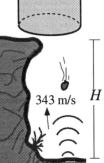

34) A ship is to make a round trip to a city 50 miles downstream along a straight river in which the water flows with a velocity of 3 miles/hour. If the ship moves with velocity v mi/hr in still water, express the time T required for the round trip as a function of v.

35) A box with an open top is to be constructed from a rectangular piece of sheet metal measuring 3 feet by 2 feet. The fabrication process involves cutting out equal squares of side x feet from each of the four corners and then folding up the sides as shown in the figure.

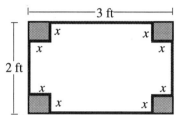

   a) Express the volume V of the box as a function of x.
   b) Sketch the graph of the volume and indicate the domain for V
   c) For what values of x is the volume equal to zero?

36) A coffee manufacturer sells its coffee in cans constructed from tin covered steel which have height h and radius r. The cylindrical part is fabricated by rolling a rectangular sheet into a cylinder and soldering the joint with tin. The material used to fabricate the cans costs x dollars per square foot for the sides and y dollars per square foot for the top and bottom. The cost of soldering the joint is z dollars per foot. Express the cost of the can as a function of radius and height.

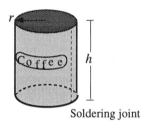

Soldering joint

37) Oil is to be piped from an offshore well to a refinery on the shore. The well is located 5 miles offshore and 10 miles up the coastline from the refinery. The cost of constructing the pipeline is estimated as $700,000 per mile on land but increases to $3,500,000 per mile for the offshore section. Assume the offshore section of the pipeline meets the shore a distance of x miles down the coast from the refinery.

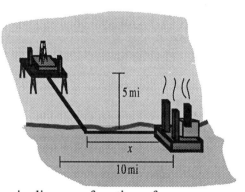

Express the total cost for construction of the pipeline as a function of x.

38) A cone can be constructed from a circle of radius R by removing a sector of angle θ from the circle and folding along the resulting seams.

a) Express the height h and radius r of the cone as functions of R and θ.

b) Express the volume V of the cone as a function of R and θ.

(Hint: the volume of the cone is $\frac{\pi r^2 h}{3}$)

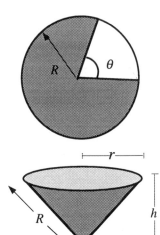

In each of the following problems assume there is no heat loss of the system to the surroundings and that the system is well mixed. The latent heat of the ice is 80 cal/g. The specific heat of water is 1 cal/g°C.

39) A jar contains 2000 grams of water at 25°C when x grams of ice at 0°C is dropped in. Express the final temperature T of the system as a function of x.

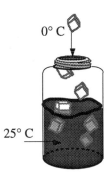

40) A thermos contains 500g of water at 25°C when x grams of water at 50°C is poured in. Express the final temperature T of the system as a function of x.

41) A tank contains 2000g of water at 25°C. Hot water at 80°C flows into the tank at the rate of 50g/sec. Express the temperature T of water in the tank as a function of time t. (in seconds)

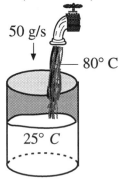

§1.5 Trigonometric Functions

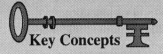
Key Concepts

- ❖ Definition of Trigonometric Functions
- ❖ Law of Sines
- ❖ Trigonometric Identities
- ❖ Law of Cosines

In this section we will first introduce the six trigonometric functions and then see how they are applied in typical geometric and engineering problems. Underlying all six trigonometric functions is a simple geometric picture - that of a right triangle inside a circle of radius one. A circle of radius one is called a <u>unit</u> circle. Applying the Pythagorean Identity to the right triangle shown in the figure, we see that every point $P = (x,y)$ on the unit circle must satisfy the identity $x^2 + y^2 = 1$.

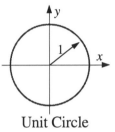
Unit Circle

Trig Machine

We can define all six trigonometric functions using the 'trig machine' shown to the right. The machine is simply an arrow of length one that pivots about the center of a unit circle. Suppose we wish to calculate the cosine of some angle θ. This angle is input into the machine by turning the arrow until it points to the angle θ, making it easy to see the corresponding horizontal component x and vertical component y as shown. Knowing these x and y components, we define:

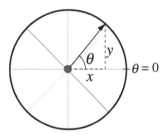
Trig Machine

$$\sin\theta = y \qquad \cos\theta = x \qquad \tan\theta = \frac{\sin\theta}{\cos\theta} = \frac{y}{x}$$

Thus the <u>cosine</u> of the angle θ is the horizontal component x and the <u>sine</u> is the vertical component y when the arrow points to the angle θ. The <u>tangent</u> is the slope of the arrow. We can tell a lot about the domains and ranges of these functions from the way the Trig Machine works. Since the slope of a vertical line is not defined, the tangent is not defined whenever the arrow points either up or down as at ±90 degrees. In contrast, the sine and cosine are defined for all real numbers. The range of either the sine or cosine consists of the closed interval [−1,1] since the projections of the arrow along either axis cannot exceed the length of the arrow.

The reciprocals of these trigonometric functions arise so often in practice that we also give them names. The reciprocal of the cosine is called the <u>secant</u>. The reciprocal of the sine is called the <u>cosecant</u> and the reciprocal of the tangent is called the <u>cotangent</u>. It is customary to use sec as an abbreviation for the secant, csc for the cosecant and cot for the tangent.

$$\csc\theta = \frac{1}{\sin\theta} = \frac{1}{y} \qquad \sec\theta = \frac{1}{\cos\theta} = \frac{1}{x} \qquad \cot\theta = \frac{1}{\tan\theta} = \frac{x}{y}$$

Following are the graphs of each of the six trigonometric functions discussed.

      Sine   Cosine   Tangent

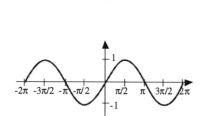

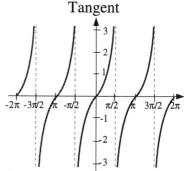

   Cosecant   Secant   Cotangent

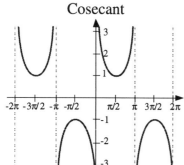
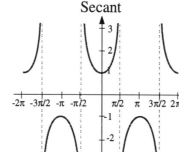
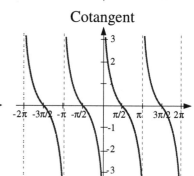

The machine shows the arrow set at 45° or π/4 radians. Both x and y are equal to $\frac{\sqrt{2}}{2}$. Thus we see that:

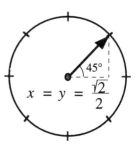

$\cos 45° = \dfrac{1}{\sqrt{2}}$ $\sin 45° = \dfrac{1}{\sqrt{2}}$ $\tan 45° = 1$

$\sec 45° = \sqrt{2}$ $\csc 45° = \sqrt{2}$ $\cot 45° = 1$

Trig Machine

§1.5 Trigonometric Functions

Proceeding in this manner we can create a table for the values of the trigonometric functions for angles that frequently arise in applications. For each angle in the table, the corresponding projections x and y were obtained by applying the Pythagorean Identity $x^2 + y^2 = 1$.

Angle	$\cos \theta$	$\sin \theta$	$\tan \theta$
0°	1	0	0
30°	$\sqrt{3}/2$	$1/2$	$1/\sqrt{3}$
45°	$1/\sqrt{2}$	$1/\sqrt{2}$	1
60°	$1/2$	$\sqrt{3}/2$	$\sqrt{3}$
90°	0	1	undefined
180°	−1	0	0

The above angles arise so frequently that it is a good idea to know the values for the sine, cosine and tangent at each angle.

Periodicity

Notice how the graphs of all six trigonometric functions are repetitive. This property arises directly from their definition using the unit circle. As we move the arrow around the circle repeatedly it returns to the same point (x, y) again and again. Since all six trigonometric functions are determined by the projections x and y they must repeat each time we return to the same point. Functions having this property are called periodic.

Definition

A function f is <u>periodic</u> if there is a positive real number p such that $f(x+p) = f(x)$ for all points x in the domain of the function. The smallest number p satisfying this property is called the <u>fundamental period</u> of the function.

Example 1

What are the fundamental periods of each of the six trigonometric functions?

Solution

By inspection of the six graphs, we see that four of the trigonometric functions have fundamental period equal to 2π. These are the sine, cosine, secant and cosecant. However, the fundamental period of the tangent and its reciprocal the cotangent is only π. Increasing the angle by π radians corresponds to reversing the arrow. When the arrow is reversed, both projections x and y change sign. However their ratio x/y stays the same! This is why the fundamental period of the tangent is only π.

Example 2

Which of the six trigonometric functions are even and which are odd?

Solution

By inspection of the six graphs, we see that the cosine and its reciprocal the secant are even functions.

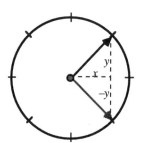

$$\cos(-\theta) = \cos\theta \qquad \sec(-\theta) = \sec\theta.$$

When the arrow is reflected in the *x*-axis, the vertical component changes sign.

The remaining four trigonometric functions are all odd functions of the angle θ.

$$\sin(-\theta) = -\sin\theta \qquad \tan(-\theta) = -\tan\theta$$

$$\csc(-\theta) = -\csc\theta \qquad \cot(-\theta) = -\cot\theta$$

These symmetry properties are easily understood by considering how the projections of the arrow change when it is reflected in the *x*-axis. Although the horizontal projection *x* does not change, the vertical component *y* changes to –*y*. Each of the last four functions contain a *y* in their definition and so change sign with it!

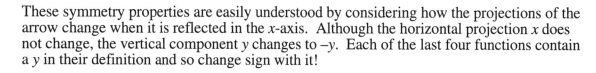

Radians or Degrees

Angles are commonly measured either in radians or degrees. In both systems the direction of the positive *x*-axis corresponds to an angle of zero and angles increase in the counterclockwise direction. The advantage of using degrees is familiarity. Everyone knows there are 360 degrees in a circle. Since four right angles make a complete circle, it is easy to see that a right angle corresponds to 90°.

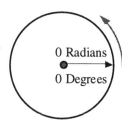

Radian Measure is Motivated by Geometry

The fact that there are 360 degrees in a circle is just an arbitrary convention. However the fact that there are 2π radians in a complete circle is a convention amply motivated in geometry. A <u>sector</u> of a circle is a portion of a circle shaped like a pizza slice as shown in the illustration. The simple formulas in the illustration for its arclength *s* and area *A* are in radian measure. These formulas would not look so nice in degrees. Notice that if $\theta = 2\pi$ we obtain the familiar formulas for the circumference and area of a circle.

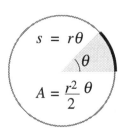

$s = r\theta$

$A = \dfrac{r^2\theta}{2}$

§1.5 Trigonometric Functions   CAFÉ   Page 65

Conversion Between Degrees and Radians

Whichever system one is working in however, it is easy to convert back and forth. Since both 360° and 2π radians correspond to a complete circle, we see that:

Conversion Between Degrees and Radians

$$1 \text{ degree} = \frac{\pi}{180} \text{radians} \quad \text{and} \quad 1 \text{ radian} = \frac{180}{\pi} \text{ degrees}$$

Application of Trigonometric Functions

Wherever angles or periodic phenomena are involved, the trigonometric functions will be important. Mechanical engineers apply trigonometric functions to such problems as controlling how data is read from and written to optical and magnetic storage drives and to such everyday problems as the design of disc brakes. An architectural engineer might apply trigonometric functions in the design of an entrance for handicapped persons. An electrical engineer uses them routinely in the study of alternating currents. The list goes on and on.

Polar Coordinates

Now that we know a little about trigonometric functions, it is time to introduce a whole new way to describe points and curves in the plane. Consider the point P shown in the illustration. Up until now we have been describing this point by giving its components x and y along the horizontal and vertical axis so that $P = (x,y)$. This method of describing points in the plane is called <u>Cartesian coordinates</u>.

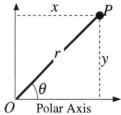

An alternative description would be to use the radius r and the angle θ shown in the illustration. In this description we would write $P = (r,\theta)$. This method of describing points is called <u>Polar coordinates</u>. The origin O is commonly called the <u>pole</u> and the positive x-axis is called the polar axis. We can convert back and forth between these systems by using the relations:

Relation Between Polar and Cartesian Coordinates

$$x = r\cos\theta \qquad \tan\theta = \frac{y}{x} \quad \Rightarrow \quad \theta = \tan^{-1}\frac{y}{x}$$
$$y = r\sin\theta \qquad r^2 = x^2 + y^2 \quad \Rightarrow \quad r = \sqrt{x^2 + y^2}$$

Just as we could describe curves in the plane by giving a relation between x and y, we can now describe curves by giving a relation between r and θ. Here are some simple examples of curves described in polar coordinates. Notice how every point on each curve satisfies the corresponding polar equation.

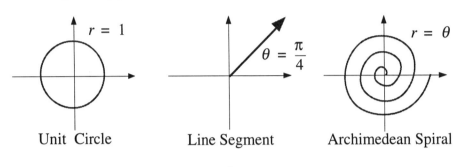

Unit Circle   Line Segment   Archimedean Spiral

Example 3

Describe the polar curve $r = 2 \cos \theta$.

Solution

Our strategy here will be to convert back into Cartesian coordinates (x and y) to see if we can recognize the curve. If we multiply through by r we obtain:

$$r^2 = 2 r \cos \theta$$

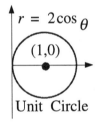

Recalling that $r^2 = x^2 + y^2$ and $x = r \cos \theta$ we find:

$$x^2 + y^2 = 2x \quad \text{or} \quad (x-1)^2 + y^2 = 1$$

The curve is a circle of radius 1 centered on the point (1,0).

Trigonometric Identities

The trigonometric functions satisfy a rich array of properties and identities. Following are some of the most useful ones.

Pythagorean Identities:

$$\cos^2 x + \sin^2 x = 1 \qquad 1 + \tan^2 x = \sec^2 x \qquad \cot^2 x + 1 = \csc^2 x$$

Double-Angle Identities:

$$\sin 2x = 2 \sin x \cos x \qquad \cos 2x = \cos^2 x - \sin^2 x$$

§1.5 Trigonometric Functions

Half-Angle Identities:

$$\cos^2(x/2) = \frac{1+\cos x}{2} \qquad \sin^2(x/2) = \frac{1-\cos x}{2}$$

Angle Sum Identities (Addition Formula):

$$\sin(x+y) = \sin(x)\cos(y) + \sin(y)\cos(x) \qquad \cos(x+y) = \cos(x)\cos(y) - \sin(x)\sin(y)$$

Reciprocal Identities:

$$\sec x = \frac{1}{\cos x} \qquad \csc x = \frac{1}{\sin x} \qquad \cot x = \frac{1}{\tan x}$$

Oblique Triangles

Many engineering problems require us to work with triangles which do not contain a right angle. Such triangles are called <u>oblique</u> triangles. The two most important formulas for working with oblique triangles are the Law of Sines and the Law of Cosines.

Law of Sines
$$\frac{\sin \alpha}{a} = \frac{\sin \beta}{b} = \frac{\sin \gamma}{c}$$

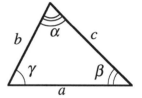

The ratio of the sine of each angle to the length of the opposite side is the same for all three angles. The Greek letters α, β and γ are called alpha, beta and gamma.

Law of Cosines
$$c^2 = a^2 + b^2 - 2ab\cos\theta$$

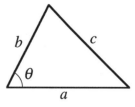

Thus knowing two sides and the angle between them we can calculate the third side. If the angle $\theta = 90°$, then the Law of Cosines reduces to the Pythagorean identity $c^2 = a^2 + b^2$.

Example 4 - Tracking The Sun

The efficiency of a solar panel can be increased by 'tracking the sun' much as the leaves of green plants do. The solar array shown in the illustration is kept perpendicular to the rays of the sun by means of a hydraulic jack DE. To what length S must the piston be extended to keep the array perpendicular to the sun's rays when the sun is at an angle of

$\phi = 45$ degrees above the horizon?

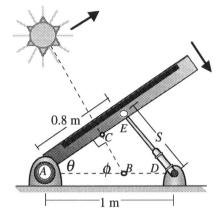

Solution

The right-angled triangle ABC in the illustration reveals that the angles of inclination of the sun and the solar array are complementary angles. That is $\theta + \phi = 90°$.

Thus the angle θ at the pivot A is also $45°$. Consider triangle ADE determined by the piston DE and the pivot at A. This is not a right triangle! However, using the law of cosines, the length S is seen to satisfy:

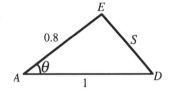

$$S^2 = 1^2 + (0.8)^2 - 2 \cdot 1 \cdot (0.8) \cos 45° = 1.64 - \frac{1.6}{\sqrt{2}} = 0.509$$

Taking square roots we find that the extension of the piston must be $\boxed{S = 0.713}$ meters.

Combining Trigonometric Functions
Sum-to-Product Formulas

In many applications of electrical engineering and signal analysis it is important to be able to convert the sum of two or more trigonometric functions into a product. The formulas for doing this are:

Sum of Sines	Sum of Cosines
$\sin A + \sin B = 2 \sin\left(\frac{A+B}{2}\right) \cos\left(\frac{A-B}{2}\right)$	$\cos A + \cos B = 2 \cos\left(\frac{A+B}{2}\right) \cos\left(\frac{A-B}{2}\right)$

We will derive only the first one involving the sum of two sines.
The technique is to rewrite A and B in the form:

$$A = \frac{A+B}{2} + \frac{A-B}{2} \qquad\qquad B = \frac{A+B}{2} - \frac{A-B}{2}$$

Thus, if we let $x = \frac{A+B}{2}$ and $y = \frac{A-B}{2}$ then $A = x + y$ and $B = x - y$.

Now we can use the angle sum identity $\sin(x+y) = \sin(x)\cos(y) + \sin(y)\cos(x)$ to find that:

$$\sin(A) = \sin(x + y) = \sin(x)\cos(y) + \sin(y)\cos(x)$$
$$\sin(B) = \sin(x - y) = \sin(x)\cos(-y) + \sin(-y)\cos(x)$$

§1.5 Trigonometric Functions   CAFÉ   Page 69

Adding together these two expressions and noting that the cosine is an even function whereas the sine is an odd function we find:

$$\sin A + \sin B = 2 \sin x \cos y = 2 \sin\left(\frac{A+B}{2}\right)\cos\left(\frac{A-B}{2}\right)$$

Example 5 - Sum to Product

Express the trigonometric function $y = \sin 2x + \sin 4x$ as a product.

Solution

Using the above sum-to-product formula we find:

$$\sin 2x + \sin 4x = 2 \sin\left(\frac{2x+4x}{2}\right)\cos\left(\frac{2x-4x}{2}\right) = 2 \sin 3x \cos x$$

Thus $\boxed{y = 2 \sin 3x \cos x}$.

Example 6 - Sum to Product

Where does the graph of trigonometric function $y = \sin x + \sin 5x$ cross the *x*-axis? That is, where is this trigonometric sum equal to zero?

Solution

Usually it is very difficult to tell when a sum is zero. However, a product of two functions is zero at a point *x* if and only if one of the functions is zero at that point. This suggests that it may be easier to solve this problem by converting the sum to a product. Using the above formula we find:

$$\sin x + \sin 5x = 2 \sin\left(\frac{x+5x}{2}\right)\cos\left(\frac{x-5x}{2}\right) = 2 \sin 3x \cos 2x$$

Thus the original function *y* is zero if and only if one of the terms $\sin 3x$ or $\cos 2x$ is zero.

Case 1: $\sin 3x = 0$ implies that $3x = n\pi$ or $x = \frac{n\pi}{3}$ where *n* is an integer.

Case 2: $\cos 2x = 0$ implies that $2x = \frac{\pi}{2} + n\pi$ or $x = \frac{\pi}{4} + n\frac{\pi}{2}$ where *n* is an integer.

Thus the original trigonometric sum *y* crosses the *x*-axis at the points $\boxed{x = \frac{\pi}{4} + n\frac{\pi}{2}}$ and $\boxed{x = \frac{n\pi}{3}}$ for every integer *n*.

Beat Phenomena

Another application of the sum-to-product rules is in the modeling of beat phenomena. Engineers who work around heavy equipment and generators are familiar with the beats or slow oscillations in loudness that can be heard over the shrill whine of metal on metal. This is particularly noticeable when large rotors or generators turn at slightly different frequencies. When the frequencies differ by a small amount, the sum of these two signals will exhibit the beat phenomena, i.e. the modulation (or amplitude of the sum) becomes a slowly varying sinusoidal function.

The following illustrates the beat phenomenon. Let the angular frequency of one source be ω and that of the other $\omega + \delta$, where δ is a small number. If both sources have the same unit strength and are originally in phase, their combined signal will be:

$$f(t) = \cos \omega t + \cos((\omega + \delta)t)$$

In this form it is not easy to see the beat phenomenon but using the sum-to-product formula, the equation can be rewritten as:

$$f(t) = 2\cos\frac{\delta t}{2} \cos\left(\frac{2\omega + \delta}{2} t\right)$$

It is now easy to see the effect of the slowly varying term $\left(2\cos\frac{\delta t}{2}\right)$ called the _modulation_. When this term is near zero, we hear virtual silence and when its magnitude is at its maximum, the sound is loudest.

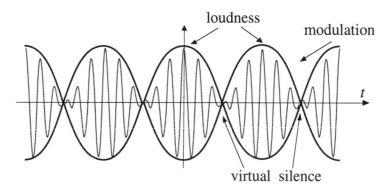

The phenomenon is easily demonstrated with two tuning forks, struck simultaneously with the same force and then held at the same distance from the observer. The pressure on the outside of the eardrum at time t may be represented by $a\cos \omega_1 t + a\cos \omega_2 t$. If ω_1 and ω_2 are very close to each other, a tone is produced that alternates between loudness and virtual silence. This is known as the beat phenomenon.

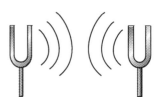

§1.5 Trigonometric Functions   CAFÉ   Page 71

1) Derive the identity $1 + \tan^2 x = \sec^2 x$ by dividing both sides of the identity $\cos^2 x + \sin^2 x = 1$ by $\cos^2 x$.

2) Derive the identity $\cot^2 x + 1 = \csc^2 x$ starting with the identity $\cos^2 x + \sin^2 x = 1$.

3) The zeroes of the sine function are located at every integer multiple of π (in radians). Letting n denote any integer we can express this symbolically as: $\sin x = 0$ if and only if $x = n\pi$.
Obtain similar expressions for the zeroes of each of the other five trigonometric functions. Indicate which of the five have no zeroes.

   a) cosine   b) tangent   c) secant
   d) cosecant   e) cotangent

4) Convert each of the following angles from radians to degrees.

   a) $\pi/4$ b) $\pi/3$ c) 2π d) θ

5) Convert each of the following angles from degrees to radians.

   a) $90°$ b) $30°$ c) $45°$ d) $d°$

6) Determine whether each of the following functions is even, odd or neither.

   a) $f(x) = \cos(1+x)$ b) $f(x) = \cos(1+x^2)$
   c) $f(x) = \sin(2x)$ d) $f(x) = \sin(x^2)$

7) Explain why it is incorrect to write the equation $\sin^2 x = \sin(x^2)$.

8) Explain why it is incorrect (a sin in fact) to write the equation $\dfrac{\sin x}{x} = \sin$.

9) The equations for each of the following lines are given in polar coordinates. Find the expression for each line in Cartesian coordinates and identify its slope and vertical intercept.
   a) $r\cos\theta = 1$ b) $r(\cos\theta - 2\sin\theta) = 1$

10) The equations for each of the following circles are given in polar coordinates. Find the equation for each circle in Cartesian coordinates and identify the center and radius of the circle.
   a) $r = 4\cos\theta$ b) $r = 2\sin\theta$ c) $r = 6\sin\theta + 8\cos\theta$

11) Two sewer lines flow into a city's water treatment plant from two towns upstream. The flow rate from the first line is $f(t) = 100 + 75\sin(2\pi t)$ cubic feet per minute and the flow rate from the second line is $g(t) = 200 + 75\sin(2\pi t + \pi/4)$. (The time t is measured in days.) The phase shift is due to the different distance to the towns. Using the sum-to-product rule, express the total rate at which water flows into the treatment plant as a constant plus a product.

12) During a drought, water flows into a town's reservoir at a rate equal to $f(t) = 100 + 75\sin(2\pi t)$ cubic feet per minute but is withdrawn at a rate equal to $g(t) = 200 + 50\sin(2\pi t)$. (The time t is measured in days.) Find a symbolic expression for the total rate at which the water is being depleted.

13) How can we measure the height of a tall building or monument? The following technique used by surveyors is an interesting application of trigonometric function. The surveyor stands a fixed distance D from the monument and measures the angle θ to its top.

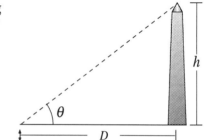

a) Express the height h of the monument in terms of the angle of elevation θ and D.

b) What is the height if $D = 400$ feet and $\theta = 30°$?

c) The height of the Washington Monument is 555 feet. What angle will be measured at a distance of 400 feet?

INTERMEDIATE EXERCISES

In the following exercises, find all solutions x that lie in the closed interval $[0, 2\pi]$.

14) $\sin 3x - \sin 5x = 0$

15) $\sin 2x - \sin 8x = 0$

16) $\cos x = \sin 2x$

17) $\cos x + \cos 2x = 0$

18) $\sin 5x - \sin x = 2\cos 3x$

19) $25\sin^2 x - \cos x - 1 = 0$

20) $\cos x - \sin x = 1$

21) $\sin x + \cos x \cot x = \csc x$

22) A circle with radius r is circumscribed in an isosceles triangle. Let θ be the angle at the top vertex of the triangle. Express the area A of the isosceles triangle as a function of θ and r.

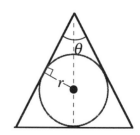

§1.5 Trigonometric Functions   CAFÉ   Page 73

23) When one circle rotates about another fixed circle of the same radius, the path traced out by a point on the moving circle was found to be given by $r = 1 + \cos 6$. This curve is called a <u>cardioid</u>. Sketch the curve.

24) **Solving triangle**
In the triangle shown, $a = 2$ and $b = 1$.
Find the length of side c for each value of γ below.
a) $\gamma = 30°$ b) $\gamma = 45°$ c) $\gamma = 90°$
d) How will the answers change if the values of a and b are interchanged so that $a = 1$ and $b = 2$?

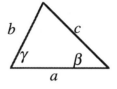

25) Use the above illustration to find the angle β for each value of γ below.
a) $\gamma = 30°$ b) $\gamma = 45°$ c) $\gamma = 90°$

26) In electrical engineering voltages and currents are frequently written in the standard form $f(t) = A \cos(Bt + C)$ where A is the <u>amplitude</u> of the signal, B is the <u>angular frequency</u> and C is the <u>phase shift</u>. Express the following signals in this standard form.
a) $f(t) = 5\cos 10t - 2\sin 10t$ b) $f(t) = \sqrt{3}\cos 4t - \sin 4t$

27) Two tuning forks are struck and generate the following combined signal:
$f(t) = \cos(1200\pi t) + \cos(1210\pi t)$. Write this sum as a product and sketch the graph. How many loudness beats will there be each second? (t is in seconds)

28) Two tuning forks are struck and generate the following combined signal:
$f(t) = \sin(1200\pi t) + \sin(1300\pi t)$. Write this sum as a product and sketch the graph. How many loudness beats will there be each second? (t is in seconds)

29) The solar array shown in the illustration is kept <u>perpendicular</u> to the rays of the sun by means of a hydraulic jack DE. To what length S must the piston be extended to keep the array perpendicular to the sun's rays when the sun is at the following angles above the horizon? (in degrees)

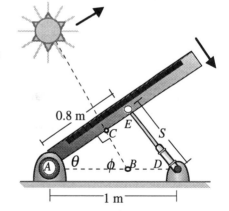

a) $\phi = 30$ b) $\phi = 60$
c) $\phi = 75$

ADVANCED EXERCISES

30) Radar Tracking
A ground radar station has spotted an aircraft flying horizontally at a height of 2 miles and initially making an angle of 30 degrees to the vertical as shown. The speed of the aircraft is 600 miles/hour. Express the angle of elevation θ as a function of the time t (in hours.)

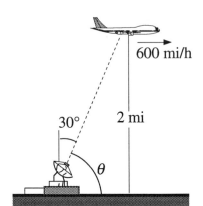

a) Assume the aircraft is flying away from the station.
b) Assume the aircraft is approaching the station. Model the situation carefully as the plane flies directly overhead.

31) Chasing smugglers
A coast guard cruiser uses radar equipment to detect a smuggling boat 3 miles east of their ship and traveling precisely northwest at a rate of 15 miles per hour. If the coast guard cruiser has a maximum speed of 20 mi/hr, answer the following:

    a) What heading should the cruiser travel in order to intercept the smugglers as quickly as possible?
    b) When will the cruiser apprehend the smugglers?

32) A small circle of radius a is tangent to a larger circle of radius b. Line OCD is tangent to both circles and line OAB passes through the centers of the two circles. Show that $\cos\theta = \dfrac{\sqrt{ab}}{\left(\dfrac{a+b}{2}\right)}$.

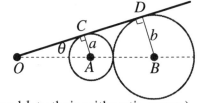

(The cosine is the ratio of the geometric mean of a and b to their arithmetic mean.)

33) From the top of an ocean cliff, Spike watches a ship sailing directly towards the cliff. If Spike is 200 feet above sea level and the angle between the horizontal and the line of sight to the ship changes from 30° to 50° during the period of observation, calculate the distance that the ship travels.

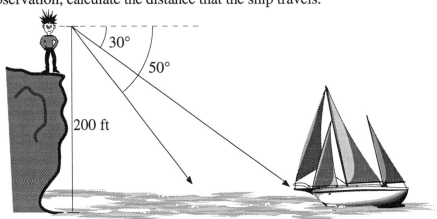

§1.5 Trigonometric Functions

34) An observer in a lighthouse 300 feet above sea level spots a ship directly off-shore. The angle of depression from the horizontal to the ship is 6.5°. Forty-five minutes later the angle of depression is measured to be 4°. Calculate the speed of the ship assuming it is constant.

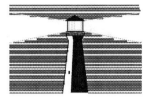

35) How can we measure the height of an object such as a mountain peak which is very distant? Using a device called theodolite surveyors are able to measure the angle of elevation to the mountain peak at two points separated by a distance d meters.

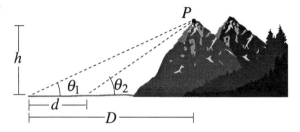

    a) Express the height h of the mountain as a function of d, θ_1 and θ_2.

    b) Express the distance D to the mountain as a function of d, θ_1 and θ_2.

36) To measure the vertical dimension of President Lincoln's profile on Mt. Rushmore, two sets of sightings were taken from ground level. The second set was taken at a point 200 feet closer to the mountain than the first. Using the following data, calculate the height of Lincoln's face.

Observation Set	Angle to top of face	Angle to bottom of face
#1	35.00°	32.00°
#2	43.03°	39.73°

37) Prove the formula for the sum of cosines using the angle sum identity $\cos(x+y) = \cos(x)\cos(y) - \sin(x)\sin(y)$.

That is, show that $\cos A + \cos B = 2\cos\left(\dfrac{A+B}{2}\right)\cos\left(\dfrac{A-B}{2}\right)$.

38) Steel is melted in a rectangular container measuring 2 ft × 2 ft × 1 ft and then poured into ingots by tilting the container at an angle θ as illustrated. In this process it is important to control the rate of turning so that a constant pour rate into the ingots is achieved.

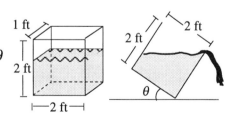

Express the total volume V poured into the ingots as a function of the angle θ. You may have to express the volume using a split or piecewise definition.

NOTES

§1.6 Exponential Functions

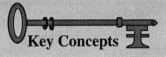

Key Concepts

- Laws of Exponents
- Properties of Logarithms
- Definition of Logarithm

The family of exponential functions arise again and again in engineering applications. The most familiar application of exponentials is to represent numbers of vastly different magnitudes. An electrical engineer needs to represent such small numbers as the charge of an electron ($Q_e = 1.602 \times 10^{-19}$ coulombs) and its rest mass ($M_e = 9.1066 \times 10^{-28}$ gram). An aerospace engineer must work with gigantic numbers such as the earth's mass ($M_{earth} = 1.315 \times 10^{25}$ pounds).

The Richter Scale

The intensity of earthquakes varies over an enormous scale. To more easily manage this vast range, geologist C. F. Richter expressed the intensity in terms of the magnitude M which is defined by the relation:

$$I = I_0 \, 10^M$$

The intensity is defined as the amplitude measured on a seismograph located 100 km from the epicenter of the earthquake. I_0 is defined as the intensity of a 'standard' earthquake and corresponds to a deflection of 1 micron. Thus each time the magnitude M increases by 1, the actual intensity of the earthquake increases by a factor of 10.

Example 1

How many times stronger is a magnitude 9 earthquake over a magnitude 7 earthquake?

Solution

The ratio of the intensities is $\dfrac{I_2}{I_1} = \dfrac{I_0 \, 10^9}{I_0 \, 10^7} = \boxed{100}$.

These examples illustrate the use of base 10 exponents. However, any positive number a can be used as a base. If n is a positive integer, we define the exponential a^n as follows:

$$a^n = \underbrace{a \times a \times a \times \cdots \times a}_{n \text{ times}}$$

Thus for example $2^3 = 2\times 2\times 2 = 8$. The other exponents satisfy the same laws as we are used to for the powers of 10.

Laws of Exponents

Product Law	Power Law	Reciprocal Law
$a^x a^y = a^{x+y}$	$(a^x)^y = a^{xy}$	$a^{-x} = \dfrac{1}{a^x}$

Example: $10^2\, 10^4 = 10^6$ Example: $(10^2)^3 = 10^6$ Example: $10^{-1} = \dfrac{1}{10}$

If the base a is greater than 1, then the function $f(x) = a^x$ increases as x increases. If the base a is less than 1, then the function $f(x) = a^x$ decreases as x increases. For example, the graph to the right shows that the function 2^x increases as we move to the right but that the graph of $\left(\dfrac{1}{2}\right)^x = \dfrac{1}{2^x} = 2^{-x}$ decreases.

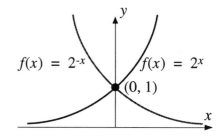

Example 2 - Bacterial growth

In the production of yogurt, bacteria (often Lactobacillus bulgaricus) are introduced into a large vat of warm milk and double in number every 30 minutes. Find an expression for the number of bacteria after t hours if one billion bacteria were originally introduced into the milk.

Solution

The bacteria quadruple in number every hour. Thus the number of bacteria after t hours is

$$\boxed{N(t) = 4^t} \text{ billion.}$$

In general, an exponential function is a function of the form $N(t) = N_0\, a^t$. If t corresponds to time, then N_0 can be interpreted as the value of the quantity at time 0. The quantity increases if $a > 1$ and decreases if $a < 1$.

§1.6 Exponential Functions

Example 3 - Radioactive Decay and Smoke Detectors

A mass of $m = 2$ grams of a radioactive isotope has been purchased by
a firm that manufactures smoke detectors. Small amounts of the
isotope are used in each detector to create ions that enhance the
detection of smoke particles. Unfortunately, the radioactive substance
decays at such a rate that only one gram is left after six months! Find
the mass $m(t)$ remaining as a function of the time t in years.

Solution

One half of the substance decays in 1/2 year so only 1/4 will be left after one year.
Thus the amount of radioactive material remaining after t years is only:

$$m(t) = m(0)\frac{1}{4^t} = \frac{2}{4^t}$$

Example 4 - Compound Interest

A city has sold one hundred million dollars in bonds to raise money for a
redevelopment project. The bonds mature in 10 years and pay 8% interest
compounded annually. What will the bill be when the bonds mature?

Solution

The original amount of the loan is one hundred million dollars or
$10^2 \, 10^6 = 10^8$ dollars. After one year, the city will owe $10^8 (1 + .08)$
dollars. Compound interest means that interest is paid on the interest. Thus
after two years the city will owe $10^8 (1 + .08)^2$ dollars.

After t years, the total debt increases to $D(t) = 10^8 (1 + .08)^t$. Since the bonds mature in
ten years, the total cost to the city of financing the redevelopment project will be:

$$D(10) = 10^8 (1 + .08)^{10} = \boxed{216 \text{ million dollars}}$$

This is more than double the original investment! Compound interest can accumulate at a
surprising rate.

The Logarithm

Many engineering problems require us to solve equations
like $y = 10^x$ for x given a value of y. From the graph of
$y = 10^x$, we can see that every positive value of y is
assumed just once and thus the corresponding x is
determined uniquely. This unique value is denoted as the
logarithm.

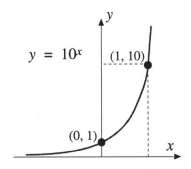

Definition

The common logarithm (or log base 10) of a number y is that unique value x satisfying $y = 10^x$ and is denoted $x = \log_{10} y$.

We will usually just write $\log x$ to denote $\log_{10} x$. Thus $x = \log y$ if and only if $y = 10^x$. From the graph of 10^x we see that the logarithm is not defined for negative values or for zero.

Several examples of logarithms are:

$\log 100 = 2$ since $100 = 10^2$ $\log 10 = 1$ since $10 = 10^1$

$\log \frac{1}{10} = -1$ since $\frac{1}{10} = 10^{-1}$ $\log 0$ is undefined

Since the logarithm is the inverse of the function 10^x, each of the laws of exponents becomes a rule for manipulating logs.

Properties of Logarithms

$\log xy = \log x + \log y$ Example: $\log(10^2 \cdot 10^4) = \log 10^2 + \log 10^4 = 6$

$\log \frac{x}{y} = \log x - \log y$ Example: $\log(10^4/10^2) = \log 10^4 - \log 10^2 = 2$

$\log x^y = y \log x$ Example: $\log 100^3 = 3 \log 100 = 6$

A simultaneous plot of 10^x and $\log x$ shows that the graph of the log is the mirror image of the graph of 10^x in the line $y = x$.

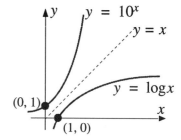

Solving Equations Containing Exponentials

Equations involving exponents arise frequently in engineering analysis and can often be solved using logarithms.

§1.6 Exponential Functions   CAFÉ   Page 81

Example 5 - Bacterial Growth Revisited

Consider the same bacterial culture introduced earlier.

a) When will the number of bacteria in the yogurt culture equal one trillion?

b) When will the number reach 4 trillion?

Note: One trillion = 10^{12} = 1000 billion.

Solution

We already saw that starting with one billion bacteria, the number of bacteria after t hours is $N(t) = 4^t$ billion.

a) To find out when the number has reached one trillion we will need to solve the equation $4^t = 1000$.

Taking the common log (log base 10) of each side we find that $\log 4^t = \log 1000$. Evaluating the logarithms gives $t \log 4 = 3$ or $t = \dfrac{3}{\log 4} = \boxed{4.98 \text{ hours}}$.

A little less than five hours makes sense since 1000 is about $1024 = 4^5$.

b) To find out when the number has reached four trillion we will need to solve the equation $4^t = 4 \times 1000$. Writing the equation in the form $4^{t-1} = 1000$ we see that this time is just one hour more than the time in part **a)**. Thus the time to reach four trillion is $\boxed{5.98 \text{ hours}}$.

Example 6 - Inflation

If inflation is running at 5% per year, how long will it take the cost of basic goods to double? (This is, how long before the dollar has lost 1/2 its present value?)

Solution

If P_0 denotes the price of a sample of goods today, then the price after t years is $P(t) = P_0(1.05)^t$.
Since we wish to know when the prices have doubled, we must solve the equation:

$$2P_0 = P_0(1.05)^t \text{ or } 2 = (1.05)^t.$$

Taking the log of each side we find that $\log 2 = t \log 1.05$. Thus:

$$t = \dfrac{\log 2}{\log 1.05} = \boxed{14.2 \text{ years}}$$

This is a lot sooner that the twenty years that would be required if the 5 % were not compounded.

Example 7 - Acids and Bases

The acidity or basicity of a solution is a measure of the hydrogen ion concentration in the solution. The concentration of hydrogen ions, denoted by $[H^+]$, varies over extreme ranges (between 10^0 and 10^{-14}) that it is measured using a <u>logarithmic</u> scale known as the pH scale. By definition, the pH of a solution is the log base 10 of the hydrogen ion concentration measured in moles per liter.

$$\text{pH of Solution}$$
$$pH = -\log[H^+]$$

Find the pH of vinegar which has a hydrogen ion concentration of about $[H^+] = 6.3 \times 10^{-3}$ moles per liter.

Solution

The pH is the negative logarithm of the hydrogen ion concentration. Therefore:

$$pH = -\log[H^+] = -\log[6.3 \times 10^{-3}] = \boxed{2.2} \qquad \text{Thus the pH of vinegar is about 2.2}$$

§1.6 Exponential Functions   CAFÉ   Page 83

Simplify the following combinations of exponentials.

1) $\dfrac{2^3 \, 4^5}{8^2}$ 2) $\dfrac{6^6}{9^2 \, 12^2}$ 3) $\sqrt{4}\,(16)^{3/2}\,8^{-1}$

Simplify the following combinations of logarithms.

4) $\log 10^6$ 5) $\log(0.00001)$

6) $10^{\log x^2}$ 7) $10^{2\log x}$

8) $(\log x - \log y) + \log \dfrac{y}{x}$ 9) $\log x^2 - 2\log xy + \log y^2$

10) Sketch the following two pair of functions on the same graph. What is the symmetry relation between the two graphs?

    i) $\begin{cases} f(x) = 3^x \\ g(x) = \left(\dfrac{1}{3}\right)^x \end{cases}$ ii) $\begin{cases} h(x) = \log x \\ l(x) = 10^x \end{cases}$

Solve the following equations.

11) $4^t = 2$ 12) $4^{t^2} = 5$ 13) $a^t = b$ 14) $2 \cdot 3^x = 5 \cdot 8^x$

15) $2^{x+1} = 5^{3x-2}$ 16) $3^{x-3} = 9^{4-x}$ 17) $\log(x+1) = 3$

18) Find the pH of carrot juice which has a hydrogen ion concentration of about 10^{-5} moles per liter.

19) Find the hydrogen ion concentration of beer which has a pH of about 4.2.

20) The pH of common solutions varies from about 1 to 14. This may not sound like a great variation, until the corresponding range of hydrogen ion concentrations is considered. Calculate the corresponding range for $[H^+]$.

21) a) How many times stronger is a magnitude 9 earthquake over a magnitude 6 earthquake?

    b) How many times weaker in concentration of hydrogen ions is a pH 10 solution over a pH 5 solution?

22) Induced Earthquakes (Koyna Dam in India)
The filling of a large reservoir behind a manmade dam can induce earthquakes. The quakes may be due to the superimposed water weight, chemical alteration of the underlying rock, or other effects due to infiltration of the water and increased pore pressure. Soon after filling the reservoir at the Koyna Dam (India), tremors began. The first two significant tremors occurred in September 1967 and had magnitudes 5.0 and 5.5 causing mild damage. In December 1967, a 6.5 magnitude earthquake occurred killing about 180 people and injuring 2200. Calculate the intensity for each of these three quakes. How many times stronger was the killer 6.5 magnitude quake than the mild 5.5 quake?

Solve the following equations.

23) $2^{-x} + 2^x = 5$ 24) $4^{x^2} = 2^{3x+2}$ 25) $2 \cdot 2^{-x} + 2^x - 3 = 0$

26) $4^x + 2 \cdot 2^x - 3 = 0$ 27) $\dfrac{10^x + 10^{-x}}{10^x - 10^{-x}} = 2$

28) $\log(x^3) = (\log x)^2$ 29) $x^{\log x} = 10$ 30) $x^{\log x} = 100x$

31) $\log \log(x + 3) = 2$ 32) $\log(x + 2) + \log(x + 5) = 1$

33) Inflation
a) How long will it take prices to double if the annual inflation rate is 100%?
b) How long will it take prices to double if the annual inflation rate is 10%?
c) How long will it take prices to double if the annual inflation rate is 1%?

34) Deflation
a) How long will it take prices to fall by 1/2 if the annual deflation rate is 10%?
b) How long will it take prices to fall by 1/2 if the annual deflation rate is 1%?

35) An initial population of one million bacteria triple in number every hour.
a) How many will be present after one <u>day</u>?
b) How many will be present after t hours?
c) When will the number of bacteria equal one trillion?

36) It is observed that an initial population of one thousand bacteria double in number every k minutes.
a) How many will be present after t hours?
b) It is also observed that after 2 hours the population of the bacteria has reached 1 million. What is the value of k here?

§1.6 Exponential Functions   CAFÉ   Page 85

37) **Radioactive decay**
A radioactive isotope has a half life of 5000 years.
   a) How much of a 1 gram sample will remain after 1000 years? After t years?
   b) How much time does it take to decay to just 1/5 of the initial amount?

38) **Bacterial Competition**
Two colonies of different bacteria (species A and B) are competing over the same food source. Initially there are 1000 members of species A and only 10 members of species B in the large food source. Species A doubles every 30 minutes whereas species B doubles every twenty minutes.
   a) Find the population of each species as a function of the time t in hours.
   b) When will the two species have the same population?
   c) What happens to the ratio of the number of members of species A to species B after a long time?

39) The number 3231 has 4 decimal digits.
   a) How many decimal digits are there in the number 9^9 ?
   b) How many decimal digits are there in the number 9^{9^9} ? Remember the number 9^{9^9} is defined as $9^{(9^9)}$, not as $(9^9)^9$.
   c) If it takes 0.1 seconds to print a digit, estimate the time it would take to print out the number 9^{9^9} in the full decimal form.

NOTES

Chapter Two

Limits and Derivatives

§ 2.1 Tangent Lines & Derivatives

Key Concepts

- Tangent Lines
- Derivatives
- Average and Instantaneous Rate of Change

Average rates of change are important in such problems as the average speed during a trip, interest accrued on a bank account and planning for increased customer demand.

❖

Definition: Average Rate of Change

The average rate of change in a quantity over a given period of time is the amount of change divided by the length of the time interval. In symbols, the average rate of change of a quantity y over the time interval from t_1 to t_2 is $\frac{\Delta y}{\Delta t} = \frac{y_2 - y_1}{t_2 - t_1}$

where $y_1 = y(t_1)$ and $y_2 = y(t_2)$.

❖

Example 1 - Thunderstorm

Shortly before a thunderstorm, the temperature often drops precipitously as a cold front collides with warm air. The following table gives the temperature as a function of time, during a particular thunderstorm.

Time	Temperature (°F)
6:00 PM	90
6:10 PM	88
6:20 PM	85
6:30 PM	80
6:40 PM	75
6:50 PM	73
7:00 PM	72

Find the average rate of change of the temperature during the storm. During which time interval was the average rate of change greatest?

Solution

During the one hour from 6:00 PM until 7:00 PM the temperature T dropped a total of 18 degrees so that the average rate of change is $\dfrac{\Delta T}{\Delta t} = \dfrac{T_2 - T_1}{t_2 - t_1} = \dfrac{72-90}{7-6} = -18\ °\text{F/hr}$.
The average rate of change is negative because the temperature is decreasing.

A glance at the table shows that over certain time intervals the temperature was falling at a much greater rate. We can make a table of the average rate of change over each ten minute interval. (Note ten minutes = 1/6 hour.)

Time Interval	Average Rate of Change in °F/hr
6:00 to 6:10	$\dfrac{88-90}{1/6} = -12$
6:10 to 6:20	$\dfrac{85-88}{1/6} = -18$
6:20 to 6:30	$\dfrac{80-85}{1/6} = -30$
6:30 to 6:40	$\dfrac{75-80}{1/6} = -30$
6:40 to 6:50	$\dfrac{73-75}{1/6} = -12$
6:50 to 7:00	$\dfrac{72-73}{1/6} = -6$

The effect of the cold front was felt most severely over the time interval from about 6:20 to 6:40 where the average rate of change is about −30 degrees per hour!

Geometric Interpretation of the Average Rate of Change
(Slope of Secant Line)

A more complete table of data during the storm was recorded and plotted. (See graph.)

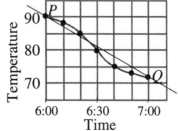

§ 2.1 - Tangent Lines & Derivatives   CAFÉ

A line through two points P and Q on a graph is called a <u>secant</u> line. In the above graph, a secant line is drawn through the two points $P = (6{:}00, 90)$ and $Q = (7{:}00, 72)$. The slope of this secant line is $Slope = \frac{rise}{run} = \frac{72-90}{7-6} = -18\,°F/hr$. This is precisely the average rate of change of the temperature during the storm.

Thus we have the following very nice geometric interpretation. The average rate of change of a function y over the time interval from t_1 to t_2 is the <u>slope of the secant line</u> through the points $P = (t_1, y_1)$ and $Q = (t_2, y_2)$.

Urban Flooding

The first graph shows the cumulative amount of rain falling over an urban area during a severe lightening storm. The total rainfall during the 24 hour period is six inches, so the average rate at which rain fell during this storm is:

$$\frac{\Delta h}{\Delta t} = \frac{6}{24} = \frac{1}{4} \text{ inches/hour}$$

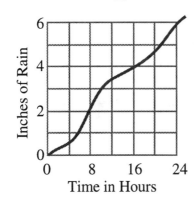
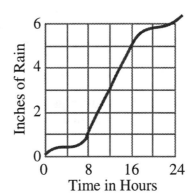

The second graph shows the amount of rain during a second storm during which approximately six inches also fell during a 24 hour period. Thus during both storm events rain fell at an average rate of 1/4 inches per hour. However the effect of each storm was very different. The first storm caused little damage and was actually beneficial to the city. The second storm caused major flooding of basements, street corners and underpasses. In addition it overwhelmed the cities combined sewer system causing raw sewage to be dumped into a nearby lake.

Given that the average rate at which the rain fell was essentially the same in each storm, why did one cause so much damage? Clearly this kind of question is important to the lives of millions of people.

Although the second storm had the same average rate of rainfall, the rain did not fall uniformly. From the graph, we can see that the rate at which the rain fell was very large at about $t = 12$. This situation reveals why engineers must consider the rate of change at a given moment instead of just average rates. The rate of change at a given moment is called the <u>instantaneous</u> rate of change.

Instantaneous Rate of Change

The following procedure is the intuitive idea underlying the concept of the rate of change of a quantity at a given moment t_1. (This is called the <u>instantaneous</u> rate of change.)

1) Find the average rate of change of the quantity between times t_1 and t_2.

$$\frac{\Delta y}{\Delta t} = \frac{y_2 - y_1}{t_2 - t_1}$$

2) Find the limit of the average rate of change as t_2 approaches t_1.

$$\lim_{t_2 \to t_1} \frac{y_2 - y_1}{t_2 - t_1}$$

❖

The graph shows in more detail, the total amount of rain that has fallen between noon and one o'clock during the previous storm. We can use the graph to estimate the instantaneous rate at which the rain is falling at noon (see point P on the graph). Select a second point Q on the rain-graph and calculate the slope of the secant line through P and Q. Then as Q approaches P, the slope of the secant line approaches the instantaneous rate of change. The instantaneous rate at noon is about one inch per hour. This is four times the average rate during the storm. This excessive rate is what caused all the flooding.

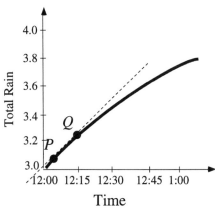

The instantaneous rate of change has an elegant geometrical interpretation as the <u>slope of the tangent line</u> to the curve at the point of interest.

❖

Procedure for Finding the Tangent Line at a Point

The concept of the tangent line is also defined using a limit procedure.
The tangent line to a curve at a point P is found by performing the following steps.

1) Find the slope of the secant line through the point P and a nearby point Q. (Average rate of change)

2) Find the limiting value of the secant slope as the point Q approaches P along the curve. If it exists, this limiting value is called the <u>slope of the curve</u> at the point P. (Instantaneous rate of change.)

3) The tangent line to the curve at the point P is the line through P with this slope.

§ 2.1 - Tangent Lines & Derivatives

As Q approaches P, the secant line PQ approaches the tangent line to the curve at P.

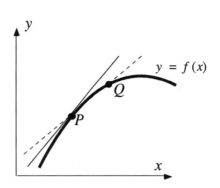

Example 2

Find the equation of the tangent line to the parabola $y = x^2$ at the point $P = (1,1)$.

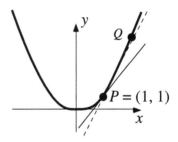

Tangent line at P.

Solution

Consider a nearby point $Q = (1+h, (1+h)^2)$ on the parabola.

1) The slope of the secant line through the points P and Q is:

$$\text{Secant Slope} = \frac{\Delta y}{\Delta x} = \frac{(1+h)^2 - 1}{(1+h) - 1} = \frac{2h + h^2}{h} = 2 + h$$

2) As Q approaches P along the parabola, h approaches 0. Thus the limit of the secant slope as Q approaches P is:

$$\lim_{h \to 0} (2 + h) = 2$$

The slope of the tangent line is 2.

3) The tangent line is the line through the point $P = (1,1)$ with slope 2. Using the point-slope formula we find $\dfrac{y-1}{x-1} = 2$ or $\boxed{y = 2x - 1}$.

Example 3

Find the slope of the quadratic $f(x) = x^2$ at an arbitrary point $P = (x, x^2)$ along its graph.

Solution

The slope of the tangent line is the limit of the secant line slopes:

$$m = \lim_{h \to 0} \frac{f(x+h) - f(x)}{h} = \lim_{h \to 0} \frac{(x+h)^2 - (x)^2}{h}$$

$$= \lim_{h \to 0} \frac{h^2 + 2xh}{h} = \lim_{h \to 0} \frac{h(h + 2x)}{h}$$

This is the crucial step since the h in the denominator can now be canceled!

$$m = \lim_{h \to 0} \frac{h(h + 2x)}{h} = \lim_{h \to 0} (h + 2x) = 2x$$

Thus the slope of the tangent line at the point $P = (x, x^2)$ is $\boxed{2x}$.

The Derivative

Intuitively, the derivative of a function f at a point x is the slope of the tangent line to the graph of the function at the point x. Since the notion of the tangent line depends on the concept of a <u>limit</u>, we will return to give a more formal treatment of the derivative after studying limits in the next section.

Definition of Derivative

Let $y = f(x)$ be a function.

The derivative of f is also a function and is denoted by $\dfrac{dy}{dx}$ or $f'(x)$ (pronounced - f prime).

The <u>derivative</u> of f at the point x is defined to be the limit:

$$\frac{dy}{dx} = \lim_{h \to 0} \frac{f(x+h) - f(x)}{h} = \lim_{x_1 \to x} \frac{f(x_1) - f(x)}{x_1 - x}$$

Geometrically, this limit corresponds to the <u>slope of the tangent line</u> at the point x.

§ 2.1 - Tangent Lines & Derivatives   CAFÉ   Page 95

The derivative of a function is another function whose value at a point is the <u>slope of the tangent line</u> to the graph of the function at that point. The illustration shows segments of each tangent line at selected points for the function $y = x^2$. The derivative at each point is the slope of the corresponding tangent line. (In this case $\frac{dy}{dx} = 2x$.)

Thus in the illustration, the derivative of the parabola is the line.

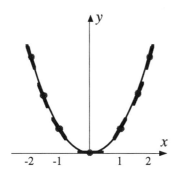
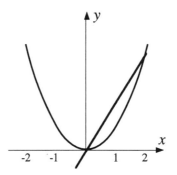

At each point a segment of the tangent line is attached. The slope of the tangent line is the derivative.

Plotting the slope of each tangent line gives a new function, the derivative. The line $y = 2x$ is the derivative of the function $y = x^2$

Example 4 - Derivative of a linear function.

In an earlier section, we noted that the slope of the line $y = mx + b$ is m. Since the tangent line at every point on the graph of our line is just the line itself, it follows that the derivative of $y = mx + b$ should be the constant function m. Show that the derivative of $y = mx + b$ is its slope m.

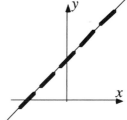

Solution

Every tangent line to a line is just the line itself!

$$\frac{dy}{dx} = \lim_{h \to 0} \frac{y(x+h) - y(x)}{h} = \lim_{h \to 0} \frac{[m(x+h) + b] - [mx + b]}{h} = \lim_{h \to 0} \frac{mh}{h} = m$$

The next sample problem uses the following algebraic result.

Combining Fractions Over Common Denominators
Fractions can be combined by bringing the numerators over a common denominator.

Don't Forget

$$\frac{a}{b} - \frac{c}{d} = \frac{ad - bc}{bd} \qquad \frac{a}{b} + \frac{c}{d} = \frac{ad + bc}{bd}$$

Example 5

Find the slope of the tangent line to the curve $f(x) = \frac{1}{x}$ at the point x.

Solution

The problem is really asking for the derivative of the function.
Using the expression for the limiting slope of secant lines we find:

$$m = \lim_{h \to 0} \frac{f(x+h) - f(x)}{h} = \lim_{h \to 0} \frac{\frac{1}{x+h} - \frac{1}{x}}{h}$$

The two fractions in the numerator can be combined by bringing the terms over a common denominator.

$$\frac{1}{x+h} - \frac{1}{x} = \frac{x - (x+h)}{x(x+h)} = -\frac{h}{x(x+h)}$$

The h in the denominator can now be canceled!

$$m = \lim_{h \to 0} \left(\frac{-h}{h \, x(x+h)} \right) = \lim_{h \to 0} \left(\frac{-1}{x(x+h)} \right) = \frac{-1}{x^2}$$

Thus the slope of the tangent line at the point x is $m = -\frac{1}{x^2}$.

In other words, $\frac{d}{dx}\left(\frac{1}{x}\right) = \boxed{-\frac{1}{x^2}}$.

Example 6 - Getting the h out!

Find the equation of the tangent line to the function $f(x) = \sqrt{x}$ at the point $x = 1$.

Solution

We will have to find the limit of the following difference quotient.

$$\frac{f(1+h) - f(1)}{h} = \frac{\sqrt{1+h} - \sqrt{1}}{h}$$

But how can we cancel out the h in the bottom when the h in the top is hiding inside the square root? What we need is some technique to get the h out of that root function. The key idea here is to recall the difference of squares identity.

$$(a-b)(a+b) = a^2 - b^2$$

So, if we multiply a difference by the corresponding sum, the result is a difference of squares and squaring is exactly what is needed to get the h out of the square root. Notice that our troublesome numerator is in fact a difference. Multiplying on top and bottom by the corresponding difference gives:

$$\left(\frac{\sqrt{1+h}-\sqrt{1}}{h}\right)\frac{\sqrt{1+h}+\sqrt{1}}{\sqrt{1+h}+\sqrt{1}} = \left(\frac{(\sqrt{1+h})^2-(\sqrt{1})^2}{h(\sqrt{1+h}+\sqrt{1})}\right) = \frac{h}{h(\sqrt{1+h}+\sqrt{1})}$$

This technique has successfully gotten the h out of the square root and the h in the denominator can now be canceled!

$$m = \lim_{h\to 0}\left(\frac{h}{h(\sqrt{1+h}+\sqrt{1})}\right) = \lim_{h\to 0}\frac{1}{(\sqrt{1+h}+\sqrt{1})} = \frac{1}{2}$$

Thus the slope of the tangent line is $\frac{1}{2}$. Since the tangent line passes through the point $(1,1)$ its equation can be written $\frac{y-1}{x-1} = \frac{1}{2}$ or $\boxed{y = \frac{x+1}{2}}$.

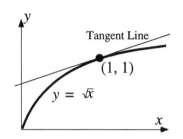

Can a Derivative Fail to Exist?

Most of the functions we have considered so far have derivatives throughout their domains. However, there are some functions for which the derivative does not exist at certain points. Since the derivative is defined as the slope of the tangent line there are two general ways in which a derivative may fail to exist. First the tangent line itself may not be defined at the point. Second, even if the tangent line exists, its slope may not. (This happens only if the tangent line is vertical.) We conclude this section with some examples that illustrate how a derivative may fail to exist.

The function on the right has a <u>jump</u> at the point P. As Q approaches P from the right, the limit of the secant lines is a horizontal tangent line. However, as Q approaches P from the left, the limit of the secant lines is a vertical tangent line. Since there is no reason to prefer either limiting line, we have to say that the tangent line is not defined here. Since the tangent line does not exist, the derivative does not exist at P.

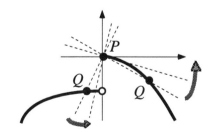

Derivative does not exist at a jump.

The function on the right has a <u>cusp</u> or <u>corner</u> at the point *P*. A different limiting line is obtained as *Q* approaches *P* from the left than that obtained as *Q* approaches *P* from the right. Since there is no reason to prefer either limiting line, we see that the tangent line and hence the derivative is not defined at *P*.

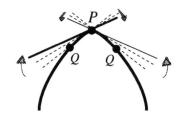

Derivative does not exist at a corner.

Even if a unique tangent line is defined at a point *P*, it is still possible that the derivative is not defined there. This happens if the <u>tangent line is vertical</u> since the slope is not defined for vertical lines.

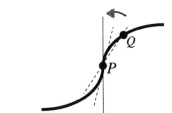

Derivative does not exist where tangent line is vertical.

Explicit examples where the derivative fails to exist are explored in the exercises.

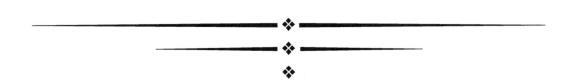

§ 2.1 - Tangent Lines & Derivatives   CAFÉ

WARMUP EXERCISES

1) The U.S. has seen some amazing average rates of change in temperature. The following data is from "The Weather Book" by Jack Williams (1992).
   a) On January 21, 1918, the temperature in Granville North Dakota went from $-33°F$ to $50°F$ in just 12 hours. What was the average rate of change in temperature during this interval?

   b) On January 19, 1892, Fort Assiniboine Montana reported a temperature increase from $-5°F$ to $37°F$ in just 15 minutes. What was the average rate of change in temperature during this interval?

2) The following world record rainfall data is from "The Weather Book" by Jack Williams. Find the average rate at which the rain was falling for each of these meteorological record holder. Express all rates in inches per hour.
   a) On Nov. 26, 1970, 1.50 inches of rain fell in one minute at Barot, Guadeloupe, West Indies.
   b) On June 22, 1947, 12 inches of rain fell in 42 minutes at Holt, Mo.
   c) On May 31, 1935, 22 inches of rain fell in 165 minutes at D'Hanis, Texas.
   d) On Aug. 1, 1977, 55.12 inches of rain fell in 10 hours at Muduocaidang, Inner Mongolia, China.

In each exercise find the derivative of the function $y = f(x)$ at the given point by evaluating the limit of the difference quotient $f'(a) = \lim_{h \to 0} \frac{f(a+h) - f(a)}{h}$.

3) $y = 1 + 2x$ at $(1,3)$ 4) $y = 1 + 2x^2$ at $(1,3)$

5) $y = 2 + x^3$ at $(1,3)$ 6) $y = 2 + \frac{1}{x}$ at $(1,3)$

7) $y = 2 + \sqrt{x}$ at $(1,3)$ 8) $y = \sqrt{10-x}$ at $(1,3)$

In each exercise find the derivative of the function $y = f(x)$ by evaluating the limit of the difference quotient $f'(x) = \lim_{h \to 0} \frac{f(x+h) - f(x)}{h}$. Then plot the original function and its derivative on the same graph.

9) $y = 1 + 2x$ 10) $y = 1 + 2x^2$

11) $y = 2 + x^3$ 12) $y = 2 + \frac{1}{x}$

13) $y = 2 + \sqrt{x}$ 14) $y = \sqrt{10-x}$

In engineering texts we often encounter many different symbols for the dependent and independent variables. This is done to make the equations more readily understood. Find the indicated derivative in each of the following exercises.

15) Find $\dfrac{dh}{dt}$ if $h = \dfrac{1}{2}gt^2$. (Distance of freefall: g is the gravitational constant.)

16) Find $\dfrac{dA}{dr}$ if $A = \pi r^2$. (Area of a circle)

17) Find $\dfrac{dx}{dy}$ if $x = y^2$. (Here x is a function of y instead of vice versa.)

18) Find $\dfrac{dC}{dV}$ if $C = \dfrac{2}{V}$. (Concentration is inversely proportional to volume.)

In each exercises find the equation of tangent line to the graph of the function at the indicated point.

19) $y = \dfrac{1}{2x+1}$ at $x = 1$ 20) $y = \sqrt{2x-1}$ at $x = 2$

21) $y = (1 - 2x)^2$ at $x = -1$ 22) $y = (x+1)^3$ at $x = 0$

INTERMEDIATE EXERCISES

23) The derivative does not exist at the points A, B, C and D in the graph. Indicate whether the failure is due to **1)** a jump in the function, **2)** a vertical tangent line or **3)** to a different limiting value as the point is approached from the left than from the right.

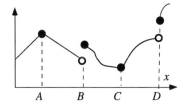

24) The derivative of each of the following functions <u>fails</u> to exist at the indicated point. Sketch the function and then indicate whether the failure is due to **1)** a jump in the function, **2)** a vertical tangent line or **3)** to a different limiting value as the point is approached from the left than from the right.

    a) $y = \sqrt[3]{x}$, at the origin.

    b) $y = |x^2 - 1|$ at $x = 1$.

    c) $f(x) = \begin{cases} x & \text{if } x \geq 0 \\ x^2 - 1 & \text{if } x < 0 \end{cases}$ at $x = 0$.

§ 2.1 - Tangent Lines & Derivatives   CAFÉ   Page 101

25) The illustration shows a graph of the function $y = f(x)$. If $f(0) = 0$ and $f'(x)$ exists, determine the limit of the ratio

$$\frac{\text{length of } PQ}{\text{length of } OQ}$$

as P approaches O along the curve.

(Hint: Interpret the limit in terms of a derivative.)

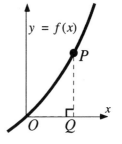

26) The graph shows the total number of cars that have crossed a bridge during morning rush hour.
   a) When is the traffic flow the greatest?
   b) When did the bridge close due to a traffic accident?
   c) When did the bridge reopen?
   d) About when does rush hour end?

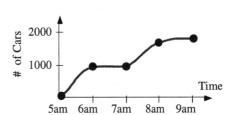

27) **Peak Load on Telephone System**
The graph shows the total number of calls handled by a local telephone company (in millions) on Mother's Day around supper time. The graph reveals that the telephone system operated at <u>maximum capacity</u> for most of the evening except for a temporary breakdown.

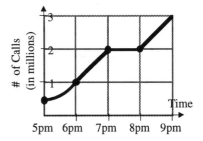

   a) At what time did the system break down and when was it repaired?
   b) What is the maximum rate at which the system can handle calls?
   c) Assuming that all the calls during the shutdown went to a competitor, how many calls were lost ? Assume the system would have been at the maximum rate.

(Engineers must design these systems so they can handle such peak loads.)

28) **Rainfall Gauges**
Networks of rainfall gauges are maintained both by the U.S. Geological Survey and the National Weather Service. Modern recording gauges are based on a small tipping bucket that tips for every 1/100th of an inch accumulated.

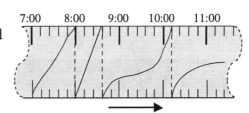

Each tip of the bucket is recorded on a strip chart that resets after each inch of rainfall (100 tips). A typical rain gauge chart is shown. Note the gauge recording resets at 8:00 PM, indicating that the first inch of rain has fallen.

- a) How many inches of rain fell in total?
- b) At what time was the intensity of rainfall the greatest?
- c) What was the greatest intensity of rainfall?
- d) Make a sketch of the accumulated total rainfall between 7 and 11 PM. Plot the derivative on the same graph. Hydrologists call the derivative of a cumulative rainfall curve a <u>hyetograph</u>.

29) The figure illustrates the accumulated cost of running a large air conditioner for four hours on a hot summer afternoon. The unit has three settings - off, medium and high.

- a) Graph the derivative of the cost function. Indicate any points where the derivative is not defined.
- b) During what time interval is the unit on high?
- c) During what time interval(s) is the unit on medium?
- d) During what time interval is the unit off?

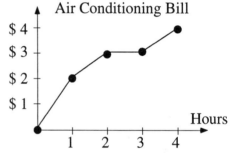

30) The illustration shows a graph of the function $y = f(x)$. If $f(0) = 0$, $f'(x)$ exists and OQ is tangent to the curve at O, determine the limit of the following ratio as P approaches O along the curve.
$$\frac{\text{length of } PS}{\text{length of } OR},$$
where S is the intersection point of OQ and PR.

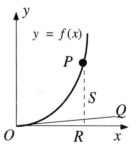

For each graph, plot the derivative on the same axes.
Points where the derivative is zero or doesn't exist are indicated by dots on the x-axis.

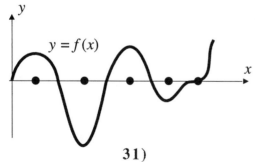
31)

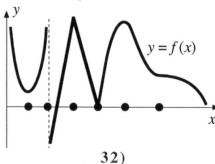
32)

For the function shown, plot its derivative on the same axes.

§ 2.1 - Tangent Lines & Derivatives   CAFÉ

Indicate all points where the derivative is zero.
Indicate any points where the derivative does not exist.

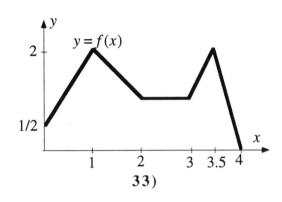

33)

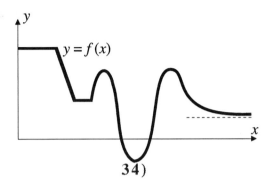

34)

NOTES

§ 2.2 Limits

Key Concepts

- Definition of a Limit
- One-Sided Limits
- Properties of Limits
- Limits at Infinity

The derivative is a crucial concept in all fields of science and engineering. Fortunately, there are rules that will enable us to calculate derivatives more efficiently than we did in the last section where we had to calculate each derivative by scratch starting with the difference quotient. To derive these differentiation rules however we will need to look a little more closely at the concept of limit. We were content to work with the limit concept in an intuitive fashion, but now it is time to nail the idea down. Thus let us set aside our discussion of derivatives for the moment and focus on the underlying concept of limit.

Intuitive Illustration of Limit

The concept of limit is actually very intuitive.
To denote that the function f has the limit L as x approaches a we write

$$\lim_{x \to a} f(x) = L$$

For the function shown on the left below it is easy to see that:

$$\lim_{x \to 1} f(x) = 1, \quad \lim_{x \to 2} f(x) = 2, \quad \lim_{x \to 3} f(x) = 3 \quad \text{and} \quad \lim_{x \to 4} f(x) = 4$$

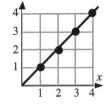
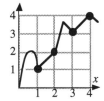
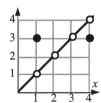

The function illustrated in the middle has the exact same limits at these four points despite the sudden changes in slope at all four points. Almost everyone can see these facts even without taking a calculus course. However, many people are surprised at first to hear that the discontinuous function shown on the right also has the same limits at these four points. In this case, the limit at 2 is 2 even though the function on the right is not defined at this point! Further, the limit at 1 is 1 even though the function on the right is has another value at this point! The counter-intuitive example of the function on the right underscores the need for a precise definition of the concept of limit.

The modern concept of limit is formalized in the following definition.

Formal Definition of Limit

The function f has the limit L as x approaches a (written $\lim\limits_{x \to a} f(x) = L$) if for every $\varepsilon > 0$ we can find a $\delta > 0$ such that $|f(x) - L| < \varepsilon$ for all points x satisfying $0 < |x - a| < \delta$.

Graphical Interpretation

The symbolic definition looks a little complicated, but the graphical interpretation is really very simple.
Suppose $\lim\limits_{x \to a} f(x) = L$ according to the above definition.

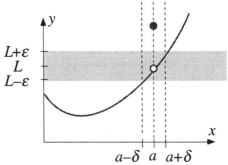

Then given **any** horizontal band centered on the limit value L, every point of the graph must lie within the band if x is chosen within a small enough vertical band centered on a. The point $x = a$ itself is exempted. At this point, the function may have any value or no value at all!

Significance of the Deleted Neighborhood

Why is the point $x = a$ excluded? Wouldn't it be more natural to insist that $f(a) = L$? It took hundreds of years after Newton and Leibnitz discovered calculus to formulate this definition of the limit concept and to realize the central importance of deleting the point $x = a$. The answer to the question as to why the point must be deleted can be seen in our working definition of the concept of instantaneous velocity in the last section. Recall how somewhat unsure of ourselves we ventured forth the idea of instantaneous velocity as:

$$v(t) = \lim_{\Delta t \to 0} \left(\frac{x(t + \Delta t) - x(t)}{\Delta t} \right)$$

At the time, we worried that this somehow might be a senseless equation because we cannot actually let $\Delta t = 0$ since division by zero is not defined. The above definition of limit nicely solves this problem by deleting the troublesome point. The notion of limit does not require that we know or even define the function at the point a, only that the output values approach some limit as x approaches a. Although the quotient $\left(\dfrac{x(t + \Delta t) - x(t)}{\Delta t} \right)$ is not defined at the point $\Delta t = 0$, the limit of the quotient as $\Delta t \to 0$ may exist since the limit does not involve the value of the function at this point!

---

§2.2 Limits

Example 1 - Volcanic Projectile

The height y in meters of a molten piece of lava ejected straight up into the air was observed to satisfy the quadratic function $y(t) = 100t - 4.9t^2$. Find the instantaneous speed at any time t. That is find the limit:

$$v(t) = \lim_{\Delta t \to 0} \left(\frac{y(t+\Delta t) - y(t)}{\Delta t} \right)$$

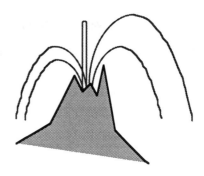

Solution

Substituting in the expression for y we find:

$$v(t) = \lim_{\Delta t \to 0} \left(\frac{\left(100(t+\Delta t) - 4.9(t+\Delta t)^2\right) - \left(100t - 4.9t^2\right)}{\Delta t} \right)$$

$$= \lim_{\Delta t \to 0} \left(\frac{100\Delta t - 9.8t\Delta t - 4.9\Delta t^2}{\Delta t} \right)$$

Notice at this stage we can cancel out the troublesome Δt in the denominator. There is no worry about division by zero because the limit point $\Delta t = 0$ is itself excluded from the definition of the limit concept. $v(t) = \lim_{\Delta t \to 0} (100 - 9.8t - 4.9\Delta t) = (100 - 9.8t)$

Thus the speed of the lava at time t is:

$$v(t) = (100 - 9.8t) \quad \text{(meters per second.)}$$

The above calculations are typical of how we will be using the limit concept. Occasionally, one must find the limit of a function using the formal definition presented at the beginning of this section. In the next problem, we see how to do one of these so-called epsilon-delta proofs.

Example 2 - Limit of the Absolute Value Function

The absolute value function has a sharp change in slope at $x = 0$. Nevertheless, as x approaches 0, the graph of the function gets very near to zero itself so it appears that the limit should exist and should equal 0. Show that:

$$\lim_{x \to 0} |x| = 0$$

Solution

Let $\varepsilon > 0$ be given and consider the horizontal band of width 2ε centered about $L = 0$. We must find a δ so that $||x| - 0| < \varepsilon$ for all points x satisfying $0 < |x - 0| < \delta$. For what value of δ does the condition $0 < |x| < \delta$ imply the condition $|x| < \varepsilon$? AHA! We can choose δ equal to ε itself.

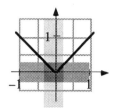

Limits by Direct Substitution

The limit of many engineering functions at a point is obtained just by evaluating the function at the same point. This notion will be made more precise in the next section when we study continuous functions. For now though let us simply note that the limit of linear functions, quadratic functions, polynomials, sines and cosines are all found just by evaluating these functions at that point. This is called the method of direct substitution. It is valid whenever the graph of the function is a single unbroken curve.

Example 3 - Limits by Direct Substitution

Find the limit as x approaches 0 for each of the following continuous functions.

a) $f(x) = x^2 + 3x + 1$ **b)** $g(x) = \sin(x)$ **c)** $h(x) = \cos(x)$

Solution

Since the graph of each of these functions is an unbroken curve we can evaluate the limits by direct substitution.

a) $\lim_{x \to 0} f(x) = f(0) = 1$ **b)** $\lim_{x \to 0} g(x) = \sin(0) = 0$ **c)** $\lim_{x \to 0} h(x) = \cos(0) = 1$

The next theorem allows us to calculate many limits with ease.

Limits Unlimited

Fortunately, knowing just a few basic limits, it is possible to find limits for almost all the functions that commonly arise in engineering applications.

Suppose we know the limit values for two functions at a certain point a.

$$\lim_{x \to a} f(x) = L_1 \qquad \lim_{x \to a} g(x) = L_2$$

§2.2 Limits   CAFÉ   Page 109

Intuitively this means for values of x near a (but excluding a) that the outputs of the function f are near L_1 and the outputs of the function g are near L_2. Clearly the outputs of the sum function $(f+g)(x) = f(x) + g(x)$ must be near the sum L_1+L_2. Similarly the outputs of the difference $(f-g)(x) = f(x) - g(x)$ must be near the difference L_1-L_2 and the outputs of the product function $(fg)(x) = f(x)g(x)$ must be near the product $L_1 L_2$. A little care is needed when we divide functions because division by zero is not defined. However, it is clear that the outputs of the quotient function $(f/g)(x) = \dfrac{f(x)}{g(x)}$ must be near the quotient $\dfrac{L_1}{L_2}$ provided the denominator $L_2 \neq 0$.

We can write these intuitive observations more formally as:

$$\lim_{x \to a} (f+g)(x) = L_1 + L_2$$

Limit of sum is sum of limits.

$$\lim_{x \to a} (f-g)(x) = L_1 - L_2$$

Limit of difference is difference of limits.

$$\lim_{x \to a} (fg)(x) = L_1 L_2$$

Limit of product is product of limits.

$$\lim_{x \to a} (f/g)(x) = L_1/L_2$$

Limit of quotient is quotient of limits. (provided $L_2 \neq 0$)

All the above statements can be rigorously proved by using the definition of limits (or equivalently the so-called epsilon-delta proofs). We omit the details.

❖

Example 4

Find the limit as x approaches 0 for the function:

$$f(x) = \sin(x)(x^2 + 3x + 2)$$

Solution

We note that $f(x)$ is a product of two functions, a sine and a quadratic polynomial. Since the limit of a product is the product of the limits we first find:

$$\lim_{x \to 0} \sin(x) = 0 \quad \text{and} \quad \lim_{x \to 0} (x^2 + 3x + 2) = 2$$

Thus the desired limit is:

$$\lim_{x \to 0} f(x) = \lim_{x \to 0} \sin(x) \cdot \lim_{x \to 0} (x^2 + 3x + 2) = 0 \times 2 = 0$$

❖

Example 5

The limits of two functions f and g at the point $x = 2$ are:

$$\lim_{x \to 2} f(x) = 100 \qquad \lim_{x \to 2} g(x) = 10$$

Find the limit for each of the following functions as x approaches 2.

a) $(f+g)(x) = f(x) + g(x)$ 　　b) $(f/g)(x) = \dfrac{f(x)}{g(x)}$

Solution

a) The limit of the sum is the sum of the limits.
$$\lim_{x \to 2} (f+g)(x) = \lim_{x \to 2} f(x) + \lim_{x \to 2} g(x) = 100 + 10 = 110$$

b) The limit of the quotient is the quotient of the limits.

$$\lim_{x \to 2} (f/g)(x) = \dfrac{\left(\lim_{x \to 2} f(x)\right)}{\left(\lim_{x \to 2} g(x)\right)} = \dfrac{100}{10} = 10$$

❖

In the last example we used the fact that the limit of the quotient is the quotient of the limits provided the limit of the denominator is not zero. In the next example, we consider a case where the limit of the denominator is zero.

Example 6

Find the limit as x approaches zero of the function $f(x) = \dfrac{x}{\sqrt{x+1} - 1}$.

Solution

Here is a case where we cannot evaluate the limit by using the fact that the limit of the quotient is the quotient of the limits because the limit of the denominator is zero!

 Division by zero is undefined!
Don't Forget

The square root in the denominator can be handled by the following technique.
$$\lim_{x \to 0} \left(\dfrac{x}{\sqrt{x+1} - 1}\right) = \lim_{x \to 0} \left(\dfrac{x}{\sqrt{x+1} - 1} \cdot \dfrac{\sqrt{x+1} + 1}{\sqrt{x+1} + 1}\right) = \lim_{x \to 0} (\sqrt{x+1} + 1) = \sqrt{0+1} + 1 = 2$$

❖

One-Sided Limits

In some applications, the two-sided limits we have been considering may not exist but the individual one-sided limits do. In the adjacent graph, we can see that the illustrated function has two-sided limits at $x = 3$ and $x = 4$.

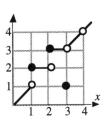

$$\lim_{x \to 3} f(x) = 3 \quad \text{and} \quad \lim_{x \to 4} f(x) = 4$$

However, the two-sided limits do not exist at $x = 1$ and $x = 2$. Nevertheless, one-sided limits clearly exist at each of these points. As we approach $x = 1$ from the left, the values of the function approach the limiting value $L = 1$. However, as we approach $x = 1$ from the right, the values of the function approach 2. To denote these one-sided limits we write:

$$\lim_{x \to 1^-} f(x) = 1 \quad \text{and} \quad \lim_{x \to 1^+} f(x) = 2$$

Similarly, the one-sided limits at the point $x = 2$ are:

$$\lim_{x \to 2^-} f(x) = 2 \quad \text{and} \quad \lim_{x \to 2^+} f(x) = 3$$

There are two important points to remember about one-sided limits.
First the two-sided limit at an interior point $x = a$ in the function's domain exists if and only if each of the one-sided limits exists at this point and have the same value.

$$\lim_{x \to a} f(x) = L \quad \text{if and only if} \quad \lim_{x \to a^+} f(x) = L \quad \text{and} \quad \lim_{x \to a^-} f(x) = L$$

Secondly, many functions are only defined on a subinterval of the real axis. For example the function $f(x) = \sqrt{x}$ is only defined on the interval $x \geq 0$. In such cases, we define the limit of the function at the endpoints of the interval by the corresponding one-sided limit. Thus, we may write $\lim_{x \to 0} \sqrt{x} = \lim_{x \to 0^+} \sqrt{x} = 0$.

Example 7 - Floor Function

Given a real input x, the floor function is defined as the greatest integer not exceeding x. Thus $\lfloor 2.9 \rfloor = 2$, $\lfloor 3 \rfloor = 3$ and $\lfloor -3.1 \rfloor = -4$.

From the graph, we can see that the two-sided limit of the floor function does not exist at any integer n. However the one-sided limits exist at all points. For any integer n the one-sided limits are:

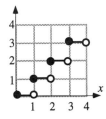

$$\lim_{x \to n^-} \lfloor x \rfloor = n - 1 \qquad \lim_{x \to n^+} \lfloor x \rfloor = n$$

Limits at Infinity

In many engineering problems, we need to know the value that a quantity approaches after a very long time. What is the limiting velocity of a marine parachuting into a trouble spot or the limiting speed of a particle being accelerated in the planned Superconducting Supercollider? These questions are different from the previous limits we have been looking at because we want to know the limit after a very long time - not at a particular fixed time.

Example 8

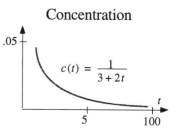

Distilled water is added to a salt water solution at a constant rate, in such a way that the concentration of the salt solution is given by the function $c(t) = \dfrac{1}{3+2t}$. If the dilution with water continues for a very long time t, then the concentration will decrease towards zero. Indeed, the graph reveals that as t becomes very large, the concentration decreases towards the t-axis and gets closer and closer to zero, although it never reaches this value. To represent a situation like this mathematicians use the notation $\lim\limits_{t \to \infty} c(t) = 0$.

Of course a function may approach other limits L and in general we have the definition.

Definition: Limits at Infinity

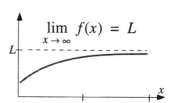

Let $f(x)$ be a function. Then we say that a number L is the limit of the function f as x approaches infinity provided the values of $f(x)$ can be made arbitrarily close to L by making x sufficiently large.

The limit at infinity is denoted by the notation: $\lim\limits_{x \to \infty} f(x) = L$

Limits at minus infinity are similarly defined and are denoted by $\lim\limits_{x \to -\infty} f(x) = L$.

For example, $\lim\limits_{x \to \infty} \dfrac{1}{2^x} = 0$ and $\lim\limits_{x \to -\infty} 2^x = 0$.

Example 9

Recall that a rational function is a quotient of two polynomials. Find the limit of the rational function $f(x) = \dfrac{2x}{1+x}$ as x approaches infinity.

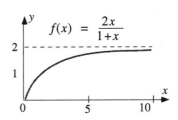

Solution

Unfortunately, the limits of both the numerator and denominator for $f(x) = \dfrac{2x}{1+x}$ are unbounded as x approaches infinity.

However, if we first divide the top and bottom by x, the resulting numerator and denominator now have finite limits at infinity. Since the limit of a quotient is the quotient of the limits, we find:

$$\lim_{x \to \infty} \left(\frac{2x}{1+x}\right) = \lim_{x \to \infty} \left(\frac{2}{\frac{1}{x}+1}\right) = \frac{\lim\limits_{x \to \infty} 2}{\lim\limits_{x \to \infty} \left(\frac{1}{x}+1\right)} = \frac{2}{1} = 2$$

Limits of Rational Functions

A similar strategy works for any rational function in which the degree n of the polynomial in the numerator does not exceed the degree m of the denominator. In this case, dividing both top and bottom by the monomial term x^m (leading monomial in the denominator) will result in finite limits for the numerator and denominator at infinity. We can then use the fact that the limit of a quotient is the quotient of the limits to find the limit at infinity of the rational function. Here is another example involving rational functions.

Example 10

Find the limit of the rational function
$f(x) = \dfrac{x^2 - 2x - 1}{3x^2 + 1}$ as x approaches infinity.

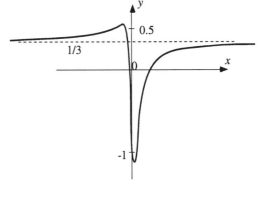

Solution

Both the numerator and the denominator have degree 2.
To find the limit at infinity, we first divide the top and bottom by x^2, the leading monomial term in the denominator. The resulting numerator and denominator now have finite limits at infinity.

$$\lim_{x \to \infty} \frac{(x^2 - 2x - 1)}{3x^2 + 1} = \lim_{x \to \infty} \frac{\left(1 - \frac{2}{x} - \frac{1}{x^2}\right)}{3 + \frac{1}{x^2}} = \frac{1 - 0 - 0}{3 + 0} = \frac{1}{3}$$

Thus, the limit as x approaches infinity is 1/3.

Infinite Limits

If a function increases without bound as x approaches infinity we write $\lim_{x \to \infty} f(x) = \infty$. Of course, ∞ is not a real number and we are only using the notation to symbolize the fact that the function's values become arbitrarily large. (A similar convention holds if a function's values increase without bound in the negative sense.)
For example:

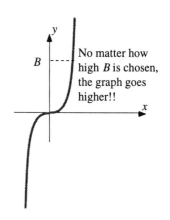

No matter how high B is chosen, the graph goes higher!!

$$\lim_{x \to \infty} x^3 = +\infty \qquad \lim_{x \to -\infty} x^3 = -\infty$$

Example 11

a) Limits of Exponentials

Let a be a positive real number and consider the function a^x.

Then,

$$\lim_{x \to \infty} a^x = 0 \text{ if } a < 1 \text{ while } \lim_{x \to \infty} a^x = \infty \text{ if } a > 1$$

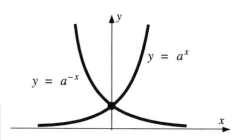

Also,

$$\lim_{x \to -\infty} a^x = \infty \text{ if } a < 1 \text{ while } \lim_{x \to -\infty} a^x = 0 \text{ if } a > 1$$

b) Limits of Powers

Let a be a positive real number and consider the function x^a.
Then $\lim_{x \to \infty} x^a = \infty$ and $\lim_{x \to \infty} \frac{1}{x^a} = 0$.

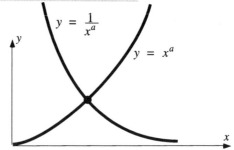

§2.2 Limits

WARMUP EXERCISES

1) Find the limit of the illustrated functions at the points 0, 1, 2 and 3.
   For the endpoints, evaluate the limit using the appropriate one-sided limit.
   At each point where a two-sided limit does not exist, state any one-sided limits that do exist.

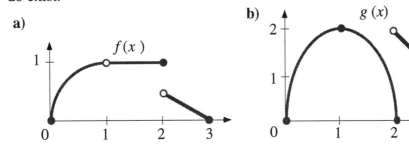

2) The graphs illustrate a function whose limit at a is L. Two "challenges" are illustrated as horizontal bands centered on the limit L. Draw the maximum possible deleted neighborhood for which the graph lies entirely within the horizontal band.

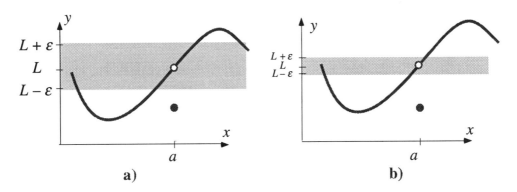

3) The graphs illustrate a function which does not have a limit at a. Two "challenges" are illustrated as horizontal bands centered on the L. Draw the maximum possible deleted neighborhood (centered on a) for which the graph lies entirely within the horizontal band if this is possible. Determine for which illustration no such deleted neighborhood can be found.

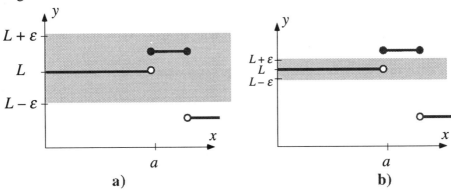

4) Given a real number x, the ceiling function $\lceil x \rceil$ is defined as the smallest integer greater than or equal to x.
Thus $\lceil 2.9 \rceil = 3$, $\lceil 3 \rceil = 3$, $\lceil -2.2 \rceil = -2$.
Sketch the graph of the ceiling function $\lceil x \rceil$, and compare it to that of floor function in the text.

5) Find the following integers.

$\lfloor 2.718 \rfloor \qquad \lfloor -\pi \rfloor \qquad \lfloor 0 \rfloor \qquad \lfloor -1.1 \rfloor$

$\lceil 2.718 \rceil \qquad \lceil -\pi \rceil \qquad \lceil 0 \rceil \qquad \lceil -1.1 \rceil$

6) Sketch the following functions on the interval from -5 to 5.

$f(x) = \lfloor x \rfloor + \lceil x \rceil \qquad g(x) = \lceil x \rceil - \lfloor x \rfloor \qquad h(x) = \lceil x \rceil - \lfloor -x \rfloor$

7) We have seen how every function can be written as the sum of an even and an odd function according to the decomposition $f(x) = \dfrac{f(x) + f(-x)}{2} + \dfrac{f(x) - f(-x)}{2}$.

Decompose both the floor function and the ceiling function into their respective even and odd parts and draw the resulting four functions. Clearly label each graph. Are any of these functions the same?

8) Suppose we know the limit values for two functions at a certain point a. That is:

$$\lim_{x \to a} f(x) = 10 \qquad \lim_{x \to a} g(x) = 4$$

Find the limit for each of the following functions as x approaches a.

a) $(f+g)(x) = f(x) + g(x)$ \qquad b) $(f-g)(x) = f(x) - g(x)$

c) $(fg)(x) = f(x)g(x)$ \qquad d) $(f/g)(x) = \dfrac{f(x)}{g(x)}$

9) The limit of a certain function f as x approaches 0 is known to be 10. That is $\lim_{x \to 0} f(x) = 10$. Let g be the function $g(x) = 2 + x^2$.
Find the limit as x approaches 0 for each of the functions
a) $f+g$, b) $f-g$, c) fg, d) $\dfrac{f}{g}$ and e) $\dfrac{g}{f}$

Evaluate the following limits by direct substitution.

10) $\lim_{x \to 0} \tan(x)$ \qquad 11) $\lim_{x \to 1} \dfrac{(x^2 + 7x + 1)}{2x + 1}$

§2.2 Limits   CAFÉ   Page 117

12) $\lim_{x \to 0} \dfrac{(x^2 + 7x + 1)}{2x + 1}$ 13) $\lim_{x \to \pi} \dfrac{\cos(x)}{2x + 1}$

14) What goes wrong if we try to find the following limits by direct substitution?

   a) $\lim_{x \to \pi} \tan\left(\dfrac{x}{2}\right)$ b) $\lim_{x \to 1} \dfrac{(x^2 - 2x + 1)}{x - 1}$

INTERMEDIATE EXERCISES

15) Recall the floor function was defined in the text. Prove that:

   $-\lceil -x \rceil = \lfloor x \rfloor$, for all real x

   Therefore, any formula expressed in terms of floor function can also be equivalently expressed in terms of the ceiling function.

16) Assume the cost of sending a parcel is 10¢ per ounce or fraction thereof. Any parcel not exceeding 20 pounds may be sent by first class mail. Express the cost C of sending a parcel as a function of its weight w in ounces and clearly state the allowed domain of the function. Note that 1 pound = 16 ounces.

17) **Taxi Fare**
   The fare charged by a Philadelphia taxi company is $1.80 for the first 1/6 mile and $.30 for each additional 1/6 mile. Express the fare $F(x)$ as a function of the miles traveled. Assume the meter favors the cab driver. That is $F(0) = \$1.80$, $F(1/6) = \$2.10$ etc.

18) A second taxi company charges three dollars for the first mile and a quarter for each succeeding quarter mile (or fraction thereof). Express the cost C as a function of the distance x traveled. Give both a piecewise definition and a definition using a combination of the unit step function and floor or ceiling functions. Assume the meter favors the company at the increment points.

19) A long distance telephone company charges customers one dollar for the first minute and then 20¢ for each additional minute (or fraction thereof) on calls between cities A and B. Express the cost C (in cents) of the call as a function of the time t (in minutes). Assume the timer favors the company, that is: $C(0) = 100¢$, $C(1) = 120¢$ etc.

Evaluate the following limits.

20) $\lim_{x \to 4} \left(\dfrac{x^2 - 16}{x - 4}\right)$ 21) $\lim_{x \to 1} \left(\dfrac{x^2 - 3x + 2}{x - 1}\right)$

22) $\lim_{h \to 0} \dfrac{\sqrt{1 + h} - 1}{h}$ 23) $\lim_{h \to 1} \dfrac{h - 1}{\sqrt{h^2 - 1}}$

24) $\lim_{x \to 0} \left(\frac{1}{x} - \frac{2}{x(x+2)} \right)$

25) $\lim_{x \to 1} \frac{\frac{1}{x} - 1}{x - 1}$

26) $\lim_{x \to 1} \frac{x - 1}{\sqrt[3]{x} - 1}$ (Hint: let $\sqrt[3]{x} = y$ and rewrite the limit in terms of y.)

Evaluate the following limits at infinity.

27) $\lim_{x \to \infty} (\sqrt{x+1} - \sqrt{x})$

28) $\lim_{x \to \infty} (\sqrt{x+k} - \sqrt{x})$

29) $\lim_{x \to \infty} \left(\frac{\sqrt{x+a} - \sqrt{x}}{\sqrt{x+b} - \sqrt{x}} \right)$

30) $\lim_{x \to \infty} \left(\frac{\sqrt{x+a} - \sqrt{x+c}}{\sqrt{x+b} - \sqrt{x+d}} \right)$, $b \ne d$

31) $\lim_{x \to \infty} \sqrt{x} (\sqrt{x+1} - \sqrt{x})$

32) $\lim_{x \to \infty} \frac{2x^2 + 1}{\sqrt{3x^4 + x^2 + 1}}$

33) $\lim_{t \to \infty} \frac{5t^3 + 1}{2t^3 - t^2 - 4}$

34) $\lim_{x \to \infty} \frac{1 + 2x + 3x^2}{3 + 2x + x^2}$

35) $\lim_{x \to \infty} \frac{1 + 2x + 3x^2}{3 + x^3}$

36) $\lim_{x \to \infty} \left| \sin \frac{x-1}{x+1} \right|$

37) $\lim_{x \to \infty} \left\lfloor \frac{x-1}{x+1} \right\rfloor$

38) $\lim_{x \to -\infty} \left\lfloor \frac{x-1}{x+1} \right\rfloor$

Evaluate the following one-sided limits.

39) $\lim_{x \to 1^+} \frac{1-x}{|1-x|}$

40) $\lim_{x \to 1^-} \frac{1-x}{|1-x|}$

41) $\lim_{x \to 0^+} \frac{|x|}{x}$

42) $\lim_{x \to 0^-} \frac{|x|}{x}$

43) $\lim_{x \to 1^+} \lceil x \rceil$

44) $\lim_{x \to 1^-} \lfloor x \rfloor$

45) $\lim_{x \to 0^+} \frac{\lfloor x \rfloor}{x}$

46) $\lim_{x \to 0^-} \frac{\lfloor x \rfloor}{x}$

47) $\lim_{x \to 0^+} 2^{-1/x}$

48) Give an example of two functions f and g such that neither has a limit as x approaches 0 but their sum does.

49) Give an epsilon-delta proof that $\lim_{x \to 1} |x-1| = 0$.

50) Give an epsilon-delta proof that $\lim_{x \to 1} x^2 = 1$.

51) A very large tank contains 1000 liters of salt water with a concentration of 20 grams of salt per liter. Pure water is then pumped into the tank at a rate of 30 liters/min.
  a) Determine the concentration of salt after t minutes.
  b) What happens to the concentration as $t \to \infty$.
The tank is assumed to have an infinite capacity.

52) A very large tank contains 1000 liters of water at 25°C.
Hot water at a temperature of 90°C pours into the tank at a rate of 30 liters/min.
    a) Determine the temperature $T(t)$ of the water after t minutes.
    b) Find $\lim_{t \to \infty} T(t)$.

The Sandwich Theorem

Suppose we know the limit values for two functions at $x = a$ are the same. That is

$$\lim_{x \to a} f(x) = \lim_{x \to a} g(x) = L$$

Furthermore we know that function $h(x)$ is sandwiched by $f(x)$ and $g(x)$ i.e.,

$$g(x) \leq h(x) \leq f(x) \text{ for } x \text{ in a neighborhood of } a..$$

then $\lim_{x \to a} h(x) = L$.

Use The Sandwich Theorem to find the following limits.

53) $\lim_{x \to 0} x u(x)$ (Recall u is the unit step function.)

54) $\lim_{x \to \infty} \dfrac{\cos x}{x + \sin x}$ **55)** $\lim_{x \to \infty} \dfrac{\lfloor x \rfloor}{x}$ **56)** $\lim_{x \to 0} x \sin \dfrac{1}{x}$

57) $\lim_{x \to 0^+} x \lfloor \dfrac{1}{x} \rfloor$ **58)** $\lim_{x \to \infty} \dfrac{\lfloor x \rfloor + 10}{x + 10}$ **59)** $\lim_{x \to \infty} \dfrac{x - 5\cos x}{x + \sin x}$

60) A right triangle has base b and height h_1. An infinite sequence of nested right triangles is constructed inside the triangle by dropping perpendiculars as shown in the diagram.

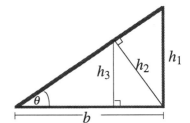

    a) Find an expression for the edge h_n.

    b) Find an expression for the infinite sum:
$$\lim_{n \to \infty} (h_1 + h_2 + \cdots + h_n)$$

(Hint: $1 + x + x^2 + \cdots + x^n = \dfrac{x^{n+1} - 1}{x - 1}$)

61) A sequence of numbers $a_1, a_2, a_3, \ldots a_n$ is generated by the following algorithm.

$a_1 = 3, \quad a_2 = \sqrt{a_1+3}, \quad a_3 = \sqrt{a_2+3}, \ldots, a_n = \sqrt{a_{n-1}+3}$

a) Find a_2, a_3 and a_4.   **b)** It is known that the limit $\lim_{n \to \infty} a_n$ exists. Determine its value.

62) Let $P = (x, x^2)$ be an arbitrary point on the parabola $y = x^2$ and let $A = (0,1)$ and $B = (1,0)$ be points along the coordinate axes. This geometry determines two triangles, $\triangle OAP$ and $\triangle OBP$.

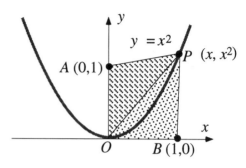

a) Determine the limit of

$\dfrac{\text{Area of } \triangle OBP}{\text{Area of } \triangle OAP}$ as point P approaches the origin at O along the parabola.

b) Determine the limit of $\dfrac{\text{Area of } \triangle OAP}{\text{Area of } \triangle OBP}$ as point P approaches infinity along the right branch of the parabola.

63) Consider the point $A = (0, a)$ located on the positive y-axis. Let $P = (x, x^2)$ be an arbitrary point on the parabola $y = x^2$ and let $B = (0, x^2)$ be the projection of P onto the y-axis. This geometry determines two triangles, $\triangle ABP$ and $\triangle BOP$.

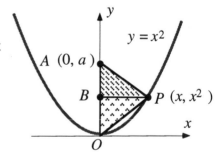

a) Determine $\lim_{x \to 0} \dfrac{\text{Area of } \triangle BOP}{\text{Area of } \triangle ABP}$.

b) Determine $\lim_{x \to \infty} \dfrac{\text{Area of } \triangle BOP}{\text{Area of } \triangle ABP}$.

64) PQ is tangent to the curve at P. Determine:

a) $\lim_{x \to 0} \dfrac{\text{area}\triangle POQ}{\text{area}\triangle POR}$

b) $\lim_{x \to 0} \dfrac{\text{length}PQ}{\text{length}OR}$

c) $\lim_{x \to 0} \dfrac{\text{length}PQ}{\text{length}PO}$

d) $\lim_{x \to 0} \dfrac{\text{length}PQ}{\text{length}PR}$

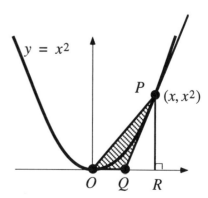

§2.2 Limits   CAFÉ   Page 121

65) A square is inscribed inside a circle of radius 1. Then a circle is inscribed inside the square and the process is continued to generate a sequence of nested squares and circles. The shaded area shown in the illustration resembles the petals of a beautiful flower. Noticing that the petals are arranged in rings we let A_n denote the area of the nth ring of petals. For example: $A_1 = \pi \cdot 1^2 - (\sqrt{2})^2 = \pi - 2$

a) Find the areas A_2 and A_3.

b) Find the total area of all the petals. That is find: $\lim_{n \to \infty} (A_1 + A_2 + \cdots + A_n)$.

66) Instead of a nested sequence of circles and squares as in the problem above, construct a triangle-circle sequence, where the first triangle is an equilateral with sides of length 1.

Find the total area of the 'petals'.

Bonus Problem

67) This problem asks you to explore a nested sequence as in the above problems but using other regular polygons instead of squares or triangles. What would the total area of the 'petals' be if a sequence of hexagons and circles were inscribed inside each other? What would the total area of the 'petals' be if a sequence of circles and regular polygons having k sides were inscribed inside each other?

NOTES

§2.3 Continuity of Functions

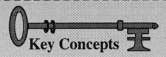
Key Concepts

- Definition of Continuity
- Max-Min Theorem
- Intermediate Value Theorem

Continuity

Intuitively, a continuous function is a function whose graph can be drawn with a single motion of the pen. The hand need not be lifted to skip or jump to new points and the graph must appear unbroken. This intuitive notion of continuity is sometimes called the 'moving pen test.' The formal definition of continuity is:

> **Definition of Continuity**
> A function f is continuous at a point a if $\lim_{x \to a} f(x) = f(a)$

Although this definition for continuity looks like a single condition, it implicitly involves checking three things.

1) The value $f(a)$ must be defined.

2) The limit $\lim_{x \to a} f(x)$ must exist.

3) The value of the function and its limit at the point a must be equal.

❖

Example 1

Test the illustrated function for continuity at the points 0, 1, 2 and 3.

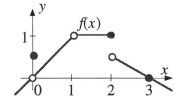

Solution

a) Discontinuous at $a = 0$.

The function is defined here $f(0) = 1/2$ so the first condition is satisfied. The limit also exists and is equal to 0. However the limit does not equal the function's value.

b) Discontinuous at $a = 1$.
The function is not defined here so the first condition is violated.

c) Discontinuous at $a = 2$.
The function is defined but the limit does not exist.

d) Continuous at $a = 3$.
The function is defined, the limit exists and both equal 0.

Applications of Continuity

Continuous functions have many important and useful properties. In this section, we will explore two of the more important applications of continuity. First we will look at the Intermediate Value Theorem and then at the Max-Min Theorem.

> **Intermediate Value Theorem (IVT)**
>
> If f is a continuous function defined on the closed interval $[a, b]$, then the outputs of f include every value between the outputs at a and b.

The word intermediate means between. The theorem guarantees that for any point y between $f(a)$ and $f(b)$, there is some point x in the closed interval $[a, b]$ such that $f(x) = y$.
The proof of the theorem may be found in more advanced calculus textbooks.

Graphical Interpretation

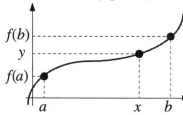

Graphically, the Intermediate Value Theorem means that every horizontal line between the lines $y = f(a)$ and $y = f(b)$ must intersect the graph of f for some x between a and b.

The necessity that the function be continuous in order for the Intermediate Value Theorem to hold is illustrated in the adjacent figure. The unit step function does not assume any of the values between its only outputs 0 and 1.

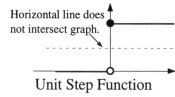

Example 2

Consider the quadratic polynomial $f(x) = 1 + x^2$. By substitution we see that $f(0) = 1$ and $f(1) = 2$. Find an input x in the interval $[0, 1]$ for which the output is $y = 3/2$.

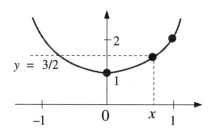

Solution

The Intermediate Value Theorem guarantees that such an input x must exist since all polynomials are continuous and $3/2$ is between the outputs 1 and 2. Solving for that input x which has output $3/2$ we find that $f(x) = 3/2$ implies $1 + x^2 = 3/2$.

Rearranging gives: $\quad\quad\quad x^2 = 1/2 \quad\quad$ so $\quad\quad x = \pm 1/\sqrt{2}$.

Of the two possible inputs, only $\boxed{x = +1/\sqrt{2}}$ lies in the required interval [0, 1].

Example 3 - Pepperoni Dilemma

Spike and his friend Cray are trying to figure out how to slice a pepperoni pizza through its center so that they each get the same amount of pepperoni. Is this possible? (Assume the four pepperonis shown are identical in size.)

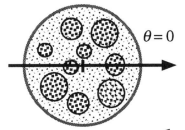

Solution

An oriented cut through the center at any angle θ divides the pizza into two slices, one to the left of the cut and one to the cut.

Let $D(\theta)$ denote the difference between the amount of pepperoni on the left minus the amount on the right. When $\theta = 0$, only one of the four pepperonis is to the left while three are to the right. Thus:

$$D(0) = 1 - 3 = -2$$

Simply reversing the direction of the cut exchanges which side is considered on the left and which is on the right. Thus:

$$D(\pi) = 3 - 1 = 2$$

We would like the difference to be 0 which is intermediate between the two values −2 and 2. Since the difference $D(\theta)$ is a continuous function of the angle θ, there is some angle where $D(\theta) = 0$ by the Intermediate Value Theorem, which proves that it is possible to divide the pizza fairly. Notice a cut at about 90 degrees divides the pizza fairly.

In the above example, our search for a fair sharing of the pepperoni pizza involved finding where the difference function $D(\theta)$ was zero. Asserting the existence of the zeroes of a function is an important application of the Intermediate Value Theorem. (A zero of a function is also called a <u>root</u> of the function.)

Existence of Zeroes   (Corollary to IVT)

If f is a continuous function such that $f(a) < 0$ and $f(b) > 0$, then there <u>exists</u> some point x between a and b such that $f(x) = 0$.

Geometrically this means that if a function's graph contains points both above and below the x-axis, it must cross this axis if it is continuous. (A function is continuous if it is continuous at all points in its domain.) This result is exploited in many computer programs that calculate the roots of functions numerically.

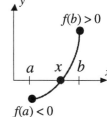

Example 4

Show that the polynomial $p(x) = x^7 + x + 1$ has a root c between -1 and $+1$. Notice it could be very hard to actually find the root in this case.

Solution

The values of the polynomial at the two endpoints are $p(-1) = -1$ and $p(1) = 3$. By the corollary to the Intermediate Value Theorem, there is some point c between -1 and 1 where $p(c) = 0$. Notice we have only shown that there <u>exists</u> a root. We have not found its value!

Another important consequence of continuity is the following theorem about the extreme values of a continuous function on a closed interval.

Max-Min Theorem

If f is a continuous function on the closed interval $[a, b]$, then f <u>assumes</u> both an absolute maximum M and an absolute minimum m on this interval. That is $m \leq f(x) \leq M$ for all x in the interval $[a, b]$ and $f(c) = m$ and $f(C) = M$ for some points c and C in this interval.

(This theorem is also called the Extreme Value Theorem.)

Graphical Interpretation

The max-min theorem means that the graph of the function lies between the lines $y = M$ and $y = m$ on the interval $[a,b]$ and further that the graph actually touches each of these lines.

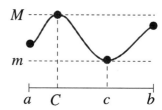

§2.3 Continuity of Functions

Counterexamples

The hypotheses of the max-min theorem require both that the function be continuous and that the interval be closed. The conclusion need not hold if either hypothesis is violated. For example, the line $y = x$ restricted to the open interval $(0,1)$ assumes neither a maximum nor a minimum on that interval (left illustration).

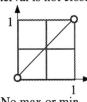
No max or min.

Interval is not closed. Function is discontinuous.
No max or min.

In the illustration on the right, the function has been given values at these endpoints so it is now defined on the closed interval, but it is discontinuous and again does not assume a maximum or a minimum.

Example 5

The max-min theorem guarantees that the quadratic polynomial $y = x^2 + 4x$ assumes a maximum M and a minimum m on the domain $[-3,3]$. Find M and m and locate the input points C and c where they occur.

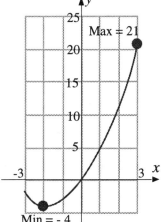

Solution:

Completing the square we find:

$$y = x^2 + 4x = (x+2)^2 - 4$$

Thus the minimum of the quadratic is $\boxed{m = -4}$ and occurs inside the interval $[-3,3]$ at the point $\boxed{c = -2}$. The maximum M will occur at one of the endpoints - but which? The value at the left endpoint is $y(-3) = -3$ and the value at the right endpoint is $y(3) = 21$. Thus the maximum is $\boxed{M = 21}$ and occurs at the right endpoint $\boxed{C = 3}$.

Combining Continuous Functions

When continuous functions are combined arithmetically using any of the operations $\{+, -, \times, /\}$, one almost always gets back a continuous function. The product, sum and difference of continuous functions is always continuous. Since division by zero is undefined, the quotient of two functions is only continuous if the denominator is not zero.

(Every family has a black sheep and the family of arithmetic operations is no exception!)

The composition of two continuous functions is also continuous.

Example 6

Consider the two continuous functions $f(x) = 1 - x^2$ and $g(x) = \sin(x)$. Discuss the continuity of the functions $f+g$, $f-g$, fg, f/g and g/f.

Solution

The original functions f and g are continuous so $f+g$, $f-g$ and fg are also continuous. Thus we only need worry about the two quotients f/g and g/f and these will be continuous wherever the denominators are not zero.

The zeroes of $f(x) = 1 - x^2$ are -1 and $+1$. The zeroes of $g(x) = \sin(x)$ are all the integer multiples of π, that is $x = n\pi$ where n is an integer.

The quotient $(g/f)(x) = \dfrac{\sin(x)}{1-x^2}$ is not defined at -1 and $+1$ where the denominator is zero, and hence cannot be continuous there.

The quotient $(f/g)(x) = \dfrac{1-x^2}{\sin(x)}$ is not defined at $x = n\pi$ where the denominator is zero, and hence cannot be continuous at any of these integer multiples of π.

Discreteness and Discontinuity

For many years there was a tendency to view discontinuous functions as "artificial" and their calculus was largely ignored. The discovery that charge comes in discrete units and the realization that energy, angular momentum and many other physical properties are also discretely packaged due to quantum mechanics has completely reversed this bias. Some cosmologists even suggest that the structure of space-time itself may be discontinuous.

Engineers were quick to exploit these scientific discoveries about the discontinuous nature of our universe. The discrete nature of charge and energy is central in the design of such machines as mass spectrometers. Charge quantization has been exploited in some ink jet and photocopier technologies and the discrete nature of atomic energy levels leads to the laser. In digital technology, the exploitation and processing of discontinuous functions are central to recent advances in computing, signal processing and transmission, data compression, sound and videophones.

§2.3 Continuity of Functions

1) Analyze the continuity of the illustrated functions at the points 0, 1, 2 and 3 using the 'moving pen test'.

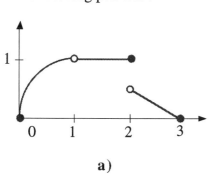

a)

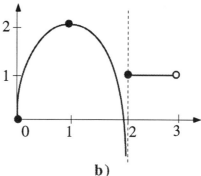

b)

2) Test the continuity of the functions illustrated above at each of the points 0, 1, 2 and 3 by formally verifying the three conditions required for continuity. Use the appropriate one-sided limits at the endpoints 0 and 3.

Identify all the discontinuities in the following functions.

3) $f(x) = 1 - x^2$

4) $f(x) = \dfrac{1}{1 - x^2}$

5) $f(x) = \dfrac{x + 1}{x^2 + 3x + 2}$

6) $f(x) = \tan(x)$

7) The function $f(x) = 1/x$ is positive for $x > 0$ and negative for $x < 0$, yet its graph never actually crosses the x-axis. Why does this not violate the Intermediate Value Theorem?

8) Draw a graph of a function satisfying $f(0) = 0$ and $f(1) = 5$ but which never assumes any of the intermediate values between 0 and 5.

Max-Min Theorem

9) Draw a graph of a function whose domain is the closed interval [0, 4] but which does not assume a maximum or a minimum on this interval. (Hint: consider a discontinuous function.)

10) Draw a graph of a continuous function whose domain is the open interval (0, 4) but which does not assume a maximum or a minimum on this interval.

11) A continuous function defined on a closed interval must assume both a maximum and a minimum there. These extreme values could be assumed in the interior of the interval or at an endpoint. Thus there are four possibilities.
    i.   Both extrema are interior to the interval.
    ii.   Both extrema are at endpoints.
    iii.   The maximum is at an endpoint but the minimum is interior.
    iv.   The minimum is at an endpoint but the maximum is interior.
    Draw functions illustrating each of these four possibilities.

INTERMEDIATE EXERCISES

For what value of the constant k are the following piecewise functions continuous?

12) $f(x) = \begin{cases} x & \text{for } x \leq 1 \\ kx+1 & \text{for } x > 1 \end{cases}$

13) $f(x) = \begin{cases} x+1 & \text{for } x \leq 0 \\ x+k & \text{for } x > 0 \end{cases}$

14) $f(x) = \begin{cases} x^2 & \text{for } x \leq 2 \\ kx & \text{for } x > 2 \end{cases}$

15) $f(x) = \begin{cases} x + \sin x & \text{for } x \leq \pi \\ kx & \text{for } x > \pi \end{cases}$

16) Consider the quadratic polynomial $f(x) = x^2 - 2x$. The outputs at -1 and 1 are $f(-1) = 3$ and $f(1) = -1$. For each part, find an input x in the interval $[-1, 1]$ whose output is equal to the given intermediate value.

    a) 0   b) 1   c) 2

17) Consider the function $f(x) = 10^{2x} + 10^x$. The outputs at 0 and 1 are $f(0) = 2$ and $f(1) = 110$. For each part, find an input x in the interval $[0, 1]$ whose output is equal to the given intermediate value.

    a) 10   b) 50   c) 100

In the following exercises show that each function has a root between -1 and 1.

18) $p(x) = x^{11} + x^7 + 1$ 19) $p(x) = x^5 - x^4 + 1$
20) $f(x) = \sin(\pi x^2) + x + 1/2$

The Max-min theorem implies that each of the following continuous functions assumes both an absolute maximum M and an absolute minimum m on the interval $[-1, 1]$. Find M and m using the method of completing the square if the function is a quadratic polynomial.

21) $y = 3x + 1$ 22) $y = \sin(x/2) + 1$ 23) $y = x^2 + 2x$

24) $y = -x^2 + 2x$ 25) $y = x^2 + x$ 26) $y = -x^2 + x$

Removable Discontinuities
(Extending the domain of a function.)

The function $f(x) = \dfrac{x^2}{|x|}$ is not defined at $x = 0$ because the denominator is zero. Thus the function is not continuous at the origin. However, the graph of the function reveals that the limit of the function at $x = 0$ exists and is equal to 0.

Removable Discontinuity

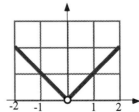

§2.3 Continuity of Functions   CAFÉ   Page 131

If we extended the domain of the function by defining $\tilde{f}(0) = 0$, the resulting function would then be continuous everywhere.

$$\tilde{f}(x) = \begin{cases} \dfrac{x^2}{|x|} & \text{for } x \neq 0 \\ 0 & \text{for } x = 0 \end{cases}$$

The tilde over the f denotes that this is a new function. It is not the same as the original function f because the domains are different. In fact, it is clear that $\tilde{f}(x) = |x|$.

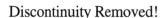
Discontinuity Removed!

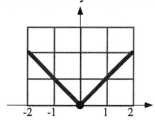
Just fill in the dot!

Each of the following rational functions is not defined wherever its denominator is zero. Thus each of these points is a point of discontinuity. However, if the rational function approaches a limit at any of these points, the discontinuity can be removed by defining the value of the function at this point to be the limit. Find all points of discontinuity and remove any discontinuities at which the function has a limit by extending the domain as in the above example.

27) $f(x) = \dfrac{x^3}{|x|}$

28) $f(x) = \dfrac{x^2 + 2x + 1}{x + 1}$

29) $f(x) = \dfrac{x^2 + 3x + 2}{x + 1}$

30) $f(x) = \dfrac{x + 1}{x^2 + x}$

31) $f(x) = \dfrac{x^2 - 4}{x^2 + 3x + 2}$

32) $f(x) = \dfrac{x + 2}{x^2 + 3x + 2}$

ADVANCED EXERCISES

33) **Calibration of Thermometer**

Due to poor quality control by the manufacturer, a mercury thermometer reads 5°F too low at the freezing point of water (32°F) and reads 5°F too high at the boiling point of water (212°F). Assume the reading on the defective thermometer is a continuous function of the actual temperature.

a) Show that there is some temperature where the reading of the defective thermometer is correct.

b) Let T denote the true temperature and let x be the false reading on the thermometer. Find an equation for the true temperature as a function of the false reading. (Assume a linear relationship.)

c) Find that temperature where the flawed thermometer gives an exact reading.

34) **Calibration of Thermometers**
Several defective thermometers are being tested by the company's quality control personnel. The readings for each thermometer at the freezing point and boiling points of water have been recorded. Assume the readings on each defective thermometer are continuous functions of the actual temperature. For which of the following thermometers can we conclude that there is some temperature where the reading is correct?

Thermometer #	Freezing	Boiling
A	37	217
B	37	200
C	30	210
D	30	214

35) The great emperor Gia Long fought many battles accompanied by a magnificent war elephant. On the death of his master, the military elephant was inconsolable and retired to the mountains where he lived by himself till his death. According to legend, once a year, always on the anniversary of his master's death, the great elephant descended the mountain to visit the grave of the emperor. Assuming that the elephant descends the mountain starting at 9:00 AM and then returns starting at 9:00 AM the following day, show that the faithful elephant travels by some point along the mountain path at the same time each day.

36) **Pizza Revisited**
In the pepperoni example problem, we assumed that there were four identical pieces of pepperoni for clarity of discussion. Show that the pepperoni on any pizza can be divided equally by a single cut through the center, no matter how many pieces there are and no matter if the pieces are different in size.

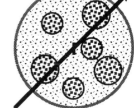

37) Show that a single vertical line can be selected that will cut a portion of steak in half, measured in terms of area.

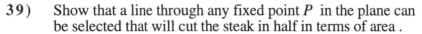

38) For a fixed slope m, show that there is a line with slope m that will cut the steak in half, measured in terms of area.

39) Show that a line through any fixed point P in the plane can be selected that will cut the steak in half in terms of area.

40) Show that a continuous function which is never zero is either always positive or always negative in its domain.

§2.3 Continuity of Functions

41) Temperature Variation
During the day the temperature rises and during the night it falls. On a certain day, the temperature is 30°F at 6:00AM and by coincidence the same temperature is measured the following morning at 6:00AM. Assume that the temperature is a continuous function in time.

   a)   Show using the <u>Intermediate Value Theorem</u> that there is some time t such that the temperature at time t in the morning is the same as the temperature at time t in the afternoon. In other words, show that there exists a twelve hour period over which the temperature variation is 0. Denote the temperature at time t by $T(t)$.

   b)   Assume that the temperature rises linearly from 30°F at 6:00AM to 60°F at 6:00PM and then returns linearly back to 30°F at 6:00AM the following morning. Find that time when the temperature is the same in the morning as it is in the afternoon.

42) Fixed Point Theorem
Let f be a continuous function defined on [0, 1] such that $0 \le f(x) \le 1$ for every x in [0, 1]. Show that there is a number z in [0, 1] such that $f(z) = z$. (Hint: Consider $g(x) = f(x) - x$.)

43) Hundred Meter Dash
A sprinter finishes the 100 meter dash in 10 seconds.
   a)   Show that he must cover exactly 50 m in some interval of 5 s. (Hint: Let $g(t)$ be the distance in meters he covers in the first t seconds. Consider $G(t) = g(t + 5) - g(t) - 50$.)
   b)   Show that the sprinter must cover exactly 10 meters in some interval of 1 second.

44) Car Racing
A race car driver finishes a one-mile circular track in 20 seconds. Show that during some 10 second interval, his car must pass through two points P and Q which lie opposite to each other on the track. Assume that the car moves with varying speed. (Hint: Let $\theta(t)$ be the angular position of the driver at time t (in seconds). For example, $\theta(0) = 0$ and $\theta(20) = 2\pi$. Then consider $G(t) = \theta(10+t) - \theta(t) - \pi$.)

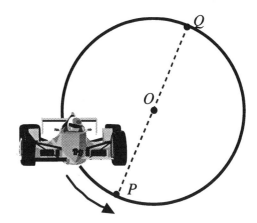

45) A car moves around a circular track. Show that the car must move with the same speed at some antipodal points (the two endpoints of a diameter) if the starting speed of the car is the same as its returning speed.

NOTES

§2.4 Differentiation Rules

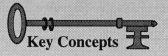
Key Concepts

- Sum Rule
- Power Rule
- Product Rule
- Quotient Rule

The concept of derivative is central in engineering analysis. So far we have had to calculate each derivative by resorting to the fundamental definition. In this section we discover useful formulas and rules that greatly speed up the calculation of derivatives. The process of calculating derivatives is called differentiation.

The derivative of a function is defined by a process involving the notion of a limit. The **derivative** of a function f is the function f' whose value at x is defined as the limit

Definition of a Derivative
$$f'(x) = \lim_{h \to 0} \frac{f(x+h) - f(x)}{h} = \lim_{x_1 \to x} \frac{f(x_1) - f(x)}{x_1 - x}$$

The quotient on the right-hand side of this equation is referred to as the **difference quotient**. If its limit exists at a point x, we say that f is **differentiable** at x.

❖

Formula for the Derivative of a Constant Function.

The derivative of a constant c is zero.

Proof
Consider the constant function $f(x) = c$. Evaluating the limit of the difference quotient we find:

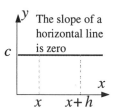
The slope of a horizontal line is zero

$$\lim_{h \to 0} \frac{f(x+h) - f(x)}{h} = \lim_{h \to 0} \frac{c - c}{h} = 0$$

The graphical interpretation of this result is that the slope of a horizontal line is zero.

If the line $y = mx + b$ is multiplied by a constant c we obtain a line of slope cm since $c(mx + b) = cmx + cb$. This simple observation has an important consequence for derivatives of any function. (Recall that the derivative is the slope of the tangent line.)

❖

Derivative of a Constant Times a Function

If a function $f(x)$ is multiplied by a constant, then its derivative is just multiplied by the same constant. That is:

$$\frac{d}{dx}(cf(x)) = cf'(x).$$

Proof

Applying the definition of the derivative we find:

$$\frac{d}{dx}(cf(x)) = \lim_{h \to 0} \frac{cf(x+h) - cf(x)}{h} = c \lim_{h \to 0} \frac{f(x+h) - f(x)}{h} = cf'(x)$$

End of Proof

❖

Formula for the Derivative of a Power

If n is a positive integer then:

$$\frac{d}{dx}(x^n) = nx^{n-1}.$$

Examples

To differentiate a power, multiply by the original exponent n and then decrease the exponent by 1.

If $n = 1$ this gives: $\quad \frac{d}{dx}(x) = \frac{d}{dx}(x^1) = 1 \cdot x^0 = 1$

If $n = 2$ this gives: $\quad \frac{d}{dx}(x^2) = 2 \cdot x^1 = 2x$

If $n = 3$ this gives: $\quad \frac{d}{dx}(x^3) = 3x^2$

Proof

To evaluate the derivative of $f(x) = x^n$ we must find the limit as h approaches 0 of the difference quotient

$$\frac{f(x+h) - f(x)}{h} = \frac{(x+h)^n - x^n}{h}$$

The numerator is a **difference of powers** so it can be factored using the identity

$$a^n - b^n = (a-b)(a^{n-1} + a^{n-2}b + \cdots + ab^{n-2} + b^{n-1})$$

§2.4 Differentiation Rules

Now we let $a = x+h$ and $b = x$. This factorization converts the numerator into a form that is divisible by h.

$$(x+h)^n - x^n = (h)\left[(x+h)^{n-1} + (x+h)^{n-2}x + \cdots + (x+h)x^{n-2} + x^{n-1}\right]$$

Dividing both sides by h gives us:

$$\frac{f(x+h) - f(x)}{h} = \left[(x+h)^{n-1} + (x+h)^{n-2}x + \cdots + (x+h)x^{n-2} + x^{n-1}\right]$$

If we take the $\lim_{h \to 0}$ of both sides, then the equation becomes:

$$\lim_{h \to 0}\frac{f(x+h) - f(x)}{h} = \lim_{h \to 0}\left[(x+h)^{n-1} + (x+h)^{n-2}x + \cdots + (x+h)x^{n-2} + x^{n-1}\right]$$

And results in:

$$\frac{d}{dx}(x^n) = n\,x^{n-1}$$

In fact the differentiation formula holds for any real number n, although we have only been able to prove it for integer values of n.

End of Proof

Example 1

At which point on the graph of the function $y = \sqrt{x}$ is the tangent line parallel to $y = x+1$?

Solution

The derivative of $\sqrt{x}$ is $\frac{1}{2\sqrt{x}}$ and the slope of $y = x+1$ is 1. Solving $\frac{1}{2\sqrt{x}} = 1$ for x, we get $x = \frac{1}{4}$.

The point in question is $\boxed{\left(\frac{1}{4}, \frac{1}{2}\right)}$

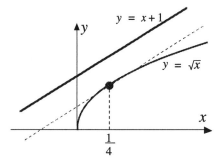

Rules of Differentiation

Next we will explore two fundamental rules from which we can derive all the other derivative formulas. These two rules are the Sum Rule and the Product Rule.

The Sum Rule

If f and g are functions, then the derivative of their sum is the sum of their derivatives at every point where f and g are both differentiable.

The Sum Rule
$$\frac{d}{dx}(f+g) = \frac{df}{dx} + \frac{dg}{dx}$$

Examples

a) Find the derivative of the sum $y = x^2 + x^3$.
$$\frac{dy}{dx} = \frac{d}{dx}(x^2) + \frac{d}{dx}(x^3) = 2x + 3x^2$$

b) Find the derivative of the sum $y = 7 + 2x + x^3$.
$$\frac{dy}{dx} = \frac{d}{dx}(7) + \frac{d}{dx}(2x) + \frac{d}{dx}(x^3) = 0 + 2 + 3x^2 = 2 + 3x^2$$

c) Find the derivative of the difference $y = x^2 - x^3$.
$$\frac{dy}{dx} = \frac{d}{dx}(x^2) + \frac{d}{dx}(-x^3) = 2x - 3x^2$$

Note that subtracting a number is the same as adding its negative. The examples point out that the sum rule applies equally well to differences and to combinations of sums and differences.

Proof of the Sum Rule

$$\begin{aligned}
\frac{d}{dx}(f+g)(x) &= \lim_{h \to 0} \frac{(f+g)(x+h) - (f+g)(x)}{h} \\
&= \lim_{h \to 0} \frac{[f(x+h) + g(x+h)] - [f(x) + g(x)]}{h} \\
&= \lim_{h \to 0} \left(\frac{f(x+h) - f(x)}{h} + \frac{g(x+h) - g(x)}{h} \right) \\
&= \lim_{h \to 0} \frac{f(x+h) - f(x)}{h} + \lim_{h \to 0} \frac{g(x+h) - g(x)}{h} \\
&= \frac{df}{dx} + \frac{dg}{dx}
\end{aligned}$$

⬅ Sum of Limits

This last crucial step in the proof exploits the fact that the limit of a sum is the sum of the limits.

𝕰𝖓𝖉 𝖔𝖋 𝕻𝖗𝖔𝖔𝖋

Derivative of a Product

What is the derivative of a product of two functions? Since we have just seen that the derivative of a sum is the sum of the derivatives, it is natural to expect that the derivative of a product should be the product of the derivatives. Is it true that $\frac{d}{dx}(fg) = \frac{df}{dx} \frac{dg}{dx}$?

§2.4 Differentiation Rules　　CAFÉ

A Case of Mistaken Identity

Whenever you suspect that some new identity might be true it is a good idea to test your conjecture. Consider two typical functions, say $f(x) = x$ and $g(x) = x^2$.

The derivative of the product is
$$\frac{d}{dx}(fg) = \frac{d}{dx}(x^3) = 3x^2$$

But the product of the derivatives is:
$$\frac{df}{dx}\frac{dg}{dx} = \frac{d}{dx}(x) \cdot \frac{d}{dx}(x^2) = 1 \cdot 2x = 2x$$

Thus it is not true that $\frac{d}{dx}(fg) = \frac{df}{dx}\frac{dg}{dx}$.

As we have just seen, the derivative of the product is **not** the product of the derivatives, but in fact the sum of **two** products, as seen in the next rule.

The Product Rule

The derivative of the product of two functions f and g satisfies

The Product Rule
$$\frac{d}{dx}(fg) = \frac{df}{dx}g + f\frac{dg}{dx}$$

wherever both functions are differentiable.

Example 2

Verify that the functions $f(x) = x$ and $g(x) = x^2$ satisfy the product rule.

Solution

The derivative of the product is
$$\frac{d}{dx}(fg) = \frac{d}{dx}(x^3) = 3x^2.$$

This is equal to the expression:
$$\frac{df}{dx}g + f\frac{dg}{dx} = 1 \cdot x^2 + x \cdot (2x) = 3x^2$$

Proof of the Product Rule

We must show that $\dfrac{d}{dx}(fg) = \dfrac{df}{dx}g + f\dfrac{dg}{dx}$.

$$\dfrac{d}{dx}(fg) = \lim_{h\to 0}\dfrac{(fg)(x+h)-(fg)(x)}{h}$$
$$= \lim_{h\to 0}\dfrac{[f(x+h)g(x+h)]-[f(x)g(x)]}{h}$$

The numerator is in the form of a difference of two products which can always be rewritten using the identity:

$$AB - ab = AB + (-Ab + Ab) - ab = A(B-b) + b(A-a)$$

$$\dfrac{d}{dx}(fg) = \lim_{h\to 0}\left[f(x+h)\left(\dfrac{g(x+h)-g(x)}{h}\right) + g(x)\left(\dfrac{f(x+h)-f(x)}{h}\right)\right]$$

$$= \lim_{h\to 0}f(x+h)\lim_{h\to 0}\left(\dfrac{g(x+h)-g(x)}{h}\right) + g(x)\lim_{h\to 0}\left(\dfrac{f(x+h)-f(x)}{h}\right)$$

The two quotients are precisely the definition of the derivatives of f and g.

Further, since f is assumed to be differentiable at x it must be continuous there and we have that: $\lim_{h\to 0}f(x+h) = f(x)$.

Putting this all together we get the desired result $\dfrac{d}{dx}(fg) = \dfrac{df}{dx}g + f\dfrac{dg}{dx}$.

End of Proof

Example 3 - The Product Rule Can Save Time and Effort

Differentiate the polynomial $y = (1 + x + x^2)^2$.

Solution

Method 1: We laboriously expand out the polynomial and then differentiate term by term.
$$\dfrac{dy}{dx} = \dfrac{d}{dx}(1 + 2x + 3x^2 + 2x^3 + x^4) = 2 + 6x + 6x^2 + 4x^3$$

Method 2: Just use the product rule! It's quick, fun and easy!
$$\dfrac{dy}{dx} = 2(1 + x + x^2)(1 + 2x)$$

§2.4 Differentiation Rules   CAFÉ   Page 141

Dimensional Aspects of the Product Rule

The mistaken product rule can easily be shown to be wrong just by considering the dimensions. Suppose the length l and width w of a rectangle are functions of time. The area A of the rectangle is the product of these two functions: $A(t) = l(t)\,w(t)$. Differentiating using the 'incorrect product rule' would give $\dfrac{dA}{dt} = \dfrac{dl}{dt}\dfrac{dw}{dt}$.

Let L denote a length dimension and let T denote a time dimension. Then the dimensions of the left side are $\dfrac{L^2}{T}$ while the dimensions on the right are $\dfrac{L^2}{T^2}$. Thus the two cannot be equal!

Visualizing the Product Rule

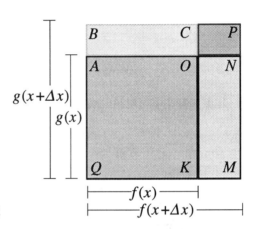

From this graph we see that $f(x+\Delta x)g(x+\Delta x) - f(x)g(x)$ is the difference of two rectangular areas.

Hence, $f(x+\Delta x)g(x+\Delta x) - f(x)g(x)$
$=$ area $ABCO$ + area $KMNO$ $=$
$(g(x+\Delta x) - g(x))f(x) + (f(x+\Delta x) - f(x))g(x)$,

Here area $CPNO$ is omitted since it is negligible in the limit as $\Delta x \to 0$. Thus, dividing the above by Δx and then letting $\Delta x \to 0$ we see the product rule holds.

Derivative of a Quotient

Many chemical and manufacturing processes involve quantities defined as quotients of other functions. Examples include concentration and density. The engineering analysis of such processes requires that we be able to readily differentiate such functions.

The derivative of the quotient of two differentiable functions is given by:

Quotient Rule
$$\left(\frac{f}{g}\right)' = \frac{f'g - fg'}{g^2}$$

Proof

Consider the quotient function $q = \dfrac{f}{g}$. Thus the numerator f of the quotient is the product
$$f(x) = q(x)\,g(x).$$

Using the product rule: $\qquad f' = q'g + qg'$

Isolating for the derivative of the quotient q we find:

$$q' = \frac{f' - qg'}{g} = \frac{f'g - fg'}{g^2}$$

End of Proof

Example 4

Find the derivative of the quotient function $q(x) = \frac{3x}{2x+1}$.

Solution

$$q'(x) = \frac{(3x)'(2x+1) - (3x)(2x+1)'}{(2x+1)^2} = \frac{3 \cdot (2x+1) - (3x) \cdot 2}{(2x+1)^2} = \boxed{\frac{3}{(2x+1)^2}}$$

Higher Order Derivatives

The derivative of a derivative is called a second or <u>double</u> derivative. The double derivative of a function $y = f(x)$ is denoted by several notations such as:

$$\frac{d^2y}{dx^2}, \frac{d^2f}{dx^2} \text{ or } f''(x)$$

Derivatives of higher order can be similarly defined. In general, we employ the notation $\frac{d^ny}{dx^n}$ or $f^{(n)}(x)$ for the nth order derivative.

Example 5

Find the double derivative of $y = x^2 + x^3$.

Solution

We begin by evaluating the first derivative:

$$\frac{dy}{dx} = \frac{d}{dx}(x^2 + x^3) = 2x + 3x^2$$

The double derivative is obtained by differentiating a second time.

$$\boxed{\frac{d^2y}{dx^2} = \frac{d}{dx}\left(\frac{dy}{dx}\right) = 2 + 6x}$$

§2.4 Differentiation Rules   CAFÉ   Page 143

Notation for the Derivative

Many different notations are used to symbolize the derivative, and each has a special advantage. The notation f' was introduced by Lagrange and has the advantage of indicating the fact that the derivative of a function is another function. The notation $\frac{dy}{dx}$ was invented by Leibnitz and has the advantage of revealing the derivative as the slope of the tangent line.

Another of the giants of calculus, Sir Isaac Newton created a completely different notation. He was principally interested in the dynamics of particles and thus evaluated the derivatives of many functions of time. For example, his notation for the velocity of a particle is $\dot{x}$. Newton's dot notation has the great advantage that it is SHORT!

When a particle moves in the plane, it has both an x coordinate and a y coordinate. To describe its location, we often use the vector notation (x,y). Its velocity is also a vector and is denoted by $\vec{v} = (v_x, v_y)$. The components of the velocity vector are calculated by differentiating each component of the position vector (x,y). Thus:

$$\vec{v} = (v_x, v_y) = \left(\frac{dx}{dt}, \frac{dy}{dt}\right)$$

The <u>speed</u> of the particle is defined as $v = \sqrt{v_x^2 + v_y^2}$. The acceleration vector $\vec{a} = (a_x, a_y)$ of the particle is defined as the derivative of the velocity vector $\vec{v} = (v_x, v_y)$ with respect to time. Thus we have:

$$\vec{a} = \left(\frac{d}{dt}(v_x), \frac{d}{dt}(v_y)\right) = \left(\frac{d^2x}{dt^2}, \frac{d^2y}{dt^2}\right)$$

Example 6

The position vector of a particle moving in the plane is $(x,y) = \left(2t, \frac{g}{2}t^2\right)$. Find the velocity and acceleration vector for this particle. What is its speed?

Solution

To find the velocity vector we just differentiate each component separately.

$$\vec{v} = (v_x, v_y) = \left(\frac{dx}{dt}, \frac{dy}{dt}\right) = \left(\frac{d}{dt}(2t), \frac{d}{dt}\left(\frac{g}{2}t^2\right)\right) = \boxed{(2, gt)}$$

To find the acceleration vector we differentiate a second time.

$$\vec{a} = (a_x, a_y) = \left(\frac{d}{dt}(2), \frac{d}{dt}(gt)\right) = \boxed{(0, g)}$$

The speed of the particle is $v = \sqrt{v_x^2 + v_y^2} = \boxed{\sqrt{4 + g^2 t^2}}$

§2.4 Differentiation Rules

WARMUP EXERCISES

Find the derivative of each of the following functions.

1) $f(x) = x^3 + 7x^5$
2) $f(x) = x^7 + a x^{11}$

3) $f(x) = 9 + \dfrac{8}{x} + \dfrac{7}{x^2}$
4) $f(x) = x^3(1-x^2)$

5) $f(x) = (1+x^2)(1-x^2)$
6) $f(x) = (1+x^2-x^3)(1-x^2+x^3)$

Find the derivative of each of the following functions.

7) $f(x) = \dfrac{1+x^2}{1-x^2}$
8) $f(x) = \dfrac{x}{1+x^2}$

9) $f(x) = \dfrac{x^3}{1+x^3}$
10) $f(x) = \dfrac{x^3}{1+2x+x^2}$

11) Let u, v and w be differentiable functions of x. Show that the derivative of the triple product uvw is
$$(uvw)' = u'vw + uv'w + uvw'.$$

Use the identity in the last problem to find the derivative of each of the following functions.

12) $g(x) = x^2(1+x^2)(2+x^2)$
13) $h(t) = (1+t^3)(2+t^2)(3+t)$

14) $g(x) = \dfrac{1}{x^3}(1+x)(1+x^2)$
15) $g(x) = f(x)(1+x^2)(2+x^2)$

INTERMEDIATE EXERCISES

In each of the following problems, a function of engineering significance is defined as a quotient. Evaluate the derivative of each.

16) The <u>concentration</u> of a substance in a solution is defined as the mass of the substance in a given volume of the solution: $C = \dfrac{M}{V}$
Thus if 20 kg of salt are dissolved in one cubic meter of water, the concentration is $C = 20$ kg/m³. In a manufacturing process, the concentration must be reduced by adding more water at a constant rate. As the water is added to the salt solution, the concentration decreases according to the equation $C(t) = \dfrac{M}{V(t)} = \dfrac{20}{1+t/10}$.
Find the rate at which the concentration decreases. Verify that the expression for the derivative is negative.

17) **Maple Syrup Production**
In the production of maple syrup, water must be removed from the sap to increase the concentration of the sugars before it can be sold as maple syrup. The sap is heated causing the excess water to evaporate. Assume the expression for the concentration is given by the formula:
$$C(t) = \frac{M}{V(t)} = \frac{2}{100-t}$$

Find the rate at which the concentration of sugars increases using the quotient rule. Verify that the rate is positive for $t < 100$.

18) The design of an electrical circuit requires a variable resistor (rheostat) in the range from 25 to 30 ohms. Unfortunately, the only variable resistor available operates in a range above 50. By combining in parallel, the expensive rheostat with an inexpensive standard 30 ohm resistor the required range can be obtained. The effective resistance R resulting from the combination of the 30 ohm resistor and the x ohm rheostat of the parallel setup satisfies $\frac{1}{R} = \frac{1}{30} + \frac{1}{x}$.

Find the rate of change of the effective resistance R as the dial on the rheostat is turned. That is, find $\frac{dR}{dx}$.

19) Van der Waals equation for a non-ideal gas is $\left(p + \frac{a}{V^2}\right)(V-b) = nRT$, where p is the pressure, V is volume, R is the ideal gas constant (8.314 J/mol-K), n is the number of moles, b is a constant representing the volume correction for the space occupied by molecules, and a is a constant representing the correction for attractive forces between molecules.
    a) Express p as a function of V.
    b) Find $\frac{dp}{dV}$.

Find the second order derivative for each of the following functions.

20) $y = x^{11}$ **21)** $y = x + x^2 + x^3$

22) $y = \dfrac{1+x}{1-x}$ **23)** $x(t) = \dfrac{1+t}{t} + t$

Consider a particle moving along the x-axis. The second derivative of the particle's position $x(t)$ with respect to time t is called its <u>acceleration</u>. Find the acceleration for a particle whose position $x(t)$ satisfies:

24) $x(t) = 1 + 2t$ **25)** $x(t) = \frac{1}{2} g t^2$

26) $x(t) = x_0 + vt$ **27)** $x(t) = x_0 + vt + \frac{1}{2} a t^2$

§2.4 Differentiation Rules

The position vector for a particle moving in the plane is given. Find the velocity vector, speed and acceleration vector for each particle.

28) $(x,y) = (2t, 3t)$

29) $(x,y) = (t^2, t^3)$

30) $(x,y) = \left(2t, \dfrac{1}{2t}\right)$

31) $(x,y) = \left(x_0 + v_x t,\; y_0 + v_y t + \dfrac{1}{2} a t^2\right)$

32) Show that the derivative of an even function is an odd function and vice versa.

Find the derivative of each of the following functions. Express your results in terms of the unit step function.

33) $f(x) = |x^2 - 4|$

34) $f(t) = (2|t| - |t-1|)^2$

35) $f(x) = x^2|x-1|$

36) $f(x) = \begin{cases} x^2 & (x \geq 0) \\ -x^2 & (x < 0) \end{cases}$

37) Recall that the product rule is not $(fg)' = f'g'$. However, one can cook up certain pairs of functions that do satisfy this mistaken identity. For example, let $f(x) = x^2$ and $g(x) = \dfrac{1}{(x-2)^2}$. Then it turns out that $(fg)' = f'g'$. Of course they always satisfy $(fg)' = fg' + f'g$.

    a) Verify that the functions given satisfy this 'incorrect' product rule.

    b) Find your own examples of two functions f and g that also satisfy this mistaken identity.

38) Let f be a function such that $f'(a)$ exists.

    a) Determine $\lim\limits_{h \to 0} \dfrac{f(a+h) - f(a-h)}{h}$ and justify your answer.

    b) Determine $\lim\limits_{h \to 0} \dfrac{f(a+3h) - f(a-2h)}{h}$ and justify your answer.

39) Assume that $f(0) = 0$, $f'(0) \neq 0$ and $f'(x)$ is continuous. The line PQ is tangent to the graph of $f(x)$ at P so that Q is the x-intercept of the tangent line PQ. Find:

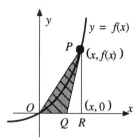

a) $\lim\limits_{x \to 0} \dfrac{\text{area } \triangle POQ}{\text{area } \triangle POR}$. (Hint: $\dfrac{\text{area } \triangle POQ}{\text{area } \triangle POR} = \dfrac{OQ}{OR}$, why?)

b) $\lim\limits_{x \to 0} \dfrac{\text{length } PQ}{\text{length } PR}$

c) $\lim\limits_{x \to 0} \dfrac{\text{length } PQ}{\text{length } PO}$

40) Assume $f(0) = 0$, $f'(0) \neq 0$. P is an arbitrary point on the graph of $f(x)$ and the abscissa of Q is $\dfrac{x}{2}$. We construct rectangles $OAPB$ and $ONQM$ as shown.

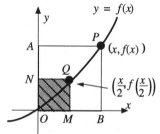

Determine $\lim\limits_{x \to 0} \dfrac{\text{area } OAPB}{\text{area } ONQM}$.

(Hint: $\dfrac{f(x)}{f(x/2)} = 2 \dfrac{(f(x) - 0)/(x - 0)}{(f(x/2) - 0)/(x/2 - 0)}$, and interpret the numerator and denominator in terms of the derivative.)

§2.5 Derivatives of Trigonometric Functions

Key Concepts

- Derivative of Sine and Cosine
- Derivative of Trigonometric Functions

The derivatives of both the sine and cosine functions are easily obtained by a very intuitive argument. Consider a particle rotating counterclockwise about a <u>unit</u> circle with angular velocity ω as shown in the figure. If at time 0 the particle is on the x-axis, then the angle θ at time t is just $\theta = \omega t$. (since $\theta(0) = 0$)

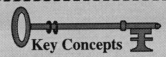

Thus the Cartesian coordinates of the particle are:
$$(x,y) = (\cos \omega t, \sin \omega t).$$

Notice that the speed of the particle is just $v = \omega$ since $r = 1$. (The speed is measured in units of length per second.) We now obtain two different equations for the particle's velocity vector. Equating the two expressions will lead to formulas for the derivatives of the sine and cosine functions.

Particle traces out a unit circle.

By definition, the velocity vector of the particle is:

$$(\dot{x},\dot{y}) = \left(\frac{d}{dt}\cos\omega t,\ \frac{d}{dt}\sin\omega t\right) \tag{1}$$

Thus if we can find the velocity vector by some means other than differentiating, we will have found expressions for the derivatives of both the cosine and sine functions.

A simple <u>geometric</u> argument leads to another expression for the velocity vector. The velocity vector has length ω since the speed of the particle is $v = \omega$. Also the velocity vector is tangent to the direction of motion as shown in the illustration and is thus at right angles to the position vector (x, y). Since the motion is counter-clockwise, the velocity vector is:

$$(\dot{x}, \dot{y}) = (-\omega \sin \omega t,\ \omega \cos \omega t) \tag{2}$$

By identifying the components of Equations 1 and 2 we can read off the derivatives of the cosine and sine functions.

$$\frac{d}{dt}\cos \omega t = -\omega \sin \omega t$$

$$\frac{d}{dt}\sin \omega t = +\omega \cos \omega t$$

Example 1 - Derivative Calculations Involving Sines and Cosines.

a) $\dfrac{d}{dx}\cos 3x = \boxed{-3\sin 3x}$

b) $\dfrac{d}{dx}(x^2 \sin 2x) = \boxed{2x \sin 2x + 2x^2 \cos 2x}$ {Product Rule}

c) $\dfrac{d}{dx}(\cos^2 x + \sin^2 x) = 2\cos x(-\sin x) + 2\sin x(\cos x) = \boxed{0}$ {Product Rule}

Since the derivative of a constant function is zero, (part c) can be viewed as another proof that $\cos^2 x + \sin^2 x = 1$. (The value of the constant being determined by evaluating the left hand side at any point such as $x = 0$.)

Knowing the derivative of the sine and the cosine, we can find the derivatives of other trigonometric functions.

Example 2 - Derivative of the Tangent

Show that the derivative of the tangent is $\dfrac{d}{dx}\tan x = \sec^2 x$.

Solution

Using the identity $\tan x = \dfrac{\sin x}{\cos x}$ the calculation of the derivative can be reduced to an application of the quotient rule.

$\dfrac{d}{dx}\tan x \quad = \dfrac{d}{dx}\left(\dfrac{\sin x}{\cos x}\right) = \dfrac{\cos x \dfrac{d}{dx}(\sin x) - \sin x \dfrac{d}{dx}(\cos x)}{\cos^2 x}$ {Quotient Rule}

$\qquad\qquad = \dfrac{\cos^2 x + \sin^2 x}{\cos^2 x} = \dfrac{1}{\cos^2 x}$ {Pythagorean Identity}

$\qquad\qquad = \sec^2 x$ {The secant is the reciprocal of cosine.}

---

§2.5 Derivatives of Trig Functions

Using the quotient rule as above, one can also find the derivatives of the remaining three trigonometric functions. The results are summarized here and will be checked in the exercises.

Summary of Trigonometric Derivatives

$$\frac{d}{dx}\sin x = \cos x \qquad \frac{d}{dx}\tan x = \sec^2 x \qquad \frac{d}{dx}\sec x = \sec x \tan x$$

$$\frac{d}{dx}\cos x = -\sin x \qquad \frac{d}{dx}\cot x = -\csc^2 x \qquad \frac{d}{dx}\csc x = -\csc x \cot x$$

Notice that the minus signs go with the 'cofunctions' - cosine, cosecant and cotangent.

Example 3 - The 100th Anniversary of the Ferris Wheel

G.W.G. Ferris's 1893 Wheel

Just over one hundred years ago, George Washington Gale Ferris was the toast of America, the engineer who had proved that Americans could outdo the Eiffel Tower. His gigantic Ferris Wheel was the star of the 1893 World's Columbian Exposition - the Chicago's World Fair.

The diameter of Ferris' Wheel was 250 feet! There were 36 gondolas and spokes. If the wheel makes a complete revolution in ten minutes, find the following quantities.

a) Find the angular speed ω and the speed v at the rim of the wheel.

b) Find the velocity vector of a gondola when at the bottom and top of the ride.

c) What is the greatest rate of ascent in the gondolas?

Solution

a) The wheel makes a complete revolution in $T = 10$ minutes. Therefore the angular speed is:

$$\omega = \frac{1}{10}\frac{\text{rev}}{\text{min}} = \boxed{\frac{\pi}{5}\frac{\text{rad}}{\text{min}}}$$

Since the diameter of the wheel is 250 feet, its radius is $r = 125$ feet and its circumference is $C = 2\pi r = 250\pi$. Therefore the speed at the rim is:

$$v = \frac{2\pi r}{T} = \frac{2\pi \cdot 125}{10} = \boxed{25\pi \frac{\text{feet}}{\text{min}}} \qquad \text{(This is about 78.5 feet/min.)}$$

b) The Cartesian coordinates of a gondola making an angle of $\theta = \omega t$ with respect to the horizontal are:
$$(x,y) = (r\cos\omega t, r\sin\omega t)$$

Thus its velocity vector is:
$$\vec{v} = (\dot{x},\dot{y}) = (-r\omega\sin\omega t, r\omega\cos\omega t)$$

At the top of the ride, the angle is 90 degrees, so the velocity vector is
$$\vec{v} = (-r\omega, 0) = \boxed{(-25\pi, 0)}$$

At the bottom of the ride, the angle is –90 degrees, so the velocity vector is
$$\vec{v} = (r\omega, 0) = \boxed{(25\pi, 0)}$$

c) The gondola is rising at the greatest rate when it is at the same height as the axle making an angle of $\theta = 0$ with respect to the horizontal. At this point, the velocity vector is $\vec{v} = (0, 25\pi)$ so that the greatest rate of ascent is $\boxed{25\pi \dfrac{\text{feet}}{\text{min}}}$.

Example 4 - Scotch Crank (Conversion of rotary motion to linear motion)

A peg on the end of a rotating crank slides freely in the vertical guide shown in the diagram. The guide is rigidly connected to a piston which moves horizontally. The crank rotates at a constant angular speed ω in a circle of radius r.

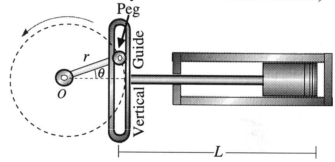

a) What is the total stroke and amplitude A of the piston's motion?
b) Determine the motion of the piston as a function of time t. Assume that the angle at time $t = 0$ is $\theta = 0$.
c) What is the maximum velocity and acceleration of the piston?

Solution

a) As the crank rotates, the vertical guide oscillates from the right to the left of the circle and back. Thus the stroke of the piston is equal to the diameter of the circle. The amplitude of the piston's motion is one half the stroke so $\boxed{A = r}$.

b) Placing the origin at the point O indicated on the diagram, then the location of the vertical guide at time t is:
$$\boxed{x = r\cos\omega t} \quad \text{(assuming it starts at } \theta = 0\text{)}$$

The location of the right face of the piston (at time t) is equal to this value plus L.

§2.5 Derivatives of Trig Functions

c) Differentiating repeatedly gives the velocity and acceleration of the piston.

$$v = \frac{dx}{dt} = -r\omega \sin(\omega t) \quad \text{and} \quad a = \frac{d^2x}{dt^2} = -r\omega^2 \cos(\omega t)$$

Since the maximum and minimum values of both the sine and cosine functions are 1 and –1 respectively, we find:

$$\boxed{v_{max} = r\omega} \quad \text{and} \quad \boxed{a_{max} = r\omega^2}$$

Example 5 - Kinematics of a Piston Engine

The crank arm AB of a piston engine has length 3 inches and connects to the piston by a rod BD of length 8 inches. If the crank arm AB rotates with a constant angular velocity of $\omega = -1500$ rpm (it is counterclockwise), determine the velocity of the piston as a function of θ. Give numerical values for the velocity when $\theta = 0, \frac{\pi}{2}$ and $\frac{\pi}{3}$ radians.

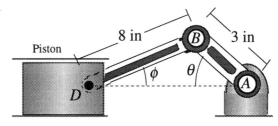

Solution

Since the angular velocity is $\omega = -1500$ rpm, we can write the angle θ of the crank arm in terms of time as $\theta = \omega t$ assuming it is initially at $\theta = 0$.

We will also convert the angular velocity to radians per second.

$$\omega = -1500 \frac{\text{rev}}{\text{min}} \cdot \left(\frac{2\pi}{60}\right) = -50\pi \frac{\text{rad}}{\text{s}}$$

Now, let $x = AD$. Applying the law of cosines to $\triangle ABD$ we have:

$$64 = 9 + x^2 - 6x\cos\theta$$

Solving the above quadratic equation for x gives:

$$x = 3\cos\theta \pm \sqrt{9\cos^2\theta + 55}$$

Only the positive answer makes sense here because x must be greater than zero. Thus:

$$x = 3\cos\theta + \sqrt{9\cos^2\theta + 55}$$

Note that both x and θ are functions of t.

Differentiating the equation $64 = 9 + x^2 - 6x\cos\theta$ using the <u>product</u> rule yields:

$$0 = 2x\dot{x} - 6(\dot{x}\cos\theta - x\omega\sin\theta) \quad \text{or} \quad \dot{x} = \frac{3x\omega\sin\theta}{3\cos\theta - x}$$

Thus:

$$\dot{x} = v = \boxed{\frac{-9\omega\sin\theta\cos\theta}{\sqrt{9\cos^2\theta + 55}} - 3\omega\sin\theta}$$

By plugging $\theta = 0, \frac{\pi}{2}$ and $\frac{\pi}{3}$ respectively into the velocity expression, we find:

$$\boxed{v = 0\,\tfrac{\text{in}}{\text{s}} \qquad v = 150\pi\,\tfrac{\text{in}}{\text{s}} \qquad v = 489.01\,\tfrac{\text{in}}{\text{s}}}$$

§2.5 Derivatives of Trig Functions   CAFÉ

Using the quotient rule, verify the following trigonometric derivatives. Recall the definitions cosecant = $\frac{1}{\text{sine}}$, secant = $\frac{1}{\text{cosine}}$ and cotangent = $\frac{1}{\text{tangent}}$.

1) $\frac{d}{dx} \csc x = -\csc x \cot x$
2) $\frac{d}{dx} \sec x = \sec x \tan x$
3) $\frac{d}{dx} \cot x = -\csc^2 x$

Find the derivatives of the following functions.

4) $y = \dfrac{1}{1 + \sin x}$ 5) $y = \dfrac{\cos x}{1 + \sin x}$ 6) $x = 2\sin \omega t + 3\cos \omega t$

7) $y = \sin^2 x$ 8) $y = \tan^3 x$ 9) $y = \tan x + \cot x$

10) $y = \dfrac{\sin x}{x}$ 11) $y = \tan x \cot x$ 12) $y = \sec^2 x - \tan^2 x$

13) $y = \tan x \cos x \csc x$ 14) $y = \sin x \sin 2x \sin 3x$

For each function find the equation of the tangent line at the indicated point.

15) $y = \sin x$, $x = 0$ 16) $y = \tan x$, $x = 0$ 17) $y = \tan x$, $x = \pi/4$

18) $y = \cos x$, $x = \pi/3$ 19) $y = \sec x$, $x = \pi/4$ 20) $y = \csc x$, $x = \pi/3$

In each of the following exercises, the position of a particle moving along the x-axis is given as a function of time. Find the velocity and acceleration of the particle.

21) $x = \sin 2t$ 22) $x = \sin 2t + \cos 7t$ 23) $x = \dfrac{g}{2} t^2 + \cos \omega t$

In the following exercises, the radius of a planar curve is given as a function of the angle theta. Find $\dfrac{dr}{d\theta}$.

24) $r = 1 - \cos\theta$ {Cardioid}   **25)** $r = \dfrac{2}{\cos\theta}$ {Line}

26) $r = \dfrac{1}{2 + \cos\theta}$ {Ellipse}   **27)** $r = \dfrac{2}{3 + \sin\theta}$ {Ellipse}

Prove the following.

28) $\dfrac{d}{dt}\cot\omega t = -\omega\csc^2\omega t$ **29)** $\dfrac{d}{dt}\csc\omega t = -\omega\csc\omega t\,\cot\omega t$

30) $\dfrac{d}{dt}\sec\omega t = \omega\sec\omega t\,\tan\omega t$

31) Which of the six trigonometric functions never have horizontal tangents?

INTERMEDIATE EXERCISES

In each of the following exercises, the <u>vector</u> position of a particle moving in the plane is given as a function of time. Sketch the particle's trajectory. Then find the velocity and acceleration vector of the particle.

32) $(x,y) = (\cos 3t, \sin 3t)$ **33)** $(x,y) = (\cos 3t, \sin 7t)$ {Lussijus curve}

34) $(x,y) = (\cos^2 t, \sin^2 t)$ **35)** $(x,y) = (t - \sin t, 1 - \cos t)$ {cycloid}

36) $(x,y) = (\sec 4t, \tan 4t)$ **37)** $(x,y) = (\cos^3 t, \sin^3 t)$ {astroid}

Differentiating known identities often leads to new identities. This is an important method for discovering new formulas.

38) The 'double angle formula for sines' is $\sin 2x = 2\sin x\cos x$. Differentiate this identity to obtain a similar double angle formula for cosines.

39) The 'triple angle formula for sines' is $\sin 3x = 3\sin x - 4\sin^3 x$. Differentiate this identity to obtain a similar triple angle formula for cosines.

§2.5 Derivatives of Trig Functions   CAFÉ   Page 157

40) The 'addition formula for sines' is $\sin(x+y) = \sin x \cos y + \cos x \sin y$. Differentiate this identity with respect to x by holding y as a constant to obtain a similar formula for cosines.

41) Find the fourth derivative of each of the following functions.

   a) $y = \sin x$ b) $y = \cos x$

42) Find an expression for the nth derivative of $y = \sin x$.
    Hint: Every integer k can be written in one of the four forms $4n, 4n+1, 4n+2$ or $4n+3$ where n is an integer.

43) The current in a 20-ohm resistor is $I(t) = 6 \sin(120\pi t)$ in amperes. How fast is the power in joules in the resistor changing at time t? (Recall $P = I^2 R$.)

44) A carrier wave of amplitude V_o and angular frequency ω_c is modulated by a sinusoidal signal of angular frequency ω_s. The resulting modulated wave is given by $V(t) = V_o(1 + m \sin \omega_s t) \sin \omega_c t$ where m is the degree of modulation. Find $\dfrac{dV}{dt}$.

45) A current I flows through a circular wire of radius r. Consider a point P located a distance x from the coil along a perpendicular axis through its center. Let θ denote the angle between the axis and a line connecting point P to any point on the circumference of the circular wire. Then the magnetic field intensity at the point P due to the current I is $\mathbf{H} = \dfrac{kI}{r} \sin^3 \theta$, where k is a constant. Find the rate of change of the magnetic field intensity with respect to θ.

Solve the following limits.

46) $\lim\limits_{x \to 0} \dfrac{\sin x}{x}$ (Hint: Write $\dfrac{\sin x}{x} = \dfrac{\sin x - \sin 0}{x - 0}$ and interpret the limit in terms of the derivative of $\sin x$.)

47) $\lim\limits_{x \to 0} \dfrac{\cos x - 1}{x}$ Hint: Interpret the limit in terms of the derivative of $\cos x$.)

Using the above results, solve the following exercises.

48) $\lim\limits_{x \to 0} \dfrac{\sin 6x}{x}$ 49) $\lim\limits_{x \to 0} \dfrac{\sin 7x}{\tan 3x}$ 50) $\lim\limits_{t \to 0} \dfrac{\sin^2 5t}{t^2}$

51) $\lim\limits_{x \to 0} \dfrac{\tan 5x}{3x}$ 52) $\lim\limits_{x \to 0} x \cot x$ 53) $\lim\limits_{x \to 0} \dfrac{x}{\sin 5x - \sin 3x}$

ADVANCED EXERCISES

54) Simple Harmonic Motion

If the position of a particle moving along the x-axis is given as a function of time t in such a way that it satisfies the differential equation $\frac{d^2x}{dt^2} = -k^2x$, where k is a constant, then the particle is said to undergo <u>simple harmonic motion</u>. For example, the motion which corresponds to that of a particle on the end of a spring (ignoring friction), is simple harmonic motion where k stands for the spring constant.

a) Show that $x(t) = A \sin kt + B \cos kt$ satisfies the differential equation for any constants A and B. The converse of the above statement is also true, and an explanation can be found in any book on differential equations. We shall not prove it here.

b) Show that $f(x) = \sin(x + y)$ satisfies $\frac{d^2f}{dx^2} = -f$, where y is a constant. Hence by the statement in part **a)** we have $\sin(x + y) = A \sin x + B \cos x$ for some constants A and B that may depend on y.

c) Determine the constants A and B in part **b)** by using special values of x. The above process is a derivation of the well–known addition formula for the sine function.

55) Assembly Line

In an assembly line to produce Hula-Hoops, a shear is used to cut a long roll of plastic tubing to the correct length needed to manufacture each hoop. The motion of the blade of the shear is approximately simple harmonic motion having an amplitude of 2 centimeters and a period of 1 second. Find the maximum acceleration and speed of the cutting blade.

56) Controlling the Frequency of a Generator

A generating plant must produce an alternating voltage with frequency as close to 60 cycles per second as possible. In the good old days, before the advent of quartz crystal technology, one method of calibrating the generator involved the use of an accurate 60-cycle tuning fork associated with a carbon-microphone button. If the maximum displacement of the tip of the tuning fork from its equilibrium position is $A = 1$ millimeter, find the maximum acceleration and speed of the tip. Assume the motion of the tip is in a straight line and that it can be treated as a simple harmonic motion.

§2.5 Derivatives of Trig Functions   CAFÉ

57) Let CB be tangent to the unit circle at B and let AD be perpendicular to OB. Determine the following limits:

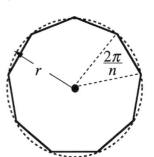

a) $\displaystyle\lim_{\theta \to 0} \frac{\text{area sector } AOB}{\text{area } \triangle AOD}$ b) $\displaystyle\lim_{\theta \to 0} \frac{\text{area sector } AOB}{\text{area } \triangle COB}$

c) $\displaystyle\lim_{\theta \to 0} \frac{\text{area sector } AOB}{\text{area } \triangle AOB}$ d) $\displaystyle\lim_{\theta \to 0} \frac{\text{length of segment } AB}{\text{length } CB}$

e) $\displaystyle\lim_{\theta \to 0} \frac{\text{length of segment } AB}{\text{length of arc } AB}$ f) $\displaystyle\lim_{\theta \to 0} \frac{\text{length of segment } AB}{\text{length } AD}$

58) Let A_n be the area and C_n the circumference of a regular n-gon (a polygon with n sides) inscribed in a circle of radius r.

a) Express A_n and C_n in terms of the trigonometric functions.

b) Determine $\displaystyle\lim_{n \to \infty} A_n$ and $\displaystyle\lim_{n \to \infty} C_n$ using the expressions in part a).

NOTES

§2.6 Chain Rule & Implicit Differentiation

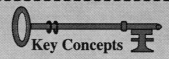
Key Concepts

❖ Chain Rule ❖ Implicit Differentiation

❖ Derivatives of Inverse Functions

We have seen how in engineering analysis, functions must often be composed as in the case $(f \circ g)(x)$. The chain rule tells us how to differentiate such composite functions. In this section we will first look at a few examples to motivate the chain rule and then we will state it formally. Finally, its application to implicit differentiation and to the derivatives of inverse functions will be considered.

Unit Conversion

Let x denote the span of a suspension bridge measured in yards. Let u denote the span of the bridge in feet and let y denote its span in inches. A precise measurement like y might be needed by the engineers who actually will build the bridge. The span of the bridge is:

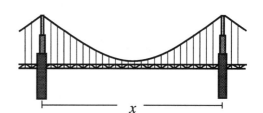

$$x \text{ yards} \quad \text{or} \quad u \text{ feet} \quad \text{or} \quad y \text{ inches}$$

Of course, given any one of these measurements we can find the others by converting the units. (one yard = 3 feet, one foot = 12 inches)

$$\begin{aligned} y &= 36x &&\text{(yards to inches)} \\ y &= 12u &&\text{(feet to inches)} \\ u &= 3x &&\text{(yards to feet)} \end{aligned}$$

Notice that the linear function $y = 36x$ (yards to inches) is the composite of the functions $u = 3x$ (yards to feet) and $y = 12u$ (feet to inches).

$$y = 12u = 12(3x) = 36x$$

How do the derivatives of these three functions compare?

$$\frac{dy}{dx} = 36 \qquad \frac{dy}{du} = 12 \qquad \frac{du}{dx} = 3 \qquad \text{Could it be that } \frac{dy}{dx} = \frac{dy}{du}\frac{du}{dx}?$$

Mechanical Analog of Chain Rule

The small, medium and large gears have 6, 12 and 18 gears respectively. Let the large gear make x turns, the small gear make u turns and the medium gear make y turns. Then:
$$u = 3x \text{ and } y = \tfrac{1}{2}u \text{ so that } y = \tfrac{3}{2}x$$

Verify that the derivatives satisfy $\dfrac{dy}{dx} = \dfrac{dy}{du}\dfrac{du}{dx}$.
Rates of change multiply!

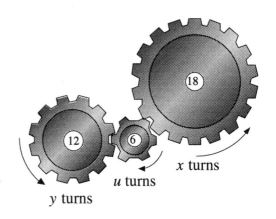

x turns
u turns
y turns

Example 1 - Crystal Growth

In a zero gravity experiment, prismatic crystals are being grown on the space shuttle. Each crystal has a square cross-section of area $A = x^2$. The edges grows at a constant rate so that their length as a function of time t is $x = c + 3t$ where c is the length of the original seed crystal. Therefore the area is the composite of the two functions $x = c + 3t$ and $A = x^2$ and can be expressed as $A = (c + 3t)^2$.

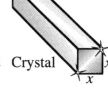

Crystal

Can you see any relation or pattern between the derivatives $\dfrac{dA}{dt}, \dfrac{dA}{dx}$ and $\dfrac{dx}{dt}$?

$$\dfrac{dA}{dt} = \dfrac{d}{dt}\left(c^2 + 6ct + 9t^2\right) = 6c + 18t = 6x, \qquad \dfrac{dA}{dx} = 2x, \qquad \dfrac{dx}{dt} = 3$$

The rates of change multiply: $\dfrac{dA}{dt} = \dfrac{dA}{dx}\dfrac{dx}{dt}$ since $6x = 2x \cdot 3$

The last example points out that the derivatives must be evaluated <u>at the appropriate places</u>. In the observation $\dfrac{dA}{dt} = \dfrac{dA}{dx}\dfrac{dx}{dt}$ we evaluate the outer derivative $\dfrac{dA}{dx}$ at the point $x(t)$.
These examples suggest that the derivative of the composite of two functions f and g is the product of their derivatives evaluated at the appropriate points. Let us now make this observation official.

Chain Rule - Heuristic Proof

The derivative of the composite of two differentiable functions f and g is:

$$\boxed{(f \circ g)'(x) = f'(g(x))g'(x)}$$

§2.6 Chain Rule & ...

Proof

Let us denote the inner function by $z = g(x)$ and let $y = f(z)$ so that $y = f(g(x))$. Note that $g(x + \Delta x) - g(x) = \Delta z$. We express the difference quotient $\dfrac{f(g(x + \Delta x)) - f(g(x))}{\Delta x}$ by using Δz. Thus, provided $\Delta z \neq 0$, we may write:

$$\frac{f(g(x + \Delta x)) - f(g(x))}{\Delta x} = \frac{f(z + \Delta z) - f(z)}{\Delta x} = \frac{f(z + \Delta z) - f(z)}{\Delta z} \cdot \frac{\Delta z}{\Delta x}$$

In the limit as $\Delta x \to 0$, Δz also approaches zero This suggests that:

$$\lim_{\Delta x \to 0} \frac{f(g(x + \Delta x)) - f(g(x))}{\Delta x} = \lim_{\Delta z \to 0} \frac{f(z + \Delta z) - f(z)}{\Delta z} \cdot \lim_{\Delta x \to 0} \frac{\Delta z}{\Delta x}$$

$$= f'(z) \frac{dz}{dx} = f'(g(x)) g'(x)$$

Hence
$$\frac{d}{dx} f(g(x)) = f'(g(x))\, g'(x)$$

End of Proof

The Outside to Inside Picture of the Chain Rule

In the composite function, $(f \circ g)(x)$, g is evaluated first, it is called the **inner** function while f is called the **outer** function. The chain rule states that the derivative of the composite $f \circ g$ equals the product of the inner and outer derivatives provided the outside derivative is evaluated at the value of the inner function.

The Chain Rule
$$(f \circ g)'(x) = f'(\,g(x)\,)\, g'(x)$$

For example, in the equation $\dfrac{d}{dx} \sqrt{1 + x^4} = \dfrac{1}{2\sqrt{1 + x^4}} \cdot 4x^3 = \dfrac{2x^3}{\sqrt{1 + x^4}}$, $\dfrac{1}{2\sqrt{1+x^4}}$ is the outer derivative (at inner value) and $4x^3$ is the inner derivative.

Generalized Power Law

A special case of the chain rule is so useful that it deserves special mention.

If u is a differentiable function of x, then:

$$\frac{d}{dx}(u^n) = n\, u^{n-1} \frac{du}{dx}$$

Several examples are:

a) $\dfrac{d}{dx}(1+\sqrt{x})^{10} = 10(1+\sqrt{x})^9 \dfrac{1}{2\sqrt{x}} = \dfrac{5}{\sqrt{x}}(1+\sqrt{x})^9$

b) $\dfrac{d}{dx}(1+x^3)^{2001} = 2001(1+x^3)^{2000} \, 3x^2 = 6003 \, x^2 (1+x^3)^{2000}$

Trigonometric Derivatives

If u is a differentiable function of x, then:

$$\dfrac{d}{dx}\sin u = \cos u \, \dfrac{du}{dx}.$$

Similar derivative expressions hold for the other trigonometric functions.

$$\dfrac{d}{dx}\sin u = \cos u \, \dfrac{du}{dx} \qquad \dfrac{d}{dx}\cos u = -\sin u \, \dfrac{du}{dx}$$

$$\dfrac{d}{dx}\tan u = \sec^2 u \, \dfrac{du}{dx} \qquad \dfrac{d}{dx}\cot u = -\csc^2 u \, \dfrac{du}{dx}$$

$$\dfrac{d}{dx}\sec u = \sec u \tan u \, \dfrac{du}{dx} \qquad \dfrac{d}{dx}\csc u = -\csc u \cot u \, \dfrac{du}{dx}$$

Example 2

Find the derivative of $f(x) = \sin(\cos x)$.

Solution

If we let $u = \cos x$, then we can write:

$$\dfrac{d}{dx}\sin(\cos x) = \dfrac{d}{dx}\sin u = \cos(u)\dfrac{du}{dx} = \cos(\cos x)\dfrac{d}{dx}\cos x$$

$$\dfrac{d}{dx}\sin(\cos x) = \boxed{-\cos(\cos x)\cdot \sin x}$$

§2.6 Chain Rule & ...   CAFÉ   Page 165

Example 3 - Multiple Compositions

When more than two functions have been composed the chain rule is used repeatedly.
Find the derivatives of: **a)** $\cos^2 \omega t$ **b)** $\sqrt{1+\sin^2 7x}$

Solution

a) $\dfrac{d}{dt}(\cos^2 \omega t) = 2(\cos \omega t)\dfrac{d}{dt}(\cos \omega t)$ {Chain rule first used}

$\qquad\qquad\qquad = 2(\cos \omega t)(-\sin \omega t)\dfrac{d}{dt}(\omega t)$ {Chain rule used again}

$\qquad\qquad\qquad = -2\omega(\cos \omega t)(\sin \omega t)$

$\qquad\qquad\qquad = \boxed{-\omega \sin 2\omega t}$ {Double Angle Formula}

b) $\dfrac{d}{dx}\sqrt{1+\sin^2 7x} = \dfrac{1}{2\sqrt{1+\sin^2 7x}}\dfrac{d}{dx}(1+\sin^2 7x)$ {Chain rule first used}

$\qquad\qquad\qquad = \dfrac{1}{2\sqrt{1+\sin^2 7x}}(2\sin 7x)\dfrac{d}{dx}(\sin 7x)$ {Chain rule used twice}

$\qquad\qquad\qquad = \dfrac{\sin 7x}{\sqrt{1+\sin^2 7x}}(\cos 7x)\dfrac{d}{dx}(7x)$ {Chain rule third time}

$\qquad\qquad\qquad = \dfrac{7 \sin 7x \cos 7x}{\sqrt{1+\sin^2 7x}} = \dfrac{7 \sin 14x}{2\sqrt{1+\sin^2 7x}}$ {Double Angle Formula}

Mechanical Interpretation of Chain Rule

The stylus or needle of a record player is an excellent mechanical illustration of the chain rule. The grooves of a record contain many little bumps which encode the music. These bumps are detected by the stylus and the resulting oscillations are converted into electrical signals and finally into your favorite music.

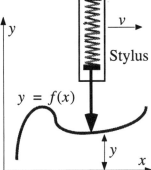

We can imagine a 'mathematical stylus' designed to trace out the "bumps of the graph" of any function $y = f(x)$. As the stylus moves along the x axis with a horizontal speed v, its tip moves up and down as it follows the contour of the graph.

The vertical velocity of the tip is $\dfrac{dy}{dt}$. In a small time interval Δt the stylus has moved along the x-axis a distance $\Delta x = v\Delta t$. Its new vertical location is $f(x+v\Delta t)$. The slope of the line connecting these points is $\dfrac{\Delta y}{\Delta t} = \dfrac{f(x+v\Delta t) - f(x)}{v\Delta t}\dfrac{v\Delta t}{\Delta t}$.

Taking the limit as Δt approaches zero, we see that: $\dfrac{dy}{dt} = \dfrac{dy}{dx}\dfrac{dx}{dt}$. The vertical velocity $\dfrac{dy}{dt}$ of the stylus is a product of two terms $\dfrac{dy}{dx}$ and $\dfrac{dx}{dt}$.

In this picture, it is clear why the derivative of the outer function $\dfrac{dy}{dx}$ must be evaluated at the inner function x because that is where the stylus is! Further, the derivative of the inner function $\dfrac{dx}{dt}$ corresponds to the speed at which the record is played.

Dimensional Check of Chain Rule

Spike's roommate has overslept as usual, so Spike has decided to wake him up by bursting a balloon with a needle. As he is inflating the balloon, (needle at the ready) he notices that this trick might provide a good check of the chain rule. The balloon is a sphere having a volume $V = \dfrac{4}{3}\pi r^3$ where r is its radius which is a function of time t. Thus by the chain rule, we should have:

$$\frac{dV}{dt} = \frac{dV}{dr}\frac{dr}{dt}$$

But are the dimensions of both sides the same? Let L denote the dimension of length and T denote the dimension of time. Thus the volume of the balloon has dimension L^3 and as it is inflated, the rate of change of volume $\dfrac{dV}{dt}$ has dimensions of $\dfrac{L^3}{T}$. The rate of change of radius $\dfrac{dr}{dt}$ has dimensions of $\dfrac{L}{T}$ and $\dfrac{dV}{dr}$ has dimensions of L^2.

Thus the chain rule satisfies the dimension check since $\dfrac{L^3}{T} = L^2 \cdot \dfrac{L}{T}$

Having proven his point, Spike proceeds to wake up his sleepy roommate.

Tangents to Planar Objects   (Implicit Differentiation)

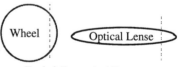

Each fails vertical line test

The graphs of many common engineering objects do not pass the vertical line test so they do not correspond to our definition of function. Such shapes as wheels, pulleys, optical lenses and gears are clearly not explicit functions of the form $y = f(x)$. However, these shapes are often describable by implicit relations such as $x^2 + y^2 = 1$ for a unit circle. There will be many times when we will have to find the tangent lines to such objects.

Example 4 - Tangent Lines to Circles

Find the equation of the tangent line to the circle $x^2 + y^2 = 1$ at the point $\left(\frac{\sqrt{2}}{2}, \frac{\sqrt{2}}{2}\right)$.

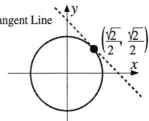

Solution

Even though we do not have an explicit relation for y as a function of x, the implicit relation $x^2 + y^2 = 1$ is enough to find the slope at any point on the circle. Simply differentiate the relation treating y as a function of x. (This is called implicit differentiation.)

$$\frac{d}{dx}(x^2) + \frac{d}{dx}(y^2) = \frac{d}{dx}(1)$$

$$2x + 2y\frac{dy}{dx} = 0 \quad \longleftarrow \text{(Use chain rule to differentiate } y^2\text{)}$$

Isolating the derivative we find:

$$\frac{dy}{dx} = -\frac{x}{y}$$

Evaluating this slope at the given point $\left(\frac{1}{\sqrt{2}}, \frac{1}{\sqrt{2}}\right)$ yields:

$$\frac{dy}{dx} = -1$$

Using the point-slope form for the equation of a line, the tangent line to the circle at this point is:

$$\frac{y - \frac{1}{\sqrt{2}}}{x - \frac{1}{\sqrt{2}}} = -1 \quad \text{or} \quad \boxed{y = -x + \sqrt{2}}.$$

Example 5

Consider the ellipse defined by the equation $\frac{x^2}{9} + \frac{y^2}{16} = 1$.

Find $\frac{dy}{dx}$ by implicit differentiation. Identify any points where the derivative is not defined.
Find the slope of the tangent line to the ellipse at the points where the ellipse intersects the line $y = x$.

Solution

Treating y as a function of x and differentiating gives:

$$\frac{d}{dx}\left(\frac{x^2}{9}\right) + \frac{d}{dx}\left(\frac{y^2}{16}\right) = \frac{d}{dx}(1) \quad \text{or} \quad \frac{2x}{9} + \frac{y}{8}\frac{dy}{dx} = 0$$

Solving for the derivative we find: $\boxed{\dfrac{dy}{dx} = -\dfrac{16x}{9y}}$.

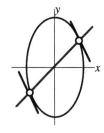

Vertical Tangent Lines

Thus the derivative is defined everywhere except where the denominator y in the above quotient is zero. From the figure it can be seen that when $y = 0$, the tangent lines to the ellipse are <u>vertical</u> and hence their slope is undefined. Thus the derivative is undefined where the ellipse intersects the x-axis ($y = 0$).

Finally, at either of the intersection points with the line $y = x$, the slope of the tangent line is $\boxed{\dfrac{dy}{dx} = -\dfrac{16}{9}}$.

Example 6

Find $\dfrac{dy}{dx}$ if y is defined implicitly as a function of x by the relation $x \sin y + y = 0$.

Solution

Treating y as a function of x and applying the chain rule gives:

$$\frac{d}{dx}(x \sin y) + \frac{dy}{dx} = 0$$

$$\left(\sin y + x \cos y \frac{dy}{dx}\right) + \frac{dy}{dx} = 0 \quad \longleftarrow \text{Note use of product rule and chain rule.}$$

Collecting together all terms containing the derivative gives:

$$\sin y + (1 + x \cos y)\frac{dy}{dx} = 0 \quad \text{or} \quad \boxed{\frac{dy}{dx} = \frac{-\sin y}{1 + x \cos y}}$$

Example 7 - Higher order derivatives of implicitly defined functions.

What is the value of the double derivative $\dfrac{d^2y}{dx^2}$ in the case of the unit circle $x^2 + y^2 = 1$?

Solution

We already know that the first derivative is given by the expression

$$\frac{dy}{dx} = -\frac{x}{y}$$

We must consider both y and $\dfrac{dy}{dx}$ as functions of x as we differentiate the above result.

$$\frac{d^2y}{dx^2} = \frac{d}{dx}\left(-\frac{x}{y}\right) = -\left(\frac{1\cdot y - x\dfrac{dy}{dx}}{y^2}\right) \quad \longleftarrow \text{Quotient rule.}$$

The derivative is already known.

$$\frac{d^2y}{dx^2} = -\left(\frac{1\cdot y - x\left(-\dfrac{x}{y}\right)}{y^2}\right) = -\frac{x^2+y^2}{y^3} = -\frac{1}{y^3}$$

Thus: $\boxed{\dfrac{d^2y}{dx^2} = -\dfrac{1}{y^3}}$.

Derivative of Inverse Functions

Finding the derivatives of the inverse of a function is an application of implicit differentiation. Recall that the inverse function $f^{-1}(x)$ of $f(x)$ is a function that satisfies

$$f\left(f^{-1}(x)\right) = x \qquad \text{(and vice versa)}$$

Using the chain rule to differentiate this functional equality we find:

$$\frac{d}{dx}\left\{f\left(f^{-1}(x)\right)\right\} = f'(f^{-1}(x)) \cdot \frac{d}{dx}f^{-1}(x) = 1 \qquad \text{or} \qquad \boxed{\frac{d}{dx}f^{-1}(x) = \frac{1}{f'\left(f^{-1}(x)\right)}}$$

Using the Leibnitz notation we have: $\boxed{\dfrac{dy}{dx} = \dfrac{1}{\dfrac{dx}{dy}}}$

Let's apply the above result to find the derivatives of $\sin^{-1} x$, $\tan^{-1} x$, and $\sec^{-1} x$.

Recall that the inverse functions are defined as follows:

$$\sin^{-1} x = \text{the number between } \frac{-\pi}{2} \text{ and } \frac{\pi}{2} \text{ (inclusive) whose sine is } x.$$

$$\tan^{-1} x = \text{the number between } \frac{-\pi}{2} \text{ and } \frac{\pi}{2} \text{ (strict) whose tangent is } x.$$

$$\sec^{-1} x = \text{the number in either } \left[0, \frac{\pi}{2}\right) \text{ or } \left[\pi, \frac{3\pi}{2}\right) \text{ whose secant is } x.$$

Examples: $\tan^{-1}(-1) = -\frac{\pi}{4}$, not $\frac{3\pi}{4}$ and $\sec^{-1}(-2) = \frac{4\pi}{3}$, not $\frac{2\pi}{3}$

By applying the above formula for the derivative of an inverse function we see:

$$\frac{d}{dx} \sin^{-1} x = \frac{1}{\cos(\sin^{-1} x)} = \frac{1}{\sqrt{1 - \sin^2(\sin^{-1} x)}} = \frac{1}{\sqrt{1 - x^2}}$$

where the positive square root was taken because of the definition of $\sin^{-1} x$.

Thus we have shown: $\quad \dfrac{d}{dx} \sin^{-1} x = \dfrac{1}{\sqrt{1 - x^2}}$

Similarly we can show that:

$$\frac{d}{dx} \tan^{-1} x = \frac{1}{1 + x^2} \qquad \frac{d}{dx} \sec^{-1} x = \frac{1}{x\sqrt{x^2 - 1}}$$

❖

Example 8

Find the derivative of each of the following expressions involving inverse functions.

a) $\tan^{-1} 2x$ **b)** $\sin^{-1} x^2$

Solution

a) $\dfrac{d}{dx} \tan^{-1} 2x = \dfrac{1}{1 + (2x)^2} \dfrac{d}{dx}(2x) = \dfrac{2}{1 + 4x^2}$

b) $\dfrac{d}{dx}\left(\sin^{-1} x^2\right) = \dfrac{1}{\sqrt{1 - (x^2)^2}} \dfrac{d}{dx}(x^2) = \dfrac{2x}{\sqrt{1 - x^4}}$

§2.6 Chain Rule & ... CAFÉ Page 171

Find the derivative of each of the following functions.

1) $y = (1 + 3x)^2$
2) $y = \cos(3 + 2x)$
3) $y = (1 + 3x^3)^2$
4) $y = 3x^2 + (1 + 3x^2)^3$
5) $f(x) = (1 + 3x + 7x^2)^2$
6) $f(x) = 1 + 3(x+3) + 7(x+3)^2$

Find the derivative of each of the following <u>tangled roots</u>.

7) $\sqrt{1+x^2}$
8) $y = \sqrt{x + \sqrt{x}}$
9) $y = \sqrt{x + \sqrt{x + \sqrt{x}}}$
10) $y = \sqrt{x + \sqrt{1+x^2}}$

Find the derivative of each of the following <u>sines of trouble</u>.

11) $y = \sin 2x$
12) $y = \sin(x^2)$
13) $y = \sin^2 x$
14) $y = \sin(\sin x)$
15) $y = \sin^2(\sin x)$
16) $y = \sin^2\left((\sin x^2)^2\right)$

Find the derivative of each of the following functions by combining the chain rule with the product or quotient rule.

17) $y = \left(\dfrac{x-1}{x+1}\right)^{1,000}$
18) $y = (x-1)^{1,000} x^{2001}$
19) $y = x^2 \tan(\cos x)$
20) $y = \dfrac{\cos(\cos x)}{1 + \sin x}$

Find the derivative for each of the following expressions that involve the inverse trig functions.

21) $\tan^{-1} x^3$
22) $\sin^{-1}(\sin x^3)$
23) $\sin^{-1}(1 + 2x)$
24) $\sin^{-1}(\cos x)$
25) $\tan^{-1}(\cot x)$
26) $(1 + 2\tan^{-1} x)^3$

27) Consider the three functions $f(x) = x^2$, $g(x) = \frac{1}{x}$ and $h(x) = 1+2x$.

These functions can be composed in six possible different orders.

    a) Write out formulas for the six multiple composites
$f \circ g \circ h$, $h \circ f \circ g$, $g \circ h \circ f$, $g \circ f \circ h$, $f \circ h \circ g$ and $h \circ g \circ f$.
    b) Are any of these six functions the same?
    c) Why are all the composites rational functions?
    d) Find the derivative of each of the composites.

28) In Einstein's theory of relativity the energy E of a particle with rest mass m_0 moving at a speed v is given by $E = mc^2 = \dfrac{m_0 c^2}{\sqrt{1-\dfrac{v^2}{c^2}}}$ where c is the speed of light. Find how the energy changes with velocity. That is, find $\dfrac{dE}{dv}$.

29) Consider a straight-wire transmitting antenna whose length is an odd number N of half-wavelengths. Let I be the antenna current and consider a point P at a distance r from the antenna measured along a line making an angle θ with respect to the axis of the antenna. The intensity $\mathbf{E}$ of the field associated with the transmitted wave at this point is $\mathbf{E} = \dfrac{60I}{r} \dfrac{\cos(2\pi N \cos\theta)}{\sin\theta}$. Find $\dfrac{d\mathbf{E}}{d\theta}$.

30) The little mouse shown has a mass of x grams. Let u be the mouse's mass in kilograms and let y be its weight in newtons. Express y as a function of u and u as a function of x. Then verify that the chain rule holds.

That is, show that $\dfrac{dy}{dx} = \dfrac{dy}{du}\dfrac{du}{dx}$.

Recall that weight W is related to the mass m by $W = mg$ where g is the acceleration of gravity.

31) The small gap D between two lengths of railroad track decreases noticeably on a hot day due to thermal expansion of the tracks. At some reference temperature T_0, each track has length L_0 and the gap measures a distance D_0. The length of each track is a linear function of temperature $L = L_0 + m(T-T_0)$ where m is a constant.

    a) Express the distance D between the tracks as a linear function of L.
    b) Calculate $\dfrac{dD}{dT}, \dfrac{dD}{dL}$, and $\dfrac{dL}{dT}$.
    c) Verify the chain rule $\dfrac{dD}{dT} = \dfrac{dD}{dL}\dfrac{dL}{dT}$.

Rate of Change of Kinetic Energy

The kinetic energy of a particle of mass m and speed v is $KE = \frac{1}{2}mv^2$.
Find the kinetic energy for each of the following types of motion. Here $x(t)$ represents the position of the particle at time t.
Then find the rate at which the kinetic energy changes in each case.

32) $x(t) = \frac{1}{2}gt^2$ Freefall (g = gravitational constant)

33) $x(t) = x_0 + v_0 t$ Motion with constant velocity.

34) $x(t) = x_0 + v_0 t + \frac{at^2}{2}$ Motion with constant acceleration.

35) $x(t) = x_0 + A\cos(\omega t)$ Simple Harmonic Motion

    Comment:
    Since energy must be conserved the change in kinetic energy requires the input or exchange of energy from some other source.

36) Consider the composite $f \circ g \circ h$ of three differentiable functions f, g and h.

    a) Given that $\frac{dh}{dx} = 0$ when $x = 0$, what is the derivative of $f \circ g \circ h$ at $x = 0$?

    b) If the derivatives of f, g and h are never zero, show that the derivatives of any of the six possible composites of f, g and h are never zero.

37) Suppose a particle moves along the x-axis in such a way that its velocity v is a function of x, that is $v = F(x)$. Show that the acceleration of the particle must satisfy the relation $a = F(x) F'(x)$.

Find $\frac{dy}{dx}$ for each of the following implicit relations between x and y using the method of implicit differentiation.

38) $x^3 + y^3 = 9xy$ Folium of Descartes

39) $y^2 = 4ax$ Parabola

40) $xy + x + y + 1 = 0$ Hyperbola

41) $x^n + y^n = nxy$ (where n is a positive integer).

42) $(x^2 + y^2)^2 = 16(x^2 - y^2)$ Lemniscate

43) The astroid is described by the implicit equation $x^4 + y^4 = 1$. Find $\frac{dy}{dx}$ by implicit differentiation. Identify any points where the derivative is not defined.

44) Find the equation of the tangent line to the curve $x^2y + xy^2 = 16$ at the point $(2,2)$.

45) Find equation of the tangent line to the curve $\sqrt{x} + \sqrt{y} = 3$ at the point $(1, 4)$.

46) Find points on the ellipse $x^2 + 4y^2 = 36$ whose slope of the tangent line is 1.

47)   **a)**   Sketch the graph of $x^3 + y^3 = 9xy$ (Notice the existence of a loop).

       b)   Determine the points on the curve where the tangent lines are horizontal.

       c)   Determine the points on the curve where the tangent lines are vertical.

48)   **a)**   Is there a point on the curve $x^2y^2 + xy - 2 = 0$ where the tangent line is horizontal?

       b)   Find all points on the curve where the slope of the tangent line is -1.

49)   **a)**   Find the equation of the tangent line to the curve $x^2 + 4x + y^2 + 6y = 12$ at the point $(3,-3)$.

       b)   Find the equation of the tangent line to the curve $x^2 + 4x + y^2 + 6y = 12$ through the point $(2,2)$. Note that the point $(2,2)$ is not on the curve.

50) The following relation is used in hydrology to describe the flow of water to wells. (k is a constant) It describes the dividing line between that portion of the water which flows to the well and that part which flows past the well.

$$y + x \tan ky = 0$$

   a)   Find $\dfrac{dy}{dx}$ by implicit differentiation.

   b)   Express $\dfrac{dy}{dx}$ as a rational function of x and y using such trigonometric identities as $\tan^2\theta + 1 = \sec^2\theta$.

   c)   The stagnation point is that point where $y = 0$. Using the small angle approximation $\tan\theta \sim \theta$ show that the stagnation point occurs at $x = -\dfrac{1}{k}$.

51) Mechanical Punch

A mechanical punch used in an assembly line is driven by an oscillating cam as shown in the illustration. For each complete oscillation of the cam, the punch will move up and down twice.

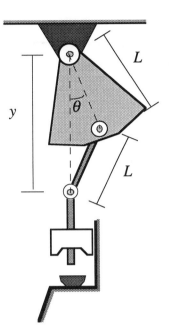

a) Express the distance y as a function of the angle θ.

b) Assuming that the time dependence of the angle theta is given by the expression

$\theta(t) = \theta_0 \cos \omega t$ express the distance y as a function of time. What is the <u>stroke</u> of the punch? Recall the stroke is the maximum linear displacement.

c) Find the velocity and acceleration of the punch as functions of time.

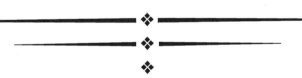

NOTES

Chapter Three

Applications of the Derivative

§3.1 Natural Exponential and Logarithm

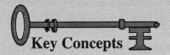

Key Concepts

- ❖ Natural Exponential Function
- ❖ Hyperbolic Functions
- ❖ Newton's Law of Cooling
- ❖ Natural Logarithmic Function
- ❖ Carbon Dating

Question: Does any function equal its own derivative?

That is, is there any function $f(x)$ satisfying $f'(x) = f(x)$.

This little query will lead us to some of the most beautiful results in modern mathematics and to what may be the most important function in engineering analysis.

What kind of function could be its own derivative?
Since the derivative of a constant is zero, the constant function $f(x) = 0$ is equal to its own derivative, but we are on the hunt for bigger game than this trivial case!

Could a polynomial equal its derivative?
No, the derivative of a polynomial of degree n is a polynomial of degree $n-1$. (Prove this.)

Could a trigonometric function equal its own derivative?
No, the derivative of a sine is a cosine and the derivative of a cosine is the negative of a sine.

Could an exponential a^x equal its own derivative?
Let us try to find the derivative of $f(x) = a^x$ using the definition of the derivative.

$$\frac{d}{dx}(a^x) = \lim_{h \to 0} \frac{a^{x+h} - a^x}{h} = \lim_{h \to 0} \frac{a^x \cdot a^h - a^x}{h} = a^x \lim_{h \to 0} \frac{a^h - 1}{h}$$

You may have noticed that we used the law of exponents $a^{x+h} = a^x \cdot a^h$.

To finish of the calculation of this derivative, we need to evaluate the limit $\lim_{h \to 0} \frac{a^h - 1}{h}$.

At this stage, we can't write an actual numerical value for this limit, but we can note that the limit is the derivative of $f(x) = a^x$ at $x = 0$.

$$\lim_{h \to 0} \frac{a^h - 1}{h} = f'(0)$$

Summarizing, we have shown that the derivative of any exponential function is _proportional_ to itself:

$$\frac{d}{dx}(a^x) = f'(0) \, a^x \quad \text{where} \quad f'(0) = \lim_{h \to 0} \frac{a^h - 1}{h}$$

This brings us very close to our goal of finding a function which is its own derivative. If the limit $\lim_{h \to 0} \frac{a^h - 1}{h} = 1$ for some choice of the base a, then $\frac{d}{dx}(a^x) = a^x$ so that the function a^x would be its own derivative! We will show that there is exactly one real number for which this is true. But how can we find this special number?

Since $f(0) = a^0 = 1$ and $f'(0) = f(0) = 1$ the tangent line through $(0, 1)$ is $y = 1 + x$. Intuitively, this tangent line should approximate the function near $x = 0$. That is, $a^x \sim 1 + x$. The approximation gets better if x gets smaller.

Letting $x = 1$, we find $a^1 \sim 1 + 1$ so $a \sim 2$.

Letting $x = 1/10$, we find $a^{1/10} \sim 1 + \frac{1}{10}$ so $a \sim \left(1 + \frac{1}{10}\right)^{10} = 2.593742$

Letting $x = 1/100$, we find $a^{1/100} \sim 1 + \frac{1}{100}$ so $a \sim \left(1 + \frac{1}{100}\right)^{100} = 2.704814$

Letting $x = 1/n$ we have $a^{1/n} \sim 1 + \frac{1}{n}$ so $a \sim \left(1 + \frac{1}{n}\right)^n$

We define:

$$e = \lim_{n \to \infty} \left(1 + \frac{1}{n}\right)^n \sim 2.718281828$$

With this number e we have:

$$\frac{d}{dx} e^x = e^x$$

Example 1 - Derivatives of functions involving e^x.

We can find derivatives for many functions involving e^x using the differentiation rules.

a) $\dfrac{d}{dx}(e^{2x}) = e^{2x} \cdot 2$ Chain Rule

b) $\dfrac{d}{dx}(e^{x^2}) = e^{x^2} \cdot 2x$ Chain Rule

c) $\dfrac{d}{dx}(x^2 e^x) = 2x \cdot e^x + x^2 e^x$ Product Rule

---

Graphical Properties of the Natural Exponential.

Since the numerical value of e is greater than one, the graph of the function $y = e^x$ increases monotonically as x increases. The exponential increases without bound as x approaches infinity and decreases towards zero as x approaches negative infinity. Thus by intermediate value theorem, every positive value y is the exponential of some real number.

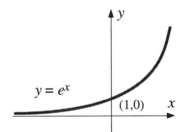

Theorem

The natural exponential function $y = e^x$ satisfies the following properties.

1) **Domain and Range**

   The exponential function is defined for all real x.
   Its range is restricted to the set of positive real numbers since its graph is entirely above the x-axis.

2) **Intercepts**

   x-axis intercepts: There are none since $e^x > 0$ for all x.
   y-axis intercept: The graph of e^x crosses the y-axis at $y = 1$ since $e^0 = 1$.

3) **Monotonicity**

   The function e^x increases monotonically over all its domain.
   It is concave up. (i.e. its derivative is also increasing)

4) **Limits at $\pm\infty$**

   $\displaystyle\lim_{x \to \infty} e^x = +\infty$ $\displaystyle\lim_{x \to -\infty} e^x = 0$

The Inverse Function

Like all the exponential functions, the natural exponential function has an inverse function. The inverse of an exponential is called a logarithm and is defined by:

$$y = \log_a(x) \quad \text{if and only if} \quad x = a^y$$

In the case of the natural logarithm we usually write $y = \ln x$ instead of $y = \log_e(x)$ to emphasize that we are using the <u>natural</u> base. In this notation we have:

$$y = \ln x \quad \text{if and only if} \quad x = e^y$$

As with any simultaneous plot of inverse functions, the graphs of the exponential and logarithm are mirror images of each other in the line $y = x$.

Because of its mirror image property and because of its definition as the inverse of the exponential function we can conclude by inspection that the natural logarithm function has the following properties.

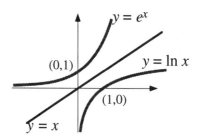

Properties of the Natural Logarithm

The natural logarithm function $y = \ln x$ satisfies the following properties.

1) Domain and Range

The logarithm is only defined for $x > 0$.
However, its range consists of the entire set of real numbers.
The domain and range are the opposite of those for the exponential function.

2) Intercepts

x-axis intercepts: The graph of $y = \ln x$ crosses the x-axis at $x = 1$ since $e^0 = 1$ implies that $\ln 1 = 0$.

y-axis intercept: There are none since the logarithm is not even defined for $x \leq 0$.

3) Monotonicity

The function $\ln x$ increases monotonically over all its domain.
It is concave down. (i.e. its derivative is decreasing)

4) Asymptotic Limits

$$\lim_{x \to \infty} \ln x = +\infty \qquad \lim_{x \to 0^+} \ln x = -\infty$$

§3.1 Natural Exponential & ... CAFÉ Page 183

Derivative of the Natural Logarithm

The definition of the natural log as the inverse function of the natural exponential states:

$$e^{(\ln x)} = x \quad \text{for } x > 0$$

Differentiating this equality and applying the chain rule gives:

$$\frac{d}{dx} e^{(\ln x)} = e^{(\ln x)} \frac{d}{dx}(\ln x) = 1 \quad \text{or} \quad x \frac{d}{dx}(\ln x) = 1$$

Isolating the unknown derivative we discover the incredible relation:

$$\boxed{\frac{d}{dx}(\ln x) = \frac{1}{x}} \quad \text{<-- Derivative of the natural logarithm.}$$

Example 2 - Derivatives of functions involving $\ln x$.

We can find derivatives for many functions involving $\ln x$ using the differentiation rules or using the properties of the logarithm.

a) i. $\quad \frac{d}{dx}(\ln 2x) = \frac{1}{2x} \cdot 2 = \frac{1}{x}$ $\qquad$ Chain rule

or

ii. $\quad \frac{d}{dx}(\ln 2x) = \frac{d}{dx}(\ln x + \ln 2) = \frac{1}{x}$ $\qquad$ Log of product is sum of logs

b) i. $\quad \frac{d}{dx}(\ln x^2) = \frac{1}{x^2} \cdot 2x = \frac{2}{x}$ $\qquad$ Chain rule

or

ii. $\quad \frac{d}{dx}(\ln x^2) = \frac{d}{dx}(2 \ln x) = \frac{2}{x}$ $\qquad$ Power rule for logs

c) $\quad \boxed{\frac{d}{dx}(\ln f(x)) = \frac{f'(x)}{f(x)}}$ $\qquad$ Chain rule

d) $\quad \boxed{\frac{d}{dx} e^{f(x)} = e^{f(x)} f'(x)}$ $\qquad$ Chain rule

---

All for one and one for all!

All the exponential functions $y = a^x$ can be written in terms of the single natural exponential function e^x. Using the inverse properties of the exponential and logarithmic functions, we can write any positive number a in the form $a = e^{\ln a}$.

Thus, $a^x = \left(e^{\ln a}\right)^x = e^{x \ln a}$.

$$a^x = e^{x\ln a} \qquad a>0$$

This 'basic' relationship enables us to express any exponential functions in terms of the natural exponential function.

Application: Calculator Microchips

Many calculators have a key for evaluating the binary function x^y.
Rather than producing circuitry for each of the infinite possible bases x, calculator manufacturers can take advantage of the above relationship to write:

$$x^y = e^{y\ln x}$$

Thus any calculator having circuitry to multiply, and calculate both the natural exponential and logarithm can also calculate exponentials in any base.

Suppose the calculator manufacturer wants to have a key for the log in any base a.

Thus given an input x, it must calculate $y = \log_a x$.
By definition of the logarithm, the unknown y is defined by $a^y = x$.
Taking the natural log of both sides we obtain $y \ln a = \ln x$ or :

$$\log_a x = \frac{\ln x}{\ln a}$$

Derivatives of exponentials and logarithms in any base.

What about the derivatives for the other exponential functions?

Method 1:
We have shown that the derivative of any exponential function is <u>proportional</u> to itself:

$$\frac{d}{dx}(a^x) = f'(0)\, a^x \quad \text{where} \quad f'(0) = \lim_{h\to 0} \frac{a^h - 1}{h}$$

To evaluate the proportionality constant we let $a = e^b$ and then take the limit.

$$\lim_{h\to 0} \frac{a^h - 1}{h} = \lim_{h\to 0} \frac{(e^b)^h - 1}{h} = \lim_{h\to 0} \frac{e^{hb} - 1}{h} = b \lim_{h\to 0} \frac{e^{hb} - 1}{hb}$$

$$= b \times (\text{derivative of } e^x \text{ at } x = 0) = b$$

Thus we can conclude:
$$\frac{d}{dx}(a^x) = (\ln a) a^x$$

Method 2:

$$\frac{d}{dx}(a^x) = \frac{d}{dx}(e^{x\ln a}) = e^{x\ln a}(\ln a) = a^x(\ln a) \qquad a > 0$$

Example 3

Find the derivatives of the exponential functions

    a) 10^x **b)** x^x **c)** $(1+x^2)^x$

    d) $(f(x))^x$,   **e)** $\log_a x$,   **f)** $\log_a f(x)$

Solution

a) $\quad \dfrac{d}{dx}(10^x) = \dfrac{d}{dx}(e^{x\ln 10}) = e^{x\ln 10}(\ln 10) = 10^x(\ln 10)$

b) $\quad \dfrac{d}{dx}(x^x) = \dfrac{d}{dx}(e^{x\ln x}) = e^{x\ln x}(\ln x + 1) = x^x(\ln x + 1) \qquad x > 0$

c) $\quad \dfrac{d}{dx}(1+x^2)^x = \dfrac{d}{dx}(e^{x\ln(1+x^2)}) = (1+x^2)^x\left(\ln(1+x^2) + \dfrac{2x^2}{1+x^2}\right)$

d) $\quad \dfrac{d}{dx}(f(x))^x = \dfrac{d}{dx}(e^{x\ln(f(x))}) = (f(x))^x\left(\ln(f(x)) + \dfrac{xf'(x)}{f(x)}\right) \qquad f(x) > 0$

e) $\quad \dfrac{d}{dx}\log_a x = \dfrac{d}{dx}\left(\dfrac{\ln x}{\ln a}\right) = \dfrac{1}{\ln a} \cdot \dfrac{1}{x}$

f) $\quad \dfrac{d}{dx}\log_a f(x) = \dfrac{1}{\ln a} \cdot \dfrac{f'(x)}{f(x)}$

Hyperbolic Functions

Any function $f(x)$ can be decomposed into the sum of an even and an odd function as follows.

$$f(x) = \underbrace{\left(\frac{f(x) + f(-x)}{2}\right)}_{\text{Even}} + \underbrace{\left(\frac{f(x) - f(-x)}{2}\right)}_{\text{Odd}}$$

Recall that a function is even if $f(-x) = f(x)$ and a function is odd if $f(-x) = -f(x)$. The even and the odd component of the exponential function arise so frequently in engineering and scientific applications that they are given names. The even component is called the hyperbolic cosine or <u>cosh</u> and the odd component is called the hyperbolic sine or <u>sinh</u>.

The even component of e^x is: $\qquad \cosh x = \dfrac{e^x + e^{-x}}{2}$

The odd component of e^x is: $\qquad \sinh x = \dfrac{e^x - e^{-x}}{2}$

The graphs of both the cosh and sinh functions are shown.

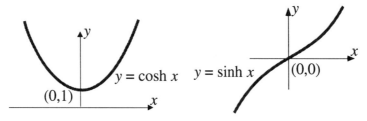

The y-intercepts of these functions are:

$$\cosh 0 = \frac{e^0 + e^{-0}}{2} = 1 \qquad \sinh 0 = \frac{e^0 - e^{-0}}{2} = 0$$

Derivatives of the Hyperbolic Functions.

The derivatives of the cosh and sinh functions are easy to calculate using the fact that the exponential function e^x is its own derivative.

$$\frac{d}{dx}\cosh x = \frac{d}{dx}\left(\frac{e^x + e^{-x}}{2}\right) = \frac{e^x - e^{-x}}{2} = \sinh x$$

$$\frac{d}{dx}\sinh x = \frac{d}{dx}\left(\frac{e^x - e^{-x}}{2}\right) = \frac{e^x + e^{-x}}{2} = \cosh x$$

Thus:

$$\boxed{\frac{d}{dx}\sinh x = \cosh x \qquad \frac{d}{dx}\cosh x = \sinh x}$$

Warning: Note the difference in sign between the derivatives of the usual trigonometric cosine and the hyperbolic cosine.

$$\frac{d}{dx}\cos x = -\sin x \qquad \frac{d}{dx}\cosh x = +\sinh x$$

Example 4 - Derivatives of the Hyperbolic Functions

a) $\quad \dfrac{d}{dx} \cosh 2x = 2 \sinh 2x$

b) $\quad \dfrac{d}{dx}\left(\cosh^2 x - \sinh^2 x\right) = 2\cosh x \sinh x - 2 \sinh x \cosh x = 0$

c) $\quad \dfrac{d}{dx}\left(\cosh x \sinh x\right) = \sinh^2 x + \cosh^2 x$

Notice that in part *b*, above the derivative was zero so that the function $\cosh^2 x - \sinh^2 x$ must be constant. To find that constant, we only need evaluate the function at a typical point such as $x = 0$. This leads to the identity:

$$\cosh^2 x - \sinh^2 x = 1$$

Note the sign difference from the usual trigonometric identity $\cos^2 x + \sin^2 x = 1$. The identity we have just proved is the motivation for calling these new functions the hyperbolic functions. The graph of the parametric curve $x = \cosh t$, $y = \sinh t$ satisfies the relation $x^2 - y^2 = 1$ which we recognize as the graph of a hyperbola.

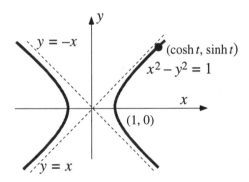

We summarize:

$$\sinh(-x) = -\sinh x, \qquad \cosh(-x) = \cosh x$$
$$\sinh(x + y) = \sinh x \cosh y + \cosh x \sinh y$$
$$\cosh(x + y) = \cosh x \cosh y + \sinh x \sinh y$$

Definitions and some elementary properties of other hyperbolic functions are included in the exercises.

Physical Applications

Example 5 - Carbon Dating

The bombardment of the upper atmosphere by cosmic rays converts nitrogen to a radioactive isotope of carbon, C-14, with a half life of about 5730 years. Vegetation absorbs carbon dioxide through the atmosphere and animal life assimilates C-14 through food chains. When a plant or animal dies it stops replacing its carbon and the amount of C-14 begins to decrease through radioactive decay. By detecting the percentage of the remaining C-14 as compared to the normal percentage in the living plant or animal, one is able to determine the age of the ancient fossil.

For example, a humanoid skull was found in a cave in South Africa along with the remains of a campfire. Archaeologists believe the age of the skull to be the same as the campfire. It is determined that only 2% of the original amount of C-14 remains in the burnt wood of the campfire. Estimate the age of the skull.

Solution

Let $y(t)$ be the amount of C-14 in the specimen of the burnt wood at time t (in years). The radioactive decay of C-14 states that $\frac{dy}{dt} = -ky$, where k is the decay constant. The solution of the above differential equation is just $y(t) = Ae^{-kt}$. To determine A, let's assume that when $t = 0$, $y(0) = y_0$. Thus:
$$y(t) = y_0 e^{-kt}.$$

Now we need to use the half-life of C-14 to determine k:
$$y(5730) = \tfrac{1}{2} y_0 = y_0 e^{-5730k}$$

Canceling y_0 from the equation leads:
$$\frac{1}{2} = e^{-5730k} \Rightarrow k = \frac{\ln 2}{5730}$$

Hence, $y(t) = y_0 e^{-\frac{\ln 2}{5730} t}$

If 2% of the original amount remains t must satisfy:
$$\frac{1}{50} y_0 = y_0 e^{-\frac{\ln 2}{5730} t}$$

By canceling y_0 and then taking the logarithm we have
$$t = \frac{\ln 50 \cdot 5730}{\ln 2} \sim \boxed{32000 \text{ years}}$$

Example 6 - Newton's Law of Cooling

This law says that the temperature of a hot object decreases at a rate proportional to the difference between the temperature of the object and the temperature of the surrounding air. A cup of hot coffee at 95°C stands in a 25°C room. After 2 minutes its temperature is 85°C.

a) Write down a differential equation for this law.
b) How long will it be before the coffee cools to 50°C?
c) What will the temperature of the coffee be after 15 minutes?

Solution

Let $y(t)$ be the temperature (in °C of the coffee) t minutes after it was 95°C. Then Newton's Law of Cooling states that

$$\frac{dy}{dt} = -k(y-25)$$

It seems that this differential equation is not quite the same as the one in our previous carbon dating problem.

Let $u(t) := y(t) - 25$. Then $\frac{du}{dt} = \frac{dy}{dt}$, so the above differential equation for $y(t)$ is converted in terms of $u(t)$.

$$u(t) = Ae^{-kt} \qquad \frac{du}{dt} = -ku$$

or equivalently

$$y(t) = 25 + Ae^{-kt}$$

We are given that $y(0) = 95$ and $y(2) = 85$.

$$\begin{cases} 95 = 25 + A \\ 85 = 25 + Ae^{-2k} \end{cases}$$

The first gives $A = 70$ and the second is reduced to $\frac{6}{7} = e^{-2k}$.

Taking logarithm, we have

$$k = \frac{1}{2}\ln\frac{7}{6}.$$

Thus,

$$y(t) = 25 + 70e^{-\left(\frac{1}{2}\ln\frac{7}{6}\right)t}$$

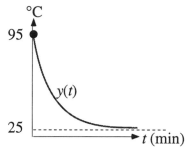

To answer part **b)**, we set $y(t) = 50$ and solve for t.

$$50 = 25 + 70e^{-\left(\frac{1}{2}\ln\frac{7}{6}\right)t} \quad \Rightarrow \quad \frac{5}{14} = e^{-\left(\frac{1}{2}\ln\frac{7}{6}\right)t}$$

$$t = \frac{2\ln\frac{14}{5}}{\ln\frac{7}{6}} = \boxed{13.359 \text{ minutes}}$$

To answer part **c)**, we calculate $y(15)$

$$y(15) = 25 + 70e^{-\left(\frac{1}{2}\ln\frac{7}{6}\right)15} = \boxed{47.029 \text{ °C}}$$

Example 7 - Carbon Monoxide Poisoning

Too often the news contains tragic stories of families overcome by the invisible menace of carbon monoxide as they slept or watched television in their homes. This odorless gas binds with the hemoglobin in the blood thus preventing oxygen from reaching the tissues of the body. The source of the carbon monoxide may be a faulty furnace or a car left running in the family garage. The problem is actually more dangerous now as the new energy efficient homes of today exchange little air with the outside enabling the CO level to rise beyond safe limits.

In this problem the car has been left running in the family garage and the CO has built up to a concentration of 4%. Toxic air from the garage is seeping into the bedroom above the garage at a rate of 1 ft³/min. Assume the volume of the bedroom is $V = 1000$ ft³, the initial concentration of CO in the bedroom is zero and that as air leaks into the bedroom from the garage, an equal amount escapes through the walls and cracks around the windows.

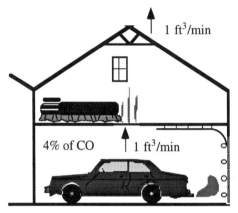

a) Find a formula for the volume $v(t)$ of carbon monoxide in the bedroom at time t.

b) Carbon monoxide is considered to be unsafe if its concentration rises above 0.015% by volume. At what time in minutes will the carbon monoxide in the room achieve this level?

c) How does the situation change if there is a fan in the bedroom window which causes 30 ft³ of fresh air to enter the room each minute? Assume an equal amount escapes through the window. Is it now safe to sleep in the room?

Solution

Let $v(t)$ be the volume (in ft³) of CO in the bedroom at time t (in minutes). Then $\frac{dv}{dt}$, the rate of change of $v(t)$, is understood as coming from two sources: one is the input rate of CO which is $1 \text{ ft}^3/\text{min} \cdot \frac{4}{100} = \frac{1}{25} \text{ ft}^3/\text{min}$, the other is the outflow rate of CO through the bedroom.

Hence,
$$\frac{dv}{dt} = \frac{1}{25} - \frac{v}{1000} \quad \text{or} \quad \frac{dv}{dt} = -\frac{1}{1000}(v - 40).$$

We employ the same technique as before.

Let $y = v - 40$ so that $\frac{dy}{dt} = -\frac{1}{1000} y$

Therefore, $y = Ae^{-\frac{t}{1000}}$ or $v = 40 + Ae^{-\frac{t}{1000}}$

Knowing $v(0) = 0$ implies $A = -40$

$$v(t) = 40\left(1-e^{-\frac{t}{1000}}\right)$$

To answer part **b)**, we solve $\dfrac{v(t)}{1000} = \dfrac{0.015}{100}$ for t.

$$1-\frac{0.015}{4} = e^{-\frac{t}{1000}} \Rightarrow t = -1000\ln\left(1-\frac{0.015}{4}\right)$$

$$\approx \boxed{3.757 \text{ minutes}}$$

To answer **c)** we notice that the modeling differential equation must be modified as:

$$\frac{du}{dt} = \frac{1}{25} - \frac{u}{1000}\cdot 30 \quad \text{or} \quad \frac{du}{dt} = -\frac{3}{100}\left(u - \frac{4}{3}\right)$$

Solving the above differential we have

$$u(t) = \frac{4}{3}\left(1 - e^{-\frac{3t}{100}}\right)$$

and the corresponding CO concentration in the bedroom after a long time is

$$\lim_{t\to\infty} \frac{\frac{4}{3}\left(1 - e^{-\frac{3t}{100}}\right)}{1000} = 0.133\% \text{ which is still higher than } 0.015\%. \text{ Sleeping in the room is}$$

probably still fatal even with the fan!

WARMUP EXERCISES

Simplify the following expressions.

1) $e^{x \ln 2}$

2) $\ln(x\, e^x)$

3) $\log_a(e^x)$

4) $\log_2\left(\dfrac{1}{\sqrt{2}}\right)$

5) $\ln(A\, e^{kt})$

6) $e^{n \ln 3}$

7) Express the function $y = 2^x$ in terms of the exponential function with base 3.

8) Express the function $y = a^x$ in terms of the exponential and log in base b, where $b > 0$, $b \neq 1$.

9) Express the function $y = \log_a x$ in terms of the log in base b, where $b > 0$, $b \neq 1$.

10) **Test the power of your calculator.**
Both compositions $e^{\ln x}$ and $\ln e^x$ should return the input value x, but some calculators have trouble if x is too large. Test your calculator and then summarize your observations.

11) Many calculators can only represent numbers in the range $10^{-100} < x < 10^{100}$. Thus the number $e^{1,000,000}$ is too large. Use your calculator to estimate the number of digits in the number $e^{1,000,000}$. Can you find the <u>leading</u> digit? (For example, the leading digit of the number 4321 is 4.)

12) To discover the number e we used the approximation $e^x \sim 1 + x$. Using a calculator, check how well the approximation works at the numbers $x = \left(\dfrac{1}{10}\right)^n$ for $n = 1, 2, 3, \cdots, 6$. Plot the function $Error(x) = e^x - (1+x)$ over the interval $\left[-\dfrac{1}{10}, \dfrac{1}{10}\right]$. What can you say about the error?

13) It can be shown that the shape of a hanging heavy cable such as a telephone wire or transmission line is described by the hyperbolic cosine. If both ends of the cable are supported at the same height, then the shape of the hanging cable is:
$y = a \cosh bx$ where a and b are constants.
What is the height of the cable when $x = 0$?
Find b if the cable is suspended from a height of h meters above the ground on poles located at $x = \pm d$.

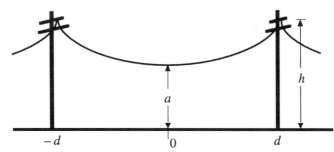

§3.1 Natural Exponential & ... CAFÉ Page 193

❖

Find the first and second derivative for each of the following functions.
Recall the second derivative is the derivative of the derivative.

14) $y = 2 \cdot 3^x + 3 \cdot 2^x$ 15) $y = 2^x + x^2$ 16) $y = x^e + e^x$

17) $y = e^{2x+3}$ 18) $y = (\ln 2)^x$ 19) $y = x^a + a^x$

20) $y = \ln(e+x)$ 21) $y = \ln(\ln x)$ 22) $N = N(0)\, e^{4t}$

23) $y = \log_2 e^x$ 24) $y = \log_2(\log_3 x)$ 25) $y = 3^{\log_2 x}$

26) $y = \ln(e^x + 1)$ 27) $y = \dfrac{\ln x}{x}$ 28) $y = \log_{10}(x^2 + 2)$

❖

The hyperbolic tangent is defined by the relation $\tanh x = \dfrac{\sinh x}{\cosh x}$. Similarly,
$\coth x = \dfrac{1}{\tanh x}$, $\operatorname{sech} x = \dfrac{1}{\cosh x}$, and $\operatorname{csch} x = \dfrac{1}{\sinh x}$.

Find the derivatives of the following functions :

29) $y = \tanh x$ 30) $y = \tanh(x^2)$

31) $y = \coth x$ 32) $y = \operatorname{sech} x$

33) $y = \operatorname{csch} x$

❖

Find the first and second derivatives for each of the following functions.

34) $y = a \cosh kx + b \sinh kx$ 35) $y = e^x \cosh x$

36) $y = \cos(\cosh kx)$ 37) $y = \sinh kx \, \cosh kx$

❖

Find the derivative for each of the following functions.

38) $y = \ln \dfrac{1+x}{1-x}$ 39) $y = \ln(x + \ln x)$ 40) $x(t) = e^{kt} \cos \omega t$

41) $y = \dfrac{1+e^x}{1-e^x}$ 42) $y = \ln(\ln(\ln x))$ 43) $y = \dfrac{e^{4t}}{\ln 4t}$

44) $y = x^{2^x}$ 45) $y = x^x$ 46) $y = (x^x)^x$

47) $y = x^{(x^x)}$ **48)** $y = (e^x)^{e^x}$ **49)** $y = x^e \cdot e^x$

50) Show that $1 = \text{sech}^2 x + \tanh^2 x$.

51) Show that $(\cosh x + \sinh x)^n = \cosh nx + \sinh nx$ for every integer n.

INTERMEDIATE EXERCISES

Using implicit differentiation, find $\dfrac{dy}{dx}$.

52) $\sin(x+y) = e^y$ **53)** $e^x \sin(y) = 2x + y$

A technique called "logarithmic differentiation" is particularly convenient for differentiation when the function to be differentiated consists of many factors. For example, suppose we need to find the derivative of the complicated function:

$$y = \sqrt{\frac{(a-x)(b-x)}{(a+x)(b+x)}}$$

By taking the logarithm of this function, it is broken up into the sum of manageable pieces.

$$\ln y = \frac{1}{2}\left(\ln(a-x) + \ln(b-x) - \ln(a+x) - \ln(b+x)\right)$$

Differentiating, we obtain:

$$\frac{dy}{dx} = -\frac{y}{2}\left(\frac{1}{a-x} + \frac{1}{b-x} + \frac{1}{a+x} + \frac{1}{b+x}\right)$$

Use "logarithmic differentiation" to find the derivatives of the following functions.

54) $y = \left(1 - \dfrac{x}{1}\right)\left(1 - \dfrac{x^2}{2}\right)\left(1 - \dfrac{x^3}{3}\right)\left(1 - \dfrac{x^4}{4}\right)$ **55)** $y = \dfrac{(x^4 - 1)^3 (x^2 + x + 1)^{\frac{3}{2}}}{(x^2 + 1)^{\frac{1}{2}}}$

Using logarithmic differentiation and implicit differentiation find the derivative of y with respect to x.

56) $x^2 y^2 e^y = 1$ **57)** $e^{xy} \ln(x+y) = e^{x^2}$

58) Find an expression for the derivative of the function $(f(x))^{g(x)}$.
(Hint: Use logarithmic differentiation)

§3.1 Natural Exponential & ...

Differentiating known identities often leads to new identities. This is an important method for discovering new formulas.

59) The 'double angle formula for hyperbolic sines' is $\sinh 2x = 2 \sinh x \cosh x$. Verify this identity by expanding both sides in terms of exponentials. Then differentiate the identity to obtain a similar formula for hyperbolic cosines.

60) The 'triple angle formula for hyperbolic sines' is $\sinh 3x = 3 \sinh x + 4 \sinh^3 x$. Verify this identity by expanding both sides in terms of exponentials. Then differentiate the identity to obtain a similar formula for hyperbolic cosines.

61) The 'angle sum identity for hyperbolic sines' is $\sinh(x+y) = \sinh x \cosh y + \cosh x \sinh y$. Verify this identity by expanding both sides in terms of exponentials. Then differentiate the identity with respect to x, i.e., treating y as a constant, to obtain a similar formula for hyperbolic cosines.

62) Find the equation of the tangent line to the curve $y = e^x$ at the point $x = a$.
   a) Show that the x-intercept of the tangent line is always one less than a.
   b) What is the value of the y-intercept of the tangent line?
   c) At what point a does the tangent line pass through the origin?

63) **a)** Find the first, second and third derivative of $y = xe^x$.
   b) A googol is the term introduced by American mathematician Edward Kasner (1878-1955) for the incredibly large number 10^{100}. What is the googolth derivative of $y = xe^x$?
   c) How many zeros are in a googol when it is expanded?
   d) A googolplex is the number $10^{(\text{googol})}$ or $10^{(10^{100})}$. What is the corresponding derivative of $x e^x$?

64) Find an approximation for 2^x near $x = 0$ similar to $e^x \sim 1 + x$.

65) Determine $\lim_{n \to \infty} n(\sqrt[n]{x} - 1)$, where $x > 0$. (Hint: Interpret the limit in terms of a derivative.)

66) Determine $\lim_{n \to \infty} n \ln\left(1 + \frac{x}{n}\right)$, (Hint: Write $n \ln\left(1 + \frac{x}{n}\right) = \dfrac{x \ln\left(1 + \frac{x}{n}\right)}{\frac{x}{n}}$ and interpret the limit in terms of a derivative.)

Physical applications

67) 300 g of a radioactive substance decays to 200 g after 5 years. How much time must pass before 10 g remains?

68) Stockpiling Radioactive Isotopes

A nuclear processing plant produces strontium-90 at the rate of 1 kg/year and stockpiles the isotope in a storage facility. The half-life of strontium-90 is 28 years.

a) How many kilograms of isotope will be in the storage facility after 28 years of production?

b) How many kilograms of isotope will be in the storage facility after a very long time?

69) Warming of Blood Plasma

Blood plasma is stored at a cool 40°F and must be warmed to at least 90°F before it can be used for a patient. When the plasma is placed in a 120°F incubator it takes 45 minutes to warm up to the required 90°F.

a) How long will it take the plasma to warm-up in a 100°F incubator?

b) How long will it take the plasma to warm-up in a 95°F incubator?

70) Suppose a brine solution contains 0.2 kg/liter of salt is poured into a tank initially filled with 500 L of water mixed with 5 kg salt. The solution enters the tank at a rate of 5 L/min, and the mixture is continually stirred. This new mixture is let out of the tank at a rate of 5 L/min.

a) Find the concentration of salt (in kg/L) in the tank after 10 minutes.

b) If the tank had a capacity of 3000 L and the outflow rate was 3 L/min. Let $A(t)$ denote the amount of salt in the tank at time t, and use the fact that the rate of increase in A = rate of input − rate of output to find a differential equation satisfied by $A(t)$.

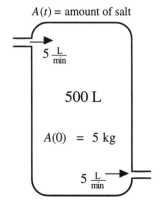

71) Cleaning a Polluted Lake

Water flows into Lake Magalene from Sweetwater Creek (an environmentally clean water source) at a rate of 300 gal/min. The lake contains about 100 million gal of water at any one time. The spraying of nearby orange groves has caused the concentration of pesticides in the lake to reach 35 parts per million. Once pesticides are banned, how much time will pass before the Lake's concentration is below 10 parts per million?

72) Criminology

It was noon on a cold December day (16°C) in Montreal. Detective Taylor arrived at the crime scene to find the sergeant leaning over the body. The sergeant said that there were several suspects. If he knew the exact time of death, the sergeant could figure out who could have murdered him. Detective Taylor took a thermometer and measured the temperature of the body. It was found that the temperature of the body was 34.5°C. He then left for lunch. Upon returning at 1:00 pm, he again measured the body temperature and found it to be 33.7°C. When did the murder occur? (Hint: Normal body temperature is 37°C. Use Newton's Law of Cooling.)

73) Coffee Debate

Two friends sit down to chat and enjoy some coffee. When the coffee is served, the impatient one of the two immediately added a teaspoon of cream to his coffee. The relaxed one waits 5 minutes before adding a teaspoon of cream (which has been kept at a constant temperature) to his coffee. The two now begin to drink their coffee. Who drinks the hotter coffee? (Hint: Assume the cream is cooler than the air, and use Newton's Law of Cooling.)

Explanation

A rope is wound around a rough circular cylinder as illustrated. Assume the coefficient of friction between the rope and the cylinder is μ, and the tension T at any point (specified by angle θ) satisfies the differential equation

$$\frac{dT}{d\theta} = \mu T$$

(Weight of the rope is neglected)
This equation can be heuristically derived as follows:

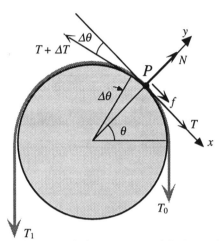

Choose a coordinate axis at point P anywhere where the rope is in contact with the cylinder so that the tension T points in the x-direction. There are four forces acting on the element of the rope indicated by angle $\Delta \ell$: the two tensions T and $T + \Delta T$, the normal force N, and the frictional force f. The fact that the sum of these four forces is zero leads us to know:

$$\left. \begin{array}{r} -(T + \Delta T) + T + f = 0 \\ (T + \Delta T)\Delta\theta = N \\ f = \mu N \end{array} \right\} \text{These equations become exact in the limit as } \Delta\theta \to 0.$$

Hence we have from the above $-\Delta T + \mu(T + \Delta T)\Delta\theta = 0$. Dividing by $\Delta \ell$ and taking the limit as $\Delta\theta \to 0$ gives:

$$\frac{dT}{d\theta} = \mu T$$

The above equation is the familiar equation of exponential growth.

74) Tug of War

A rope tied to a boat is wrapped around a piling n times. The boat then attempts to motor out of the bay, but is restricted by Spike holding the other end of the rope. If the boat is pulling with a force of 5 tons (T_1), and Spike pulls with a force of 50 lbs (T_0), what is the smallest n (an integer) that will keep the boat stationary as long as the rope does not break? (Assume 1 ton = 2000 lbs and coefficient of friction $\mu = 0.3$.)

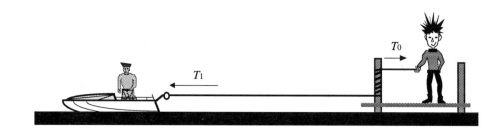

§ 3.2 Linear Approximation & Newton's Method

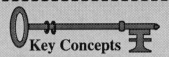
Key Concepts

❖ Zooming Theorem
❖ Linearization
❖ Newton-Raphson Method

Under the Microscope

Key advances in science and technology have often come from careful examination of material or specimens under the microscope. Discoveries made this way include the cell theory, chromosomes and DNA, neurons, and many advances in material sciences. Recently, scientists have even imaged individual atoms.

It should be no surprise then that when we look closely at functions 'under the microscope' a number of great discoveries can be made. In mathematics, placing a function under the microscope means zooming in on its graph at some point. By adjusting the scale, we can magnify the graph of a function without limit.

The illustration on the left shows the graph of the function $y = \sin x$ at normal magnification. The small dotted circle in the illustration has one half the radius of the full field of view. Thus when the magnification is doubled, the portion of the graph inside the small circle now occupies the entire field of view. (right illustration)

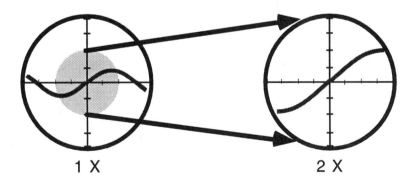

1 X   2 X

In the next figure, the graph of the sine function is shown at magnifications of 1×, 5×, 25× about the origin. Each 5× magnification corresponds to enlarging the portion of the image within a small circle of radius 1/5 so that it fills the complete field of view.

Graphs of $f(x) = \sin x$ seen at magnifications of 1×, 5×, 25×.

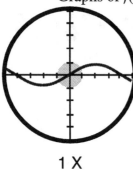
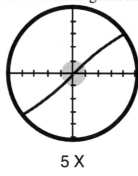
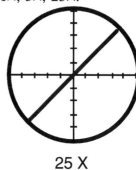

1 X   5 X   25 X

At five times the magnification, the graph appears very similar to a line; at 25 times the graph is indistinguishable from a line of slope 1. This kind of crucial observation reveals an innate simplicity of functions when viewed up close. Could it be that all functions look like lines when viewed closely enough?

In the spirit of discovery let us examine some further 'specimens'. Below we show two more functions each examined at 1×, 5× and 25× magnification. Examine each carefully. Do they both look like lines under great magnification?

Graphs of $f(x) = \tan x$ seen at magnifications of 1×, 5×, 25×.

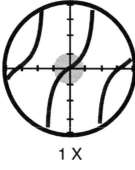
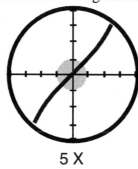
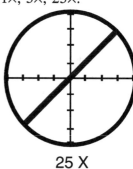

1 X   5 X   25 X

Graphs of $f(x) = |x|$ seen at magnifications of 1×, 5×, 25×.

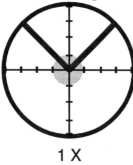
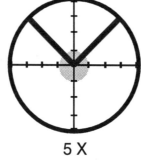
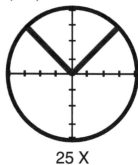

1 X   5 X   25 X

§3.2 Linear Approximation & ...

What can we conclude from this graphical experiment? Notice that at high magnification, the sine and the tangent resemble straight lines but the absolute value function does not! The absolute value function has a cusp at the origin and is not differentiable there. It turns out that under high magnification about a point every <u>differentiable</u> function will look exactly like a line. This critical observation is the most important fact about differentiable functions. Its importance to the foundations of calculus is comparable to the importance of the cell theory to biology or to the atomic nature of matter to physics. Let's try to formalize what we have discovered.

When we look under a microscope at the graph of a function $f(x)$, the image we see is a function of the tick marks t in our field of view. Since the tick marks are on the ocular lens (eyepiece) they are fixed and do not change as the magnification is varied.

Mathematical Expression for the Image Seen under the Microscope.

The mathematical expression for the magnified image of the function depends on where we center the scope and what magnification is used. Let us first center the scope on the graph of the function at a point x_0. At normal magnification $M = 1$, the image seen as a function of the tick marks is:

$$\text{Scope}(t) = f(x_0+t) - f(x_0).$$

The subtraction corresponds to adjusting the microscope view until the image is centered.

At double magnification $M = 2$, the image seen as a function of the tick marks is:

$$\text{Scope}(2\times, t) = 2 \times [f(x_0+t/2) - f(x_0)]$$

Notice the magnification factor $2\times$ appears not only in front of this expression but also in the tick mark variable as $t/2$. When we magnify, both the vertical and horizontal axes must be scaled by the magnification factor.

At magnification M, the image seen is: $\quad \text{Scope}(M\times, t) = M \times [f(x_0+t/M) - f(x_0)]$

Now we will be able to express our fundamental observation about differentiable functions under extreme magnification in a precise manner.

Zooming Theorem

Let $f(x)$ be differentiable at the point x_0. Under extreme magnification about the point x_0, the magnified image of its graph approaches a straight line with slope equal to $f'(x_0)$. That is:

$$\lim_{M \to \infty} \text{Scope}(M\times, t) = f'(x_0)\, t$$

Proof

At magnification M, the image of the function under the scope is given by the function

$$\text{Scope}(M \times, t) = M \times [f(x_0 + t/M) - f(x_0)].$$

Let h denote the reciprocal of the magnification, that is $h = \frac{1}{M}$.
As M approaches infinity, h approaches zero. Thus we can write:

$$\lim_{M \to \infty} \text{Scope}(M \times, t) = \lim_{h \to 0} \frac{f(x_0 + ht) - f(x_0)}{h} = \lim_{h \to 0} \frac{f(x_0 + ht) - f(x_0)}{ht} t = f'(x_0)\, t$$

End of Proof

An Equivalence Relation

Thus if we zoom in on the graph of any differentiable function at a point, its magnified graph will look like a straight line with slope equal to its derivative! This is an example of an <u>equivalence relation</u>. Any two functions that intersect at a point and that have the same slope there will appear the same in the limit of extreme magnification. As far as the microscope is concerned, both functions will be indistinguishable from their common <u>tangent line</u> in this limit. As a special case, we see that any function is equivalent under extreme magnification to its tangent line since they both have the same slope at the point of intersection.

Example 1

Verify that the graphs of the functions
$f(x) = 1 + \sin 2x$ and $g(x) = (1+x)^2$
appear the same at the point $x = 0$ under extreme magnification. That is show that these two functions have the same tangent line at $x = 0$.

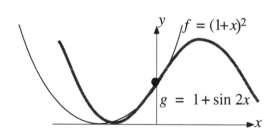

Solution

First we verify that the two graphs actually intersect when $x = 0$.
Indeed $f(0) = 1$ and $g(0) = 1$.

Then we show that their slopes are also equal at $x = 0$.
From the derivatives $\frac{df}{dx} = 2\cos 2x$ and $\frac{dg}{dx} = 2(1+x)$, we see that both functions have slope 2 at $x = 0$. The common tangent line for both functions is $\boxed{y = 1 + 2x}$.

Linear Approximation

Often we only need know the values of a function near a certain point. Since the tangent line at a point is the best linear approximation to the function at that point, we can often replace the actual values of the function in our calculations with the values assumed by the tangent line. This 'linear approximation' is often used if the simplicity gained in the numerical calculations is more important than the slight errors introduced by the approximation.

Nomenclature: **Linear Approximation**

Let $f(x)$ be a differentiable function. The equation of the tangent line to the graph of $f(x)$ at the point $x = a$ is $y = f(a) + f'(a)(x-a)$. In numerical work, this tangent line is called the <u>linearization</u> or <u>linear approximation</u> to the function $f(x)$ at the point $x = a$.

Example 2

a) Find the linear approximation to $f(x) = \tan x$ at the point $x = 0$ and then obtain a numerical estimate for $\tan\left(\frac{1}{100}\right)$.

b) Find the linear approximation to $f(x) = \tan x$ at the point $x = \pi/4$ and then obtain a numerical estimate for $\tan\left(\frac{3}{4}\right)$.

Solution

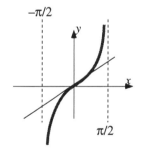

a) The linear approximation at a point is just the tangent line at this point.
The derivative of the tangent function is given by

$f'(x) = \sec^2 x$.

At the point $x = 0$, $f(0) = \tan 0 = 0$ and

$f'(0) = \sec^2 0 = 1$.

Thus the equation of the tangent line at $x = 0$ is:

$$y = f(0) + f'(0)(x-0) = 0 + 1\cdot(x-0) = x$$

This shows that $\tan x \sim x$ if x is near zero.

Therefore $\tan\left(\frac{1}{100}\right) \sim \boxed{0.01}$.

The actual value computed on a calculator is 0.01000033335, so the approximation is good to six digits!!

b) At the point $x = \pi/4$, $f(\pi/4) = \tan(\pi/4) = 1$ and
$f'(\pi/4) = \sec^2(\pi/4) = 2$.
The equation of the tangent line is therefore:
$$y = f(\pi/4) + f'(\pi/4)(x - \pi/4) = 1 + 2(x - \pi/4)$$

This shows that $\tan x \sim 1 + 2(x - \pi/4)$ if x is near $\pi/4$.
Therefore,
$$\tan\left(\tfrac{3}{4}\right) \sim 1 + 2(3/4 - \pi/4) = 1 + \frac{3 - \pi}{2} = \boxed{0.9292036732}$$
The actual value computed on a calculator is 0.9315964599.

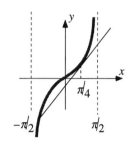

Example 3

Find the linear approximation to $f(x) = \sqrt{x}$ about $x = 100$ then estimate $\sqrt{101}$.

Solution

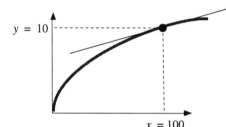

From the derivative $f'(x) = \dfrac{1}{2\sqrt{x}}$ the slope of the tangent line is $f'(100) = \dfrac{1}{2\sqrt{100}} = \dfrac{1}{20}$.

Thus the equation of the tangent line is $y = 10 + \dfrac{1}{20}(x - 100) = 5 + \dfrac{x}{20}$.

Thus $\sqrt{x} \sim 5 + \dfrac{x}{20}$ for x near 100.

Since 101 is near 100 we can estimate $\sqrt{101} \sim 5 + \dfrac{101}{20} = \boxed{10.05}$

Compare this with the value displayed by your calculator.

Example 4 - An Important Linear Approximation

Show that for small x, the linearization of the function $f(x) = (1+x)^k$ gives
$(1+x)^k \sim 1 + kx$

Solution

The derivative of the function is $\dfrac{d}{dx}(1+x)^k = k(1+x)^{k-1}$.

Thus $f'(0) = k$, so the equation of the tangent line at the origin is $y = 1 + kx$.
Therefore $(1+x)^k \sim 1 + kx$ for small x.

Conversion of Mass to Energy

Perhaps the most celebrated formula in modern times is Einstein's equation $E = mc^2$ relating mass to energy. The constant c represents the speed of light - about 300,000 miles/second. Einstein further showed that if a particle is moving with a speed v, then its mass is given by $m = \dfrac{m_0}{\sqrt{1-\frac{v^2}{c^2}}}$ where m_0 is the particle's rest mass. Linearizing the root term we have $\dfrac{1}{\sqrt{1-\frac{v^2}{c^2}}} \sim 1 + \dfrac{v^2}{2c^2}$ if the speed of the particle is small compared with the speed of light. Inserting this linearization into Einstein's equation leads to:

$$E \sim m_0 c^2 \left(1 + \frac{v^2}{2c^2}\right) = m_0 c^2 + \frac{1}{2} m_0 v^2.$$

The second term is just the classical kinetic energy of a particle moving with speed v. Thus in some sense, classical Newtonian physics is just a 'linearization' of Einstein's theory of relativity.

Root Finding (Newton-Raphson method)

Recall that a root or zero of a function $f(x)$ is a point where $f(x) = 0$. Geometrically, a root is a point where the function touches the x-axis. The illustration shows a function having roots at -2 and $+1$. The positive root illustrates that a function need not cross the x-axis at a root. (It only need touch the axis.) Although we have formulas for the roots or zeroes of such functions as lines and parabolas, there are many important engineering functions for which no known formula exists. However, we can use the fact that a differentiable function looks like a line under 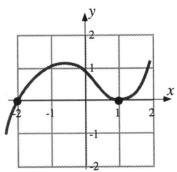 high magnification to create an algorithm that in most cases homes in on the roots of a function to any degree of accuracy. In fact given a differentiable function $f(x)$ we will define a function $Z(x)$ which finds the zeroes of f. Let us first see how to apply the root-finding algorithm and then we will see why it works.

Locating Zeroes of Functions   (Newton-Raphson Method)

To locate the zeroes or roots of a differentiable function f, first construct the associated zero-finding function $Z(x) = x - \dfrac{f(x)}{f'(x)}$ and make an educated guess x_0 for the location of each root. Apply Z repeatedly to the original guess x_0 to get better and better estimates of each root. The sequence of iterates, $x_1 = Z(x_0)$, $x_2 = Z(x_1)$, ... , $x_{n+1} = Z(x_n)$, will in most cases rapidly converge to the desired root. This method is called the Newton-Raphson Method or just Newton's Method.

Example 5

The square root of two is the positive zero of the function $f(x) = x^2 - 2$. Estimate $\sqrt{2}$ using three iterations of the Newton-Raphson formula and the guess $x_0 = 1$.

Solution

First construct the zero-finding function $Z(x)$ using the above formula.

$$Z(x) = x - \frac{f(x)}{f'(x)} = x - \frac{x^2-2}{2x} = \frac{x^2+2}{2x}$$

Starting with the initial guess $x_0 = 1$ and repeatedly applying $Z(x)$ we find:

$$x_1 = Z(1) = \frac{1^2+2}{2 \cdot 1} = \frac{3}{2} = 1.5$$

$$x_2 = Z(1.5) = \frac{(1.5)^2+2}{2 \cdot (1.5)} = \frac{4.25}{3} = 1.41\dot{6}$$

The dot over the six denotes that the six is repeated over and over in the decimal expansion of this number. Continuing:

$$x_3 = Z(1.41\dot{6}) = \frac{(1.41\dot{6})^2+2}{2 \cdot (1.41\dot{6})} = 1.414215686$$

Using our calculators we see that the fist ten digits of $\sqrt{2}$ are in fact 1.414213562. Thus after only three iterations of the zero-finding function, we have been able to get very close to the actual value of $\sqrt{2}$.

How The Root-Finder Works.

We would like to find the root of the function $y = f(x)$ illustrated in the figure. Unfortunately the actual root is often too difficult to calculate directly. By graphical or other means, it may be possible to find a rough estimate x_0 of the root. The goal of Newton's Method is to improve on this rough estimate. Consider the tangent line to the graph of the function $y = f(x)$ at the point x_0. (See left illustration.) The equation of this tangent line is $y = f(x_0) + f'(x_0)(x-x_0)$. Now here is the main point. Because the tangent line is the best linear approximation to the original function, its root will be very close to the actual root.

§3.2 Linear Approximation & ...

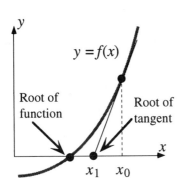
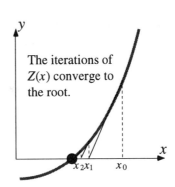

But unlike the actual root, the root of the tangent line is easy to calculate and is given by:

$$x_1 = x_0 - \frac{f(x_0)}{f'(x_0)} = Z(x_0)$$

We can repeat the process using x_1 as the new guess. Thus each iterate is obtained by applying the zero-finding function $Z(x) = x - \frac{f(x)}{f'(x)}$ to the previous estimate. As shown in the right illustration, the iterates usually converge very rapidly to the actual root of $y = f(x)$.

Example 6 - The value of π.

The number π is the zero of the sine function closest to $x = 3$. Estimate π by iterating the corresponding zero-finding function starting with an initial guess of $x_0 = 3$.

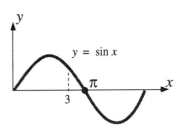

Solution

Let us first simplify the expression for the zero finding function $Z(x)$ associated with the function $f(x) = \sin(x)$.

$$Z(x) = x - \frac{f(x)}{f'(x)} = x - \frac{\sin x}{\cos x} = x - \tan x$$

Starting with $x_0 = 3$ and iterating this zero function we find:

$$\begin{aligned} x_1 &= Z(3) = 3 - \tan 3 = 3.1425465430742778053 \\ x_2 &= Z(x_1) = 3.1415926533004768155 \\ x_3 &= Z(x_2) = 3.1415926535897932385 \\ x_4 &= Z(x_3) = 3.1415926535897932385 \end{aligned}$$

Amazingly, the value of x_4 and all subsequent iterates is the same as that of x_3 to the twenty digits shown. Thus to twenty digits, the value of π is 3.1415926535897932385.

In many applications of Newton's Method, the equation whose roots are to be found is not just given to us but has to be found by modeling a real problem. This is the case in the next sample problem.

Example 7 - FIDO' Fog Dispersal

Many methods have been tried to disperse the intensive fog that often collects over airport runways. In one method, the fog is cleared by igniting long troughs filled with fuel that have been placed along the sides of the runway. A trough for holding the fuel is to be fabricated from a rectangular sheet of metal having the dimensions shown. In cross-section, the trough is an arc of a circle having radius r and subtending an angle θ.

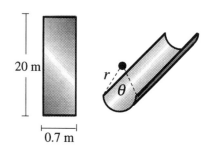

FIDO - Fuel Trough

The manufacturer has specified that each trough should hold 0.85 m³ of fuel. Determine the required central angle using Newton's Method.

Solution

The cross-sectional area of the trough is $A = \frac{1}{2}r^2\theta - \frac{r^2}{2}\sin\theta = \frac{r^2}{2}(\theta - \sin\theta)$.

Thus the volume of the 20 meter trough is $V = 20A = 10r^2(\theta - \sin\theta)$.

Recall that the length of a circular arc of radius r subtending an angle θ is $r\theta$.

Since the width of the original rectangular sheet is 0.7 meters, we have that $r = \frac{0.7}{\theta}$.

Thus the volume can be written $V = 10r^2(\theta - \sin\theta) = 4.9\frac{(\theta - \sin\theta)}{\theta^2}$.

Since each trough should hold a volume $V = 0.85$ m³ of fuel we are faced with solving the monstrous equation $0.85 = 4.9\frac{(\theta - \sin\theta)}{\theta^2}$. After a little rearranging this 'simplifies' to:

$$\frac{17}{98}\theta^2 + \sin\theta - \theta = 0$$

The desired angle is the nonzero root of this truly miserable equation. We will find the root using Newton's Method. Plotting the function, we can see the desired root is slightly larger than one radian. As an initial guess we choose $\theta_0 = 1$.

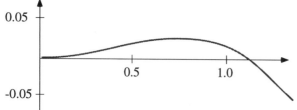

Applying Newton's Method repeatedly we find:

$\theta_1 = 1.132498367$
$\theta_2 = 1.107651049$
$\theta_3 = 1.106637031$
$\theta_4 = 1.106635356$
$\theta_5 = 1.106635356$

Inspecting the iterates to the right, we notice that the fourth iteration gives the root to 10 decimal places!

Thus the sheet metal should be bent until its edges subtend an angle of about 63 degrees.

§3.2 Linear Approximation & ...   CAFÉ   Page 209

Functions under the microscope

When we look at a function $f(x)$ under the microscope, the image we see is a function of the tick marks t in our field of view. The mathematical expression for the image depends on where we center the scope and what magnification is used. If we focus on the point x_0 at magnification M, the image seen as a function of the tick marks is:

$$\text{Scope}(M \times, t) = M \times [f(x_0 + t/M) - f(x_0)].$$

Use the scope function to help solve the following.

1) Let $f(x) = 1 + x + x^2$. Find expressions for the image seen if we focus on the point $x_0 = 0$ with magnifications 1×, 5× and 10×. Plot all three images on the same graph with the tick marks ranging from –1 to +1.

2) Let $f(x) = \sin x$. Find expressions for the image seen if we focus on the point $x_0 = 0$ with magnifications 1×, 5× and 10×. Plot all three images on the same graph with the tick marks ranging from –1 to +1.

3) Let $f(x) = \sin x$. Find expressions for the image seen if we focus on the point $x_0 = \pi/2$ with magnifications 1×, 5× and 10×. Plot all three images on the same graph with the tick marks ranging from 0 to π.

4) Let $f(x) = |x|$. Find expressions for the image seen if we focus on the point $x_0 = 0$ with magnifications 1×, 5× and 10×. Plot all three images on the same graph with the tick marks ranging from –1 to +1.

Under extreme magnification at $x = 0$, which functions in the following groups will be equivalent to $\sin x$ in the sense introduced in the text. Recall two equivalent functions should pass through the same point.

5) a) $(1+x)^{1/3} - 1$ b) $\dfrac{(1+x)^{1/2}}{\sqrt{1+x^5}} - 1$

    c) $e^x - 1$ d) $1 - \cos x$

6) a) $2\sin \dfrac{x}{2}$ b) $\tan x$

    c) $x^2 + 2x$ d) $x^2 - \dfrac{x}{x+1}$

7) a) $2\sqrt{x+1} - 2$ b) $x^{1/3}$

    c) $\sqrt[3]{1+3x} - 1$ d) $x^{2/3}$

8) a) $e^x \sin x$ b) $\frac{1}{2}\sin 2x$
    c) $\ln(x+1)$ d) $x \cos x$

Linearization

9) Find a linearization of $f(x) = \sqrt{x}$ about the point $x = 16$ and use it to estimate $\sqrt{17}$. Compare the estimation with the result from your calculator.

Estimate each of the following roots by linearizing an appropriate function about a suitable integer

10) $\sqrt{24}$ 11) $\sqrt[3]{65}$ 12) $\sqrt[3]{1001}$

13) Which of the six trigonometric functions do not have linearizations at $x = 0$?

Verify each of the following small angle formulas for x near 0.

14) $\sin x \sim x$ 15) $\tan x \sim x$ 16) $\cos x \sim 1$

Newton's Method

17) The zero-finding function $Z(x)$ is based on the notion of approximating a function by its tangent line. Thus it should work very well if applied to an actual line $y = mx + b$.
    a) What is the root of the generic line $y = mx + b$? (assume $m \neq 0$)
    b) What is the zero-finding function $Z(x)$ corresponding to this line?
    c) Show that the zero-finding function finds the zero on the first try no matter what guess we start with!
    (Of course, one would never use Newton's method to find the root of a linear function. This exercise is only intended as a check of Newton's Method.)

18) Using the quadratic formula, find both roots of $f(x) = x^2 - x - 1 = 0$. Then estimate both roots using three iterations of Newton's method. Start with initial guesses +1 and –1. (The positive root is known as the golden section. It often appears in the design of Greek architecture.)

19) Graphically estimate each root of the cubic equation $x^3 - x - 1 = 0$. Then improve each estimate using three iterations of Newton's method.

20) Show that there is a real root of the quintic equation $x^5 - x - 1 = 0$ in the interval [1,2]. Then estimate using three iterations of Newton's method.

§3.2 Linear Approximation & ...   CAFÉ

21) Approximate the cube root of 3 to an accuracy of four decimal places. Using Newton's method, start with an initial guess of 1.

What polynomial equation with integer coefficients would you solve when applying Newton's method for approximating the following numbers?

22) $\sqrt[3]{5}$ **23)** $\sqrt{2} + \sqrt{3}$ **24)** $3 + \sqrt[7]{11}$ **25)** $\dfrac{1}{\sqrt[3]{2}} + \sqrt{3}$

26) Find a quadratic polynomial $f(x)$ having $\sqrt{5}$ as a root. Calculate the zero-finding function $Z(x)$ for this polynomial and then estimate $\sqrt{5}$ starting with a guess $x_0 = 2$ and iterating the zero-finding function three times.

27) Using Newton's method, find all nonzero roots of the equation $f(x) = 2\sin x - x$ to an accuracy of four decimal places. (Be sure to graph the function so you can estimate the roots.)

28) In the sample problem involving 'Fog Control' we were lead to find the nonzero solution of the equation: $\dfrac{17}{98}\theta^2 + \sin\theta - \theta = 0$.

   a) Find an expression for the zero-finding function $Z(\theta)$.

   b) Solve for the root using an initial guess of $\theta_0 = 1.1$ and four iterations of Newton's method.

29) Occasionally, the method of approximating the actual root by the roots of successive tangent lines can do unexpected things. Consider the function $f(x) = x^{1/3}$ which has the single root $x = 0$. The zero-finding function $Z(x)$ for this monomial is $Z(x) = x - \dfrac{f(x)}{f'(x)} = x - \dfrac{x^{1/3}}{\left(\frac{1}{3}x^{-2/3}\right)} = -2x$. Thus starting with any nonzero guess, each successive iterate is twice as far <u>away</u> from the root $x = 0$ as the previous estimate. The method fails miserably!

Consider the general monomial $f(x) = x^r$ ($r > 0$ and $r = \dfrac{p}{q}$ when p and q are integers with q odd) which has the single root $x = 0$.

   a) For which values of the rational number r will the method fail?

   b) For which value of r does the method find the root $x = 0$ on the first try for all possible guesses.

30) Newton's method can also fail if the tangent line is horizontal at any of the iterates. The expression $Z(x) = x - \dfrac{f(x)}{f'(x)}$ is then undefined since a horizontal tangent has zero slope and we cannot divide by zero. Give an example of a function $f(x)$ and an initial guess x_0 such that $Z(x_0)$ is undefined.

ADVANCED EXERCISES

31) A barrel is to be fabricated from sheet metal. The manufacturer has specified that it hold a volume $V = 10{,}000\,\text{cm}^3$ and have a total surface area (including bottom) of $A = 2400\text{ cm}^2$. The area of the barrel does not include the lid however.

a) Show that the radius r of the barrel must satisfy the equation $2V + \pi r^3 - Ar = 0$ where V and A have the values given above.

b) Using Newton's Method, find the radius of the barrel to the accuracy of four decimal places. (Is the answer unique?)

32) Determine the central angle θ to an accuracy of four decimal places of a unit circle so that the arc length subtended is two times as big as the length of the chord. (i.e., $\widehat{AB} = 2AB$)

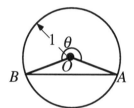

33) A semi-cylindrical boat is made of iron sheet of density 150 kg/m^2. The length is 6 m and the radius of the cylinder is 1 m. Assume the density of water is 1000 kg/m^2. Calculate the draft (the depth that is submerged) of the whole structure to an accuracy of four decimal places. (Hint: Use the Archimedes Principle.)

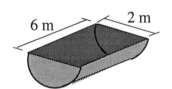

Front View:

34) A flat belt of length 2 m is used to transmit a torque from pulley A to pulley B. The radius of pulley A is 0.25 m and the radius of pulley B is 0.1 m The belt must be kept taut. Calculate the angle θ of subtention as shown to an accuracy of four decimal places.

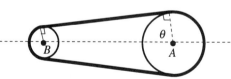

§ 3.3 Related Rates

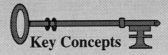

Key Concepts

❖ Relation and Implicit Differentiation

What are Related Rates Problems?

In related rates problems, one must find the rate of change of some quantity, knowing the rate of change of another. The following example is a typical related rates problem.

❖

Example 1 - Failure of Boiler Pipe

The inner radius of a boiler pipe is decreasing at a steady rate due to deposits of minerals from the water supply. When new, the inner radius of the pipe was $r = 2$ inches, but measurements show that the radius of the pipe is decreasing at the rate of 1/5 inch per year. How fast is the cross-sectional area of the pipe decreasing when the radius is 1/2 its original value?

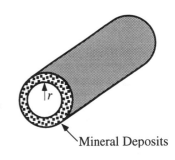

Mineral Deposits

Solution #1

The cross-sectional area is <u>related</u> to the radius by $A = \pi r^2$.

To find a <u>relation between their rates of change</u> we differentiate. This gives:

$$\frac{dA}{dt} = 2\pi r \frac{dr}{dt}$$

At the half way mark the radius is 1 inch. Since the radius is decreasing at a rate of 1/5 inch/year, the area is decreasing at the rate $\frac{dA}{dt} = -2\pi \cdot 1 \cdot \frac{1}{5} = \boxed{-\frac{2\pi}{5} \frac{\text{in}^2}{\text{yr}}}$.

Thus from the rate of change of the radius, we easily obtain the rate of change of the area. This is why such problems are called <u>related rates problems</u>.

Solution #2 (Not recommended)

An alternative approach would be to write the area A explicitly as a function of time by eliminating the radius r. Since r is a linear function which begins at 2 inches and decreases at the rate of 1/5 inch per year its value is:

$$r(t) = 2 - \frac{t}{5}$$

Thus the area as a function of t is
$$A(t) = \pi\left(2 - \frac{t}{5}\right)^2$$

Differentiating we find:
$$\frac{dA}{dt} = -\frac{2\pi}{5}\left(2 - \frac{t}{5}\right)$$

At what time has the radius decreased to half its value? Solving $r(t) = r(0)/2$ leads to $1 = 2 - \frac{t}{5}$ or $t = 5$ years. Evaluating $\frac{dA}{dt}$ at $t = 5$ years we find the same answer as above.

Notice however that the second method involved significantly more work! This is because in the second approach, we failed to exploit the simple connection between the rates of change of the radius and the area.

Example 2 - Crystal Growth

In a space shuttle experiment, the length x of each edge of a cubic crystal was observed to increase at the rate of 1 millimeter per day.
a) At what rate is the volume of the crystal increasing when $x = 10$ mm?
b) At what rate is the volume of the crystal increasing when $V = 8$ cm³?

Solution   The length and the volume are related by the equation:

$$V = x^3$$

The <u>rates of change</u> of the volume and sides can be <u>related</u> by differentiating this expression.

$$\frac{dV}{dt} = 3x^2 \frac{dx}{dt}$$

a)   When $x = 10$ mm, the rate of change of the crystal's volume is:

$$\frac{dV}{dt} = 3 \cdot 10^2 \cdot 1 = \boxed{300 \frac{\text{mm}^3}{\text{day}}}$$

b)   To solve part **b)**, we need to first find x given that $V = 8$ cm³. From the relation $V = x^3$ we see that $x = 2$ cm $= 20$ mm. Thus:

$$\frac{dV}{dt} = 3x^2 \frac{dx}{dt} = 3 \cdot 20^2 \cdot 1 = \boxed{1200 \frac{\text{mm}^3}{\text{day}}}.$$

§3.3 Related Rates — CAFÉ — Page 215

The next problem involves finding a relation between the rates of change of the height and the volume of a concrete pier as it is being poured. As is typical with related rates problems, we first find a relation between these variables and then differentiate.

Example 3 - Pouring a Concrete Pier

A concrete pier is being constructed which will later support a railway bridge. The cross-section is square with the base measuring 3 meters on each side and the top 2 meters on each side. The total height is 2 meters.

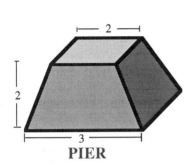
PIER

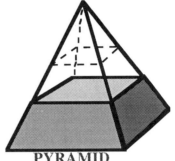
PYRAMID

If the concrete is poured at a constant rate of $\dfrac{dV}{dt} = 0.01 \dfrac{m^3}{sec}$, at what rate is the height h of the poured concrete increasing when $h = 1$ (half way up)?

Solution

The shape of the pier is referred to as a **frustum** of a pyramid. The word frustum is Latin for 'piece'. If the top of the pier was extended to form a complete pyramid, its height would be six meters. (Check this.)

When the concrete has been poured to a height of h meters, it can be considered as a pyramid with the top $(6-h)$ meters removed. The volume of a pyramid is $V = \dfrac{1}{3} A H$ where A is the area of its base and H is its height.

AH-ha !   Thus the volume of concrete poured as a function of height is:

$$V = \frac{1}{3}(AH - ha) = \frac{1}{3}\left[9 \cdot 6 - (3 - h/2)^2 (6-h)\right] = 18 - \frac{(6-h)^3}{12}$$

The volume is a cubic function: $V = 18 - \dfrac{(6-h)^3}{12}$.
Now that we have found a relation between the variables V and h we can differentiate to find a relation between their rates of change.

$$\frac{dV}{dt} = \frac{(6-h)^2}{4} \frac{dh}{dt}$$

Thus $\dfrac{1}{100} = \dfrac{(6-1)^2}{4} \dfrac{dh}{dt}$ which implies that $\boxed{\dfrac{dh}{dt} = \dfrac{1}{625}}$ meters per second.

When the concrete had reached a height of one meter (half way up) the foreman supervising the job noticed it had taken eight loads of concrete so far and ordered that eight more be delivered. Was this wise? How many would you have ordered?

Exampe 4 - 'Tracking the Sun'

The efficiency of a solar panel can be increased by 'tracking the sun' much as the leaves of green plants have been doing for hundreds of millions of years. Since the Earth makes a complete revolution of 360° about its axis each day, the rate of change of the angle of the sun's inclination ϕ is:

$$\frac{d\phi}{dt} = \frac{360°}{24 \text{ hr}} = 15 \frac{\text{deg}}{\text{hr}}.$$

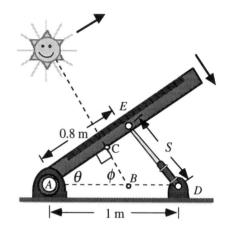

a) The solar panel's angle of inclination is θ. Find the rate of change of θ, if the solar array is kept perpendicular to the rays of the sun by means of a hydraulic jack.

b) At what rate must the length of the hydraulic jack be decreased when $\theta = 45°$ to maintain the panels perpendicular to the sun's rays? That is find $\frac{dS}{dt}$ where S is the length of the hydraulic jack.

Solution

a) Triangle *ABC* in the illustration reveals that the angles of inclination of the sun and the solar array are <u>complementary</u> angles. That is $\theta + \phi = 90°$. Differentiating this relation we see that $\dfrac{d\theta}{dt} = -\dfrac{d\phi}{dt} = \boxed{-15 \dfrac{\text{deg}}{\text{hr}}}$. Thus the angle of the solar array must be <u>decreased</u> at the same rate at which the angle of the sun <u>increases</u>.

b) A relation between the angle of inclination of the array θ and the length S of the hydraulic jack can be found from the law of cosines.

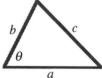

Law of Cosines: $c^2 = a^2 + b^2 - 2ab \cos \theta$

Applying this identity to the triangle *ADC* in the illustration we see that:

$$S^2 = 1^2 + (0.8)^2 - 2 \cdot 1 \cdot (0.8) \cos \theta = 1.64 - 1.6 \cos \theta$$

Thus the rate of change of the hydraulic jack's length S is: $\dfrac{dS}{dt} = 0.8 \dfrac{\sin\theta}{S} \dfrac{d\theta}{dt}$ where we must be careful to express $\dfrac{d\theta}{dt}$ in <u>radians</u> per hour.

$$\dfrac{d\theta}{dt} = -15 \dfrac{\deg}{\text{hr}} \dfrac{\pi \text{ rad}}{180 \deg} = -0.262 \dfrac{\text{rad}}{\text{hr}}$$

Finally, when the arrays are inclined at an angle of 45° we find:

$$\dfrac{dS}{dt} = 0.8 \dfrac{1/\sqrt{2}}{\sqrt{1.64 - 1.6/\sqrt{2}}} \cdot (-0.262) = \boxed{-0.2078 \dfrac{\text{m}}{\text{hr}}}$$

WARMUP EXERCISES

Summary

All related rates problems involve finding the rate of change of some quantity (which we cannot usually measure directly), knowing the rate of change of some other quantity. This is achieved in two steps.
  1) Find a relation between the two quantities.
  2) Differentiate to find a relation between their rates of change.

Step two is always a simple matter of differentiation and can be easily carried out by computer programs. However, the search for relationships between the variables of interest (step 1) often requires all the higher level thinking skills that one can muster. Each of the following related rates problems has been designed to exercise your ability to visualize geometric situations, uncover hidden patterns, explore unexpected relationships and to develop successful modeling strategies. These are the essential skills for the discovery of new knowledge.

1) The radius of a heated metal disc is increasing at the rate $\frac{dr}{dt} = 1$ mm/hr due to thermal expansion. Find the rate of change of its area A at the instant when $r = 10$ centimeters. That is find $\frac{dA}{dt}$.

2) The volume of a cone of height h and radius r is $V = \frac{1}{3}\pi r^2 h$. At a given instant, the radius and height of the cone are both 2 inches.

   a) Find the rate at which the volume is changing if the height of the cone remains constant and the radius is increasing at the rate of 1 inch/sec.
   b) Find the rate at which the height is changing if the radius of the cone remains constant and the volume is increasing at the rate of 1 in³/sec.
   c) Find the rate at which the radius is changing if the height of the cone remains constant and the volume is increasing at the rate of 1 in³/sec.

3) A particle is moving along a curve $xy = 1$ in such a way that the horizontal velocity is always 1 cm/sec. Find its vertical velocity when the particle is at the point (1, 1).

4) If the two sides of a right triangle are increasing at the rate of 1 cm/sec and 2 cm/sec respectively, how fast is the hypotenuse increasing when it forms an isosceles right triangle of side 10 cm?

5) Helium is being pumped into a spherical balloon at a rate of 50 cm³/sec. How fast is the radius of the balloon increasing when the radius is 60 cm? Recall that the volume of a sphere is $V = \frac{4}{3}\pi r^3$.

§3.3 Related Rates   CAFÉ   Page 219

6) A conical pile of gravel is growing in such a way that the shape is always a right circular cone with height equal to its diameter. Gravel is being dumped from a conveyor belt at a rate of 1 m³/min. How fast is the height increasing when the pile is 5 m high? Recall that the volume of a circular cone is $\dfrac{\pi r^2 h}{3}$.

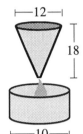

7) A solution is filtering through a conical filter 18 inches deep and 12 inches across into a cylindrical vessel whose diameter is 10 inches. When the depth of the solution in the filter is 12 inches, its level is falling at the rate of 1 inch per minute. At what rate is the level in the cylinder rising?

8) In a crystal growing experiment, two cubical crystals are placed side by side and act as a single seed crystal. Each crystal originally has edges of length 1 millimeter. Thus the length, width and height of the twin crystal are 2, 1 and 1 millimeters respectively. Each dimension increases at the rate of 1 mm/hour during the crystal growing experiment.
   a) At what rate does the volume of the crystal originally begin to increase?
   b) At what rate does the volume of the crystal increase at the instant its length l is 1 cm?

9) A log is lifted into a lumber mill for processing by a power winch which rotates with an angular velocity $\omega = 0.1$ rad/sec. The winch has a radius $R = 0.45$ meters.

   a) What is the speed of the log while it is being pulled by the winch?
   b) At what rate does the height of the log increase?

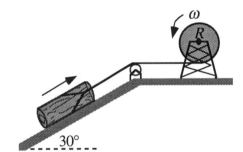

10) A department store escalator has two rotating drums of radius $R = 2$ feet, each rotating at an angular velocity of $\omega = 0.5$ rad/sec. The escalator rises at an angle of 30° from the horizontal.
   a) What is the speed of a person standing on the escalators?
   b) At what rate does the height of a person from the ground increase?

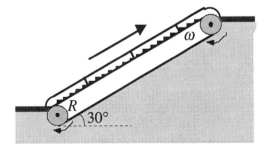

11) The inner radius of a boiler pipe was originally 1 inch but is decreasing at the rate of 1/10 of an inch per year due to mineralization. At what rate is its cross-sectional area of the opening decreasing at the instant when the radius is 1/2 inch?

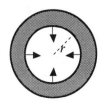

12) During a thunder storm, rain <u>suddenly</u> begins to fall at the rate of one inch/hour. The precipitation is monitored by a conical rain gauge having height 6 inches and radius 3 inches.

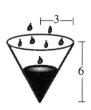

  a) At what rate does the height of rain water in the gauge increase at the instant the rain starts? Is it finite? The gauge is originally empty.
  b) At what rate is the height of rain water in the gauge increasing just before it begins to overflow?

INTERMEDIATE EXERCISES

13) A lazy tongs is made from metal segments each of length 4 inches. The gap between points A and B is decreasing at the rate of 1 inch per second at the instant the metal segments align to form 2×2 squares.

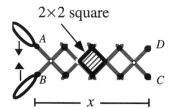

  a) At what rate are the tongs extending?
  b) At what rate is the area of the rectangle $ABCD$ decreasing?

14) A tractor and pulley are used to hoist hay bales onto transport trucks. If the tractor has a velocity of 1 meter/sec, and $h = 10$ meters, determine the upward velocity of the bale when $x = 5$ meters.

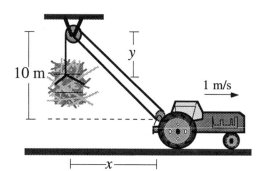

15) The height of the automobile jack shown is controlled by rotating a screw ABC which is single-threaded at each end (right-handed thread at A, left-handed thread at C). The pitch is 2.5 mm.

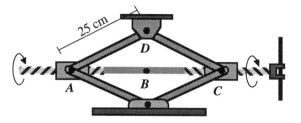

If the screw ABC is rotated at a rate of 10 rpm (revolutions per minute), how fast is the car moving upward when $BC = 20$ cm? (B is the midpoint of AC)

§3.3 Related Rates

16) A van is lifted into the air by a hydraulic jack for repairs. If the jack's piston extends at a rate of 0.3 ft/sec, what is the upward velocity of the van when $\theta = 30°$?

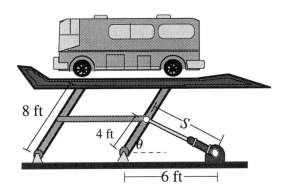

17) A collar slides along a rod at a constant speed v. A thin wire is attached to the collar as shown. The motion of the collar causes additional wire to unravel from the reel located at point O.

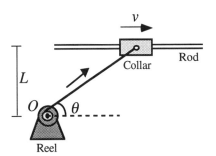

a) Show that $\dfrac{d\theta}{dt}$, the rate of change of the angle θ, satisfies $\dfrac{d\theta}{dt} = -\dfrac{v}{L}\sin^2\theta$.

b) Show that the angular velocity ω of the reel satisfies $\omega = \dfrac{v}{R}\cos\theta$ if its radius is R.

18) A hockey puck slides along the ice with a velocity of 30 meters/sec towards the goalie located 15 meters to its right. A second player skates at 8 meters/sec along a line that makes a 60° angle with the path of the puck in an effort to intercept the puck before the opposing team scores. Determine the rate of change of the distance between the puck and the second player if that player is 4 meters from the goalie at the instant shown.

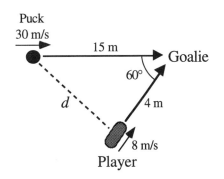

19) Before it is filled, an empty water tank contains air at a temperature of 27°C and at one atmosphere pressure (10^5 Newton/m²). The tank is completely sealed except for an intake valve at the bottom. Water is then pumped through the intake valve at a rate of 0.5 m³/min. What is the rate of increase of air pressure in the tank when the water level is 1 meter high? (Assume the temperature of air remains unchanged so that $PV = $ constant applies.)

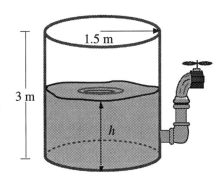

20) Filling the Oil Tank
An empty underground oil tank of dimension shown was laid horizontally. At the beginning of the winter, the heating oil company fills the tank in preparation of the cold weather. Assume the oil enters at a rate of 0.4 m³/min. Determine how fast the oil level is rising when the depth is 0.25 m.

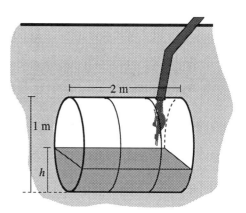

21) Beer is poured into a pitcher in the shape of a frustrum at a rate of $50 \frac{cm^3}{sec}$.

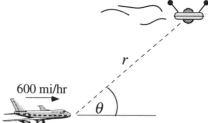

a) Ignoring any foam, how fast is the beer level rising when the pitcher is filled to one half its height?
b) How fast is the beer level rising when the pitcher is one half full?
c) How fast is the beer level rising just before the pitcher begins to overflow?

22) A radar equipped aircraft flies at a constant altitude with a speed of 600 miles/hour. Its radar detects a UFO moving in the same vertical plane. The computerized radar records the distance r and angle θ to the UFO as a function of time and then numerically computes $\dot{r}$ and $\dot{\theta}$.

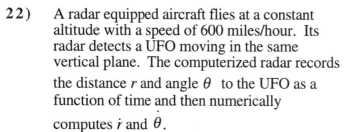

a) Determine the ground speed of the UFO knowing from the computerized radar that $\theta = 20°$, $r = 5$ miles, $\dot{\theta} = 1$ °/sec and $\dot{r} = 1000$ mi/hr.

b) Is the UFO flying at constant altitude at the instant the above data was recorded?

§3.3 Related Rates   CAFÉ   Page 223

23) The pouring of molten steel into ingots is controlled by means of a hydraulic jack. The steel has been melted in a box-shaped container having dimensions 2 x 2 x 1 meters.
   a) If the hydraulic jack extends at a rate of 0.01 m/sec, determine the rate at which the molten steel pours from the container (in m^3/sec) when it has been tipped by an angle $\theta = 30°$.
   b) At what rate should the jack be extended (in m/sec) so that the steel pours at the constant rate of 0.1 m^3/sec? (Express the answer in terms of θ.)

This is an nice example of using calculus to control the rate of a process.

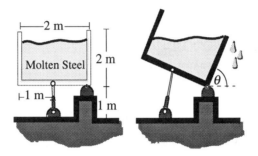

24) The molten contents of a gigantic crucible can be emptied by a cable arrangement. When cable AB is pulled, the crucible will pivot at C and tilt. The distance BD is 0.8 meters and the distance CD is 1 meter. The angle BDC is $120°$. If cable AB is pulled at a rate of 0.05 meters/sec, how fast is the angle of tilt θ increasing at the instant when cable AB has a length of 2 meters?

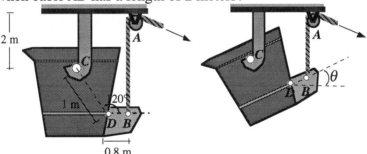

25) A circular swimming pool, having the cross-section shown, is being filled with water. Each horizontal cross-section of the pool is a circle. At the top, the radius of the pool is 10 meters. At the bottom, the radius is only one meter. The pool is being filled at a rate of 5 m^3/min. Determine the rate at which the water is rising in the pool when its height is: (a) 0.25 meters   (b) 1.25 meters

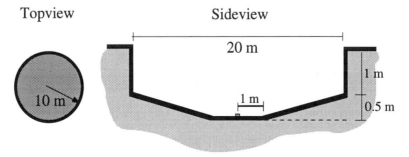

26) A crane is hoisting a tractor by decreasing the length of the cable *AB* at a rate of 0.1 meters/sec. Find the horizontal and vertical components of the tractor's velocity when the boom of the crane is raised at an angle $\theta = 60°$.

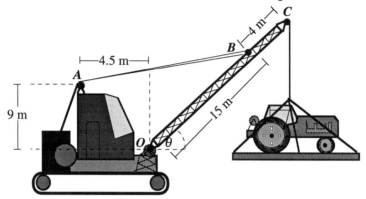

27) Satellites do not last forever, and many eventually come crashing back to Earth. This is a particularly nasty problem when the satellite is nuclear powered. Radar stations must track the reentry of these satellites to predict where they will crash and help prevent loss of life. The following table shows the distance and angle of elevation of a satellite as tracked by a radar station minutes before it crashed.

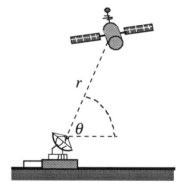

a) Using the approximation,

$$f'(t) \sim \frac{f(t+\Delta t) - f(t)}{\Delta t}$$

estimate the derivative of r and θ at the time $t = 20$ seconds. (Let $\Delta t = 5$).

b) Estimate the velocity vector $\vec{v} = (v_x, v_y)$ of the satellite at this time. (We have assumed the motion takes place in a plane.)

Time (sec) t	Radius (km) r	Angle(°) θ
0	36.4	110.5
5	29.9	100.0
10	26.2	91.0
15	24.1	83.7
20	22.7	77.7
25	21.8	72.7
30	20.9	67.7
35	20.5	63.2
40	20.1	58.6
45	19.7	55.3
50	19.3	52.0

§3.3 Related Rates   CAFÉ   Page 225

28) The telescoping arm is used to raise a worker (Spike) so he can repair damaged power lines. As the hydraulic jack extends, θ increases. If rate of extension is 0.1 ft/sec, determine the rate at which the worker's height increases at the instant when $\theta = 60°$.

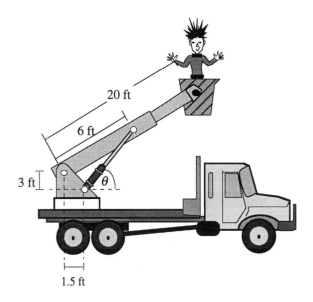

29) **Quick Return Mechanism**
The slotted lever pivots back and forth about the point O. Its motion is driven by a peg D attached to the drive wheel and which is free to slide in the lower slot of the lever. The peg traces out a circle of radius a at a constant angular speed ω. In the upper slot of the lever is a collar C which is constrained to glide horizontally in a guide.

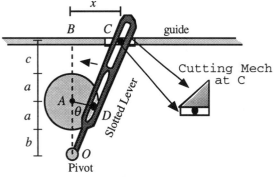

The object of the mechanism is to attach to C a cutting tool that slices one way only and is idle during the return stroke. Therefore the return stroke must be made as quick as possible.)

a) Express the distance x between points B and C as a function of the angle $\theta = \angle DAO$.

b) Determine the velocity of C in terms of θ.

c) Graph the velocity of C as a function of θ for the case $a = 1$, $\omega = 1$, $b = c = 2$ to verify the quick return property.

30) The figure illustrates the drive mechanism attached to the front wheel on a locomotive which is moving to the left with speed v. The drive rod (of length L) connects at one end to a sliding block at A that moves back and forth along a guide. At the other end, the drive rod connects to a crank pin, located at point B on the front driving wheel. As the wheel of radius R rotates, the crank pin traces out a circle of radius r. The center line of the guide is tangent to the circle traced by the crank pin. Determine the velocity of the sliding block A (relative to its guide) as a function of the angle ϕ.

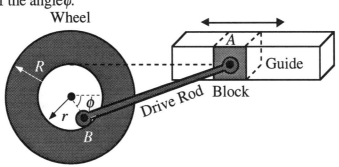

31) In a chemical extraction process, a cupful of methyl alcohol in the conical form shown was left open to the air for 8 hours. During this time, 1/3 of the alcohol evaporated. Assume the evaporation rate is proportional to the area exposed to the atmosphere. How much time would it take before the next 1/3 (of the full cup) evaporates?

§ 3.4 Optimazation

Key Concepts

- ❖ Critical Points
- ❖ First Derivative Test
- ❖ Fermat's Theorem

Local Extrema and Critical Points

The graph below has properties inherent in typical engineering functions.

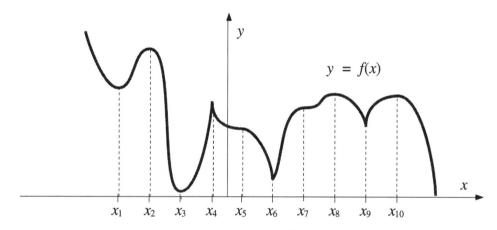

The labeled points separate the graph into monotonic pieces, that is, pieces which are always rising or always falling when we move along the graph from left to right. For example, the graph in the intervals (x_1, x_2), (x_3, x_4), (x_6, x_7), (x_7, x_8) and (x_9, x_{10}) is increasing and the derivative is positive. Similarly, the graph is decreasing in the intervals (x_2, x_3), (x_4, x_5), (x_5, x_6), (x_8, x_9), and (x_{10}, ∞) and its derivative is negative. These intervals are separated by points such as $x_1, x_2, x_3, \ldots, x_{10}$. Notice that the derivative of these points is zero except for x_4, x_6 and x_9, where the derivative does not exist. Such points are called critical points.

> **Definition**
> A point x in the domain of a function is called a critical point if either the derivative is zero or does not exist at the point.

Critical points not only determine intervals of monotonicity, but also determine the locations of valleys (local minima) and peaks (local maxima) of a function.

Definition
A point x in the domain of a function is called a local extremum if it is either a local maximum or a local minimum.

In the example above, the points x_1, x_3, x_6 and x_9 are all local minima (valleys), while x_2, x_4, x_8 and x_{10} are all local maxima (peaks). Points x_5 and x_7 are not local extrema although they are critical points because the derivative is zero. Although not all critical points correspond to a local extremum, the converse is true and is called Fermat's Theorem.

Fermat's Theorem

If $f(x)$ attains a peak (local maximum) or a valley (local minimum) at a point c, then either

a) $\qquad f'(c) = 0 \qquad$ (Tangent line is horizontal)

or

b) $\qquad f'(c)$ does not exist.

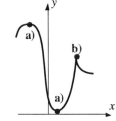

As we saw above, the converse of the statement is not true. Some critical points are neither local maxima or minima. The <u>First Derivative Test</u> helps to decide whether a critical point c is a local maximum, local minimum, or neither.

First Derivative Test

Let c be a critical point of a function $f(x)$.
If a and b are numbers such that $a < c < b$, f is continuous in $[a, b]$ and c is the only critical point of f in $[a, b]$, then:

1) $\quad f(c)$ is a local maximum of f if $f'(a) > 0$ and $f'(b) < 0$.

2) $\quad f(c)$ is a local minimum of f if $f'(a) < 0$ and $f'(b) > 0$.

3) $\quad f(c)$ is not a local extremum otherwise.

The above theorem is illustrated in the following examples.

§3.4 Optimazation — CAFÉ — Page 229

Example 1

Find the local extrema of the cubic polynomial $f(x) = x^3 + 3x^2 - 2$.

Solution

First we differentiate to find the critical points of the function.

$$f'(x) = 3x^2 + 6x = 3x(x+2)$$

Since the derivative is defined at all points, the only critical points are where the derivative is zero. Thus -2 and 0 are the critical points of f.

Let us first test the critical point -2.
Note that $f'(-3) = 9 > 0$ and $f'(-1) = -3 < 0$.
Therefore by the First Derivative Test, the point $x = -2$ is a local maximum. The value at this point is $f(-2) = 2$.

Now consider the critical point $x = 0$.
Note that $f'(-1) = -3 < 0$ and $f'(1) = 9 > 0$.
Therefore by the First Derivative Test, the point $x = 0$ is a local minimum. The value at this point is $f(0) = -2$. The graph of $f(x)$ is sketched as shown. To obtain a qualitative picture of the graph of the function f, the location of the critical points must be first determined, then the extremum characteristics for each of them may then be checked.

Example 2

Find the local extrema of the function $f(x) = \sqrt[3]{x}\,(x-14)^2$.

Solution

First we differentiate to locate <u>all</u> the critical points.

$$f'(x) = \tfrac{1}{3}x^{-2/3}(x-14)^2 + 2x^{1/3}(x-14) = \frac{7(x-2)(x-14)}{3x^{2/3}}$$

There are three critical points.
One is at 0 where $f'(x)$ is undefined and the others are at 2 and 4 where $f'(x) = 0$.
Next we test the sign of the derivative at sample points on either side of each critical point.

$$f'(-1) = 105 > 0 \qquad f'(1) = \frac{91}{3} > 0$$

$$f'(3) = -\frac{77}{3^{5/3}} < 0 \qquad f'(15) = \frac{91}{3 \cdot 15^{2/3}} > 0$$

These four sample points are enough to apply the First Derivative Test to all three critical points. Checking the sign of the derivatives about each critical point we find:

1) $f(0) = 0$ is not a local extremum
2) $f(2) = 144 \cdot 2^{\frac{1}{3}}$ is a local maximum
3) $f(14) = 0$ is a local minimum

The graph of f is shown to the right:
Notice that f has a vertical tangent line at the origin which explains the critical point there.

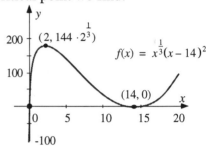

Applied Optimization Problems

Optimization, a concept used in everyday life constantly, is particularly important in engineering.

Example 3

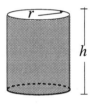

A plastic waste basket is designed in the form of a cylinder without the top lid. The designer has specified that it hold a volume $V = 12000$ cm³ and have a total surface area of $A = 2200$ cm². (The area does not include the lid).

Is this design of the basket feasible?
What dimensions r and h will minimize the amount of plastic used?

Solution

One way to see the feasibility of the design is to find the minimum surface area for the given volume $V = 12000$ cm³, then compare the minimum area with the specified surface area. The volume of the cylinder if $V = \pi r^2 h$. Hence, we have $12000 = \pi r^2 h$. The function we have to minimize is the surface area of the basket.

$$A = 2\pi r h + \pi r^2$$

First solve for h in terms of r using the volume relation: $\quad h = \dfrac{12000}{\pi r^2}$

Now we can express A completely in terms of r:

$$A = 2\pi r \left(\frac{12000}{\pi r^2}\right) + \pi r^2 \quad \text{or} \quad A = \left(\frac{24000}{r}\right) + \pi r^2 \quad \text{with } r > 0$$

Thus the surface area A is a <u>rational</u> function of the radius r.
To find the minimum value of A, we first find its derivative.

$$A' = -\frac{24000}{r^2} + 2\pi r$$

The only critical point is just $r = \left(\dfrac{12000}{\pi}\right)^{1/3}$.
(Note that $r = 0$ is not a critical point because the domain of the function is $r > 0$.)
The minimum surface area is $A\left(\left(\dfrac{12000}{\pi}\right)^{1/3}\right) = 3\pi^{1/3}(12000)^{2/3} = 2302.986$ cm².

§3.4 Optimazation CAFÉ Page 231

The graph of $A(r)$ is sketched as illustrated.
We see that the minimum surface area is greater than the specified value 2200 cm² so the initial specifications for the area and volume of the basket are impossible.

Finally we show that for the optimal design, the height and the radius of the waste basket are the same.

$$\frac{h}{r} = \frac{\left(\frac{12000}{\pi r^2}\right)}{r} = \frac{12000}{\pi r^3} = \frac{12000}{\pi\left(\frac{12000}{\pi}\right)} = 1$$

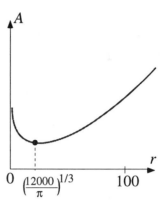

Example 4 - Rescue Mission

Spike has a summer job working as a lifeguard. What path should Spike follow to reach the drowning victim shown in the illustration in the least amount of time?

(Spike runs at a speed of 8 meters per second and swims at a speed of 2 meters per second. The radius of the pool is 20 meters.)

Solution

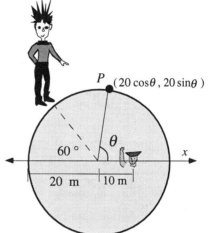

Assume that Spike runs along the edge of the pool to the point P, where he jumps into the water swimming directly toward the drowning victim.
The time required is:

$$T(\theta) = \frac{20\left(\frac{2\pi}{3}-\theta\right)}{8} + \frac{\sqrt{(20\cos\theta-10)^2+(20\sin\theta)^2}}{2} = \frac{5\left(\frac{2\pi}{3}-\theta\right)}{2} + \frac{\sqrt{500-400\cos\theta}}{2}$$

where $0 \le \theta \le \frac{2\pi}{3}$

Next we find the derivative of $T(\theta)$ to locate the critical points.

$$T'(\theta) = -\frac{5}{2} + \frac{100\sin\theta}{\sqrt{500-400\cos\theta}}$$

Since the derivative is defined everywhere, the only critical point comes from solving $T'(\theta) = 0$. Thus $40\sin\theta = \sqrt{500-400\cos\theta}$.
Squaring both sides of the equation and using the identity $\sin^2\theta + \cos^2\theta = 1$ yields:

$$16\cos^2\theta - 4\cos\theta - 11 = 0$$

This leaves: $\cos\theta = \dfrac{1 \pm 3\sqrt{5}}{8}$.

The negative root gives $\theta = 135.52°$ which is irrelevant here.

Thus $\theta = \cos^{-1}\dfrac{1+3\sqrt{5}}{8} = 15.522°$.

So the best strategy for Spike to save the victim is to run around the pool to the point specified by $\theta = \cos^{-1}\dfrac{1+3\sqrt{5}}{8}$ and jump into the water, then swimming directly toward the victim.

The optimal time to reach the victim is $T(15.522°) = 9.9110$ seconds. The graph of T versus θ is shown in the illustration.

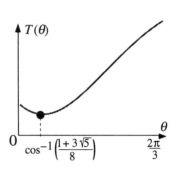

Example 5

A snow-making machine provides artificial snow for a ski resort. Let $C(x)$ denote the cost per hour of running the machine on a setting that produces x kilograms per hour. (See picture.)

Resurfacing the 'Killer Trail' requires a total of 10,000 kg of snow. Illustrate graphically the most economical production rate x (in kg/hr).

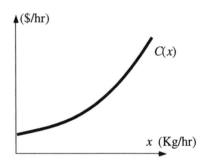

Solution

Suppose the machine is set for a production rate of $x\,\dfrac{\text{kg}}{\text{hr}}$. It will take $\dfrac{10000}{x}$ hours to produce the required 10,000 kg of snow. Therefore, the total production cost P satisfies:

$$P(x) = \dfrac{10000\,C(x)}{x}, \quad \text{where } x > 0.$$

In order to minimize $P(x)$, the derivative is taken:

$$P'(x) = 10000\,\dfrac{xC'(x) - C(x)}{x^2}$$

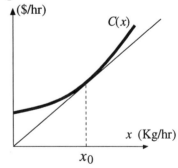

Setting the above derivative equal to zero, we get

$$10000\,\dfrac{xC'(x) - C(x)}{x^2} = 0 \quad \Rightarrow \quad xC'(x) - C(x) = 0$$

We can now see that the critical point satisfies $C'(x) = \dfrac{C(x)}{x}$.

§3.4 Optimazation

The above equation has an interesting graphical interpretation
Draw a line tangent to the curve $C(x)$ from the origin.
The point of tangency x_0 is the optimal production rate.

WARMUP EXERCISES

Find the extrema of the following functions and sketch.

1) $f(x) = x^2 - 2x + 5$ on $[0,5]$
2) $f(x) = x^3 - 6x^2 + 9x$
3) $f(x) = (x+1)^2(x-2)^3$
4) $f(x) = (x+1)(x-2)^2$
5) $f(x) = x^2 + \dfrac{1}{x^2}$
6) $f(x) = x^4 - 4x^3 + 10$
7) $f(x) = x^{\frac{4}{3}}(x-1)$
8) $f(x) = e^{-x} - e^{-2x}$
9) $f(x) = xe^{-x}$
10) $f(x) = e^{-x}\cos x$
11) $f(x) = x\ln x$
12) $f(x) = \dfrac{\ln x}{x}$
13) $f(x) = \dfrac{e^x - e^{-x}}{e^x + e^{-x}}$
14) $f(x) = \ln(9 - x^2)$
15) $f(x) = x^x$ (Hint: Write $x^x = e^{x \ln x}$)
16) $f(x) = x^{\frac{1}{x}}$ (Hint: Write $x^{\frac{1}{x}} = e^{\frac{1}{x}\ln x}$)
17) $f(x) = \sqrt{x}\ln x$
18) $f(x) = x - \ln x$
19) $f(x) = \sin x + \cos 2x$ on $[0, 2\pi]$

INTERMEDIATE EXERCISES

20) a) Is there a real number x so that $\dfrac{1}{2}x = \ln x$?
    b) Is there a real number x so that $\dfrac{1}{3}x = \ln x$? If there is, find all possible x's that satisfy the equation. Use Newton's Method to determine the values to an accuracy of 4 decimal places.
    c) Find a number a so that $ax = \ln x$ has exactly one solution. What is the corresponding unique solution for x?

§3.4 Optimazation CAFÉ Page 235

21) The common use of a superscript to indicate both exponential and power functions can lead to some confusion. Yet the power function $y = x^a$ is very different from the exponential function $y = a^x$.
   a) Consider just $x > 0$. Show that if $a > 1$ and $a \neq e$, we have two values of x so that $a^x = x^a$.
   b) What happens to part a) if $0 < a < 1$?
   c) What happens in part a) if $a = e$?

22) Show that of all the rectangles with a given perimeter, the one with the greatest area is a square.

23) Among all isosceles triangles with perimeter L, find the length of the sides of the triangle with the largest possible area.

24) A right circular cylinder is inscribed in a sphere of radius r. Find the largest possible surface area of such a cylinder.

25) A rectangle of perimeter P is rotated about one of its sides so that it generates a cylinder. Among all such possible rectangles find the dimensions of the one that generates a cylinder of maximum volume.

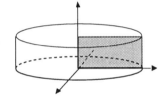

26) Find the dimensions of a right circular cone of maximum volume inscribed in a sphere of radius r.
   Recall the volume of a cone is $\dfrac{\pi r^2 h}{3}$.

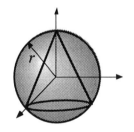

27) Find the rectangle of maximum area inscribed in a semicircle of radius r.

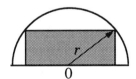

28) Of all the right circular cylinders that can be inscribed in a given right circular cone, prove that the one of greatest volume has height one-third of the cone.

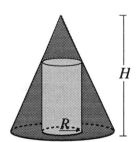

29) In our calculus course we have learned how to maximize and minimize a function $f(x)$ by calculating the derivative $f'(x)$ and solving $f'(x) = 0$ for critical points. Show that this technique is not always correct for the following problem. A rectangle is to be inscribed in a regular trapezoid so as to have the largest area. Assume that the base length is greater than 4. (Hint: The answer depends on a and 8 is an important number for a)

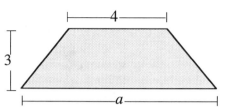

30) **Snell's Law**
According to Fermat's Principle, a ray of light will always travel along a path from a point A to another point B that will minimize the time needed. Assume point A is in air, where the speed of light is denoted by v_a and point B is in water, where the speed of light is v_w. Show that the path of the light ray satisfies

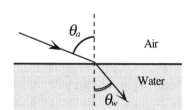

$$\frac{\sin \theta_a}{\sin \theta_w} = \frac{v_a}{v_w},$$

where θ_a is the angle of incidence and θ_w is the angle of refraction. This equation is known as Snell's Law.

31) Find the most advantageous length of a lever to raise a weight of 500 pounds, if the distance of the weight from the fulcrum is one foot and the lever weighs 3 pounds per foot.

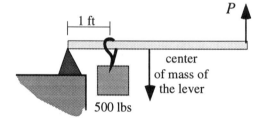

32) **Design of Green House**
A green house is to be designed in the form shown so as to hold 3000 cubic meters of volume. In winter time, because of the material used in its construction, three times as much heat per square meter is lost through the walls as through the roof.

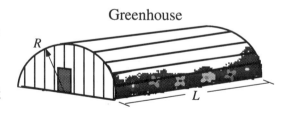

(There are two walls and one circular roof). No heat is lost through the floor. What shape should the green house be designed so as to minimize the heat loss? What is the ratio of optimal length, L, to the optimal radius.

§3.4 Optimazation — CAFÉ — Page 237

33) **Design of Garden**
A garden of rectangular shape must be designed into the form consisting of 10 identical rectangles of planting areas arranged in two rows and five columns (see illustrationa). The width of the sidewalk is 2 meters so that tourists can have enough room to walk through the garden. The total area of the garden (planting area and area of sidewalk) is 2500 square meters. How should the length and width of the garden be designed so that it has the largest planting area.

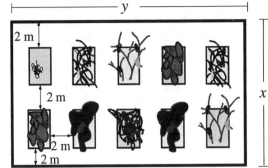

34) **Rescue Mission Revisited**
Spike was a lifeguard of a standard swimming pool. One day he saw a child drowning at a location as shown. Spike runs at a speed of 8 meters per second and swims at a speed of 2 meters per second. He must reach the drowning child as soon as possible. What path should he follow? What is the least amount of time for him to reach the drowning child? The illustration shows two types of paths that Spike might have to consider before he rushes into water.

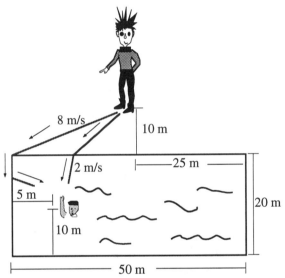

35) A man is trapped in a swamp as shown at $(3/4, 0)$. The rescue team must stay as close to him as possible so that they can pass a rope to him. Find the best location of the rescue team. Also find the minimum length of rope required to save him. The swamp area is unshaded.

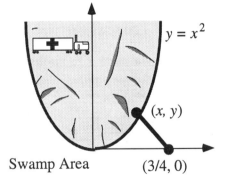

Swamp Area $(3/4, 0)$

36) This problem is a generalization of a sample problem shown in the text above. Here, we assume that the drowning child is a distance a meters from the center. What path should Spike follow to reach the drowning victim in the least amount of time in this general case? And how long, in terms of a, does it take him? Assume the Spike runs at a speed of 8 m/s and swims at a speed of 2 m/s.

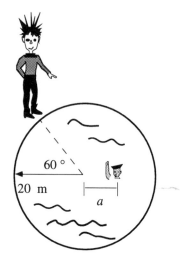

37) **Safari**
A team of jeep-riding zoologists with a dart gun are in a savanna when they spot a cheetah. When the cheetah notices the team, the animal is crossing the road 1/3 mi away. It promptly runs west at a speed of 40 mi/hr, as the jeep approaches at 45 mi/hr due north. The terrain is rough enough that the jeep cannot travel on it to follow the cheetah. To have an accurate shot, the team must be within within 0.2 mi of the cheetah. Will they be able to catch the animal?

38) A man is on the bank of a river that is one mile wide. He wants to travel to a town on the opposite bank, 6 miles upstream. The river flows at a speed of 2 miles per hour and he rows at a speed of 4 miles per hour in still water. Because he has an emergency in the town, he must reach the town as soon as possible.

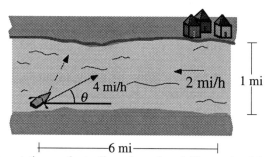

In order to reach his destination in the least time what direction should he point his boat to if he rows first and walks 5 miles per hour along the river bank.

39) A publisher wants to publish a book using the least amount of paper. He agrees that the page area is 300 square centimeters. The margin widths are shown to the right. In order to use the least amount of paper, the printing area must be maximized. What is the optimal dimension of a page.

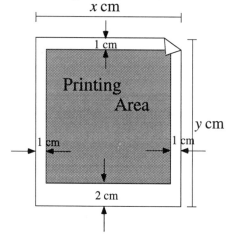

§3.4 Optimazation

40) Design of Buoy
A factory must produce can-buoys in the form of a double cone. The buoy is hollow and must be made to have the largest volume possible so that it can provide the biggest buoyant force for floating purpose. Suppose the amount of material that makes the buoy is fixed. Find the optimal design. The area of a cone is $\pi r L$, where L is the slant height.

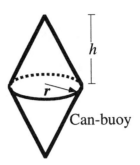
Can-buoy

41)
In a remote area a clinic center is to be built somewhere along a straight road as shown so that patients from two nearby villages A and B may toil least as a whole. The population of village A is three times as large as the population of village B. A reasonable indication of the toil for the patients to reach the clinic center is $3AC + BC$ where AC is the distance between A and C etc,.

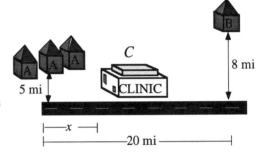

(Hint: Use Newton's Method to calculate the critical point to an accuracy of four decimal places.)

42)
A commercial cake case is a clear hemispherical cap of radius 15 inches. How big (in terms of its volume) can a circular cylinder cake be if it is stored inside the case without getting squeezed by the case.

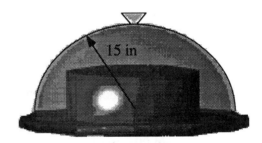

43) Design of Oil Tank
A huge stationary oil tank is to be constructed in the form shown. It has an underground hemispherical cement bottom and a cylindrical outlook with a circular lid. The cylindrical part is made with metal which costs 600 dollars per square meter. The cement bottom costs 500 dollars per square meter. This huge tank is to hold 10000 cubic meters of crude oil. The designer wants to find the minimum construction cost and the optimal dimension of the tank.

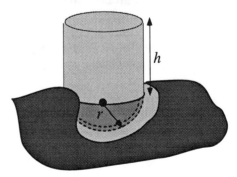

44) Maximizing Profit
A gas station manager wants to maximize the profit by adjusting the gas price. The current price is 1 dollar per gallon and he sells an average of 4000 gallons per day. He pays 70 cents per gallon to the oil company. From survey he knows that for every penny hike in gas price, there will be 80 gallons less sold each day. For every penny drop there will be 200 gallons more sold per day. How should he set the gas price to maximize the profit.

45) A real estate office handles 400 apartment units. When the rent of each unit is 100 dollars per month, all the units are occupied. However, from past experience the office knows, in general, for each 5 dollars increase in rent, two of the units become vacant. Each occupied unit requires, on average, 10 dollars per month for service and repairs. What rent should be set to achieve the most profit? Also find how many units are vacant when the profit is at the maximum.

46) A rug company is under contract to manufacture a rectangular rug to cover the maximum floor area of an elliptical shaped room. The major axis of the ellipse is 20 meters and the minor axis is 15 meters. The price for the rug is 100 dollars per square meter and the installation fee is 10 dollars per square meter. How much should the total charge be? What are the dimensions of the carpet.

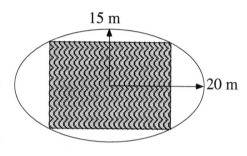

47) An apple grower has 35 trees per acre, and the average yield is 380 apples per tree. For each additional tree planted per acre, the average yield per tree is reduced by approximately 8 apples. How many trees per acre will give the largest crop of apples? (Remember to give the number of trees in whole numbers)

48) **Trout Harvesting**
A fish farmer wants to decide the best timing for harvesting his trouts in the pond. At first he introduced P trouts in the pond. He then feeds the trouts and takes care of them so that the trout population grows according to the curve shown to the right. Assume that feeding and maintenance cost for the trout pond is C_1 dollars per week and the market price for trout is C_0 dollars per trout.
For how long should he wait before harvesting so that he can maximize his profit per unit time? Show the results graphically. <u>No numerical calculation is required</u>. N denotes the vertical axis as the number of trouts.

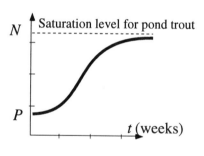

49) **Gas Mileage**
U.S. Air wants to maximize the mileage per gallon of jet liners. An imperical curve for gas consumption rate (in gallons/hour) vs. speed (in miles/hour) of the jet liner is given to the right. At what speed should the jet liner fly so that the maximum gas mileage is achieved? Show the result graphically. <u>No numerical calculation is required</u>.

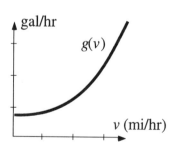

§3.4 Optimazation   CAFÉ   Page 241
                                              ❖

50) Taxation
For a given administration, the effective civil service delivered vs. total tax collected is given to the right. The initial flat portion of the graph represents the use of tax money for basic administrative set up so that only a small portion of the tax collected can be used for civil service. The slow increase of service delivered is viewed as an indication of corruption, inefficiency in the administration. From a tax payer's point of view, the effective civil service per tax dollar collect is relevant.

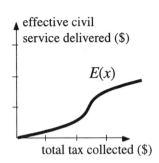

The higher the number is, the better the tax money is spent. Our problem here is to find the optimal total tax collected so that effective civil service per tax dollar is maximized. This problem is a mathematical model indicating that taxing too much does not pay off. People actually get less of what is worth. Show the result graphically. No numerical calculation is required.

51) Two highways intersect at a right angle as shown. For the convenience of people living in the north part to visit the shopping mall then going east, a third high way is to be constructed that connects the two highways and passes through the shopping mall. For economic reasons this highway must be the shortest. What should the design be?

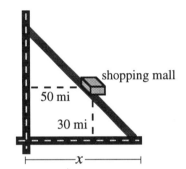

52) Design of irrigation channel
An irrigation channel is to be made so that its cross section is a trapezoid with equally sloping sides. The channel is 10 miles long. It is important that we use the least amount of concrete (for economic reasons) and also make the channel the biggest capacity. What should the optimal design be? How much is the total volume of concrete that must be used. 1 mile = 5280 ft.

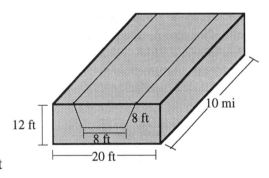

53) A sun porch in front of a house extends 8 feet into a garden. The porch roof is flat and is 10 feet above ground level. Above the porch are bedroom windows to be cleaned. The window cleaner attempts to place his 25 foot ladder against the house wall, resting on the ground and passing over the sun porch. Can the window cleaner make it to the windows? What is the minimum ladder length that can just pass over the porch and still rest against the house.

54) A 30 foot long girder is in a hallway 10 feet wide. What must be the minimum width of a second hallway at right angle to the first to allow the girder to be carried through?

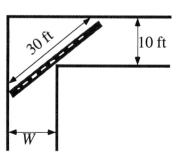

55) A cardboard rectangle 3 feet by 8 feet is to be cut at the corners as shown so that it can be folded into an enclosed rectangular box (six sided with a top lid included). What is the largest volume of rectangular box? How should we design it?

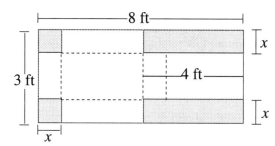

56) A parcel box is the shape shown. This box has a square base. When completely folded the two lids will become double layered. If there is 24 square feet of cardboard available to make such a box. Find the largest possible volume of the box.

57) Design of Planting Pot
The design of a planting pot is of the form of a circular frustum with upper base area equal to twice the lower base area. The clay that is used to make the pot is kept fixed (or the total surface area of the pot is a constant 10 ft^2). One needs to make the pot in such a way to allow the maximum amount of soil to be held. Find the optimal design and the maximum volume of the pot.

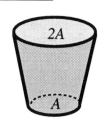

58) A plastic trash can is designed in the form of a circular frustum so that its top lid has an area 3/2 times the area of the bottom. The plastic material used in its construction has a fixed total area 20 ft^2. Find the optimal design that has the maximum volume.

Chapter 4
Integration

§ 4.1 The Area Problem

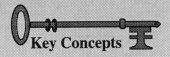

- ❖ Rate of Change of Area
- ❖ Antiderivative and Indefinite Integrals

Area of Polygons

Before the invention of calculus, it was only possible to find expressions for the area of simple shapes such as polygons whose edges are all straight lines.

$A = l^2$

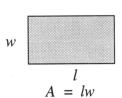
$A = lw$

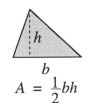
$A = \frac{1}{2}bh$

For example, the area of a square is the square of the length of the sides. The area of a rectangle is defined as the product of the length and the width. The area of a triangle is half the base multiplied by the height.

More complicated polygons can be broken into smaller, simpler shapes whose areas are easily found. To find the area of a polygon we can divide it into triangles and add the areas of the triangles.

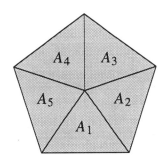

$A = A_1 + A_2 + A_3 + A_4 + A_5$

However, many engineering objects such as the cross-section of an airplane wing or propeller are not shaped liked polygons. How can the areas of these shapes be obtained?

❖

Discovery Exercise: Snowplow

A snow plow is clearing snow from a long, straight section of highway. The area cleared by the plow is a rectangle that increases in size.
At what rate does the plowed area increase as the snow plow progresses down the highway?

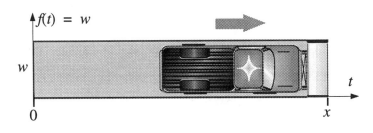

The width of the snow plow is a constant w and the length of the road plowed is x. Thus the area $A(x)$ that has been cleared of snow is a function of the distance x the plow has traveled along the highway.

$$A(x) = wx$$

The area function $A(x) = wx$ corresponds to the area under the graph of the function $f(t) = w$ over the variable interval $[0, x]$. The graph of $f(t) = w$ has a constant <u>height</u> w which corresponds to the <u>width</u> of the plow.

At what rate does the area plowed increase as the length of road x increases?

Note that $\dfrac{dA}{dx} = w$ so that the derivative of the area is equal to the height. Also note that $A(0) = 0$ since when $x = 0$ no snow has been plowed.

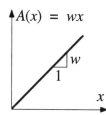

Discovery Exercise: **Trapezoid**

What is the area $A(x)$ under the line $y = f(t) = kt$ over the variable interval $[a, x]$? At what rate does the area increase as x increases?

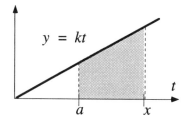

Notice that the area under the curve is a trapezoid. You may recall that the area of a trapezoid is the product of its base and its average height. The base of the trapezoid is $x - a$. The left side has height ka and the right side has height kx. Thus the average height is:

$$h_{ave} = \frac{ka + kx}{2} = k\left(\frac{a + x}{2}\right)$$

Therefore the area of the trapezoid is:

$$A(x) = (x - a) \cdot k\left(\frac{a + x}{2}\right) = \frac{k}{2}\left(x^2 - a^2\right)$$

What is the rate of change of the area of the trapezoid as x increases?

$$\frac{dA}{dx} = \frac{d}{dx}\left(\frac{k}{2}x^2\right) - \frac{d}{dx}\left(\frac{k}{2}a^2\right) = kx - 0 = f(x)$$

In this case, the height of the function $y = kt$ is not constant, but we still have $\dfrac{dA}{dx} = f(x)$. That is, the derivative of the area is the height. However, the height must be evaluated at the right endpoint x of the interval $[a, x]$. Also note that $A(a) = 0$ since the trapezoid degenerates to a vertical line!

§4.1 The Area Problem

Example 1 - Area of a Parabolic Region

In this example we will find the area $A(x)$ under the parabola $f(t) = t^2$ over the variable interval $[1, x]$. This problem was considered as a difficult problem in ancient times because the "curved" boundary of the region defied all known formulae.

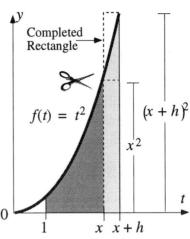

Solution

Rather than find the area $A(x)$ directly, we will first show that the area function has the following two properties.

1) $A(1) = 0$ 2) $\dfrac{dA}{dx} = x^2$

Later we shall see that these two properties will uniquely determine the area $A(x)$.

The first property is obvious. As x approaches 1, the area under the curve becomes smaller and smaller. When $x = 1$, the parabolic region shrinks to a vertical line at 1 which has obviously zero area.

Now we will show that $\dfrac{dA}{dx} = x^2$. Choose a point a little to the right of x, say $x + h$. Consider the difference $A(x + h) - A(x)$. Since $A(x + h)$ is the area of the parabola over the interval $[1, x + h]$, the difference $A(x + h) - A(x)$ is just the area of the thin vertical strip of width h adjacent to the point x (see illustration). Notice the top of the strip forms a curved cap. Removing this cap from the strip, we obtain a rectangle of height x^2 and width h. The area of this rectangle is $x^2 h$ and is clearly less than the area of the strip. Thus,

$$x^2 h \leq A(x + h) - A(x)$$

Similarly, by completing the curved cap we obtain a larger rectangle of height $(x + h)^2$ and width h. The area of the completed rectangle is thus $(x + h)^2 h$ and is clearly larger than the area of the strip. Combining this with the above inequality we have:

$$x^2 h \leq A(x + h) - A(x) \leq (x + h)^2 h$$

Dividing each term in the inequality by h, we obtain:

$$x^2 \leq \frac{A(x + h) - A(x)}{h} \leq (x + h)^2 \qquad \text{(Recall } h > 0\text{)}$$

Recall that the derivative $\dfrac{dA}{dx}$ is defined as the limit $\dfrac{dA}{dx} = \lim_{h \to 0} \dfrac{A(x + h) - A(x)}{h}$.
Since our inequality is only true for $h > 0$, we must take a one-sided limit $\left(\lim_{h \to 0^+}\right)$.
Taking the one-sided limit as $h \to 0^+$ of our inequality we obtain:

$$\lim_{h \to 0^+} x^2 \quad \leq \quad \lim_{h \to 0^+} \frac{A(x + h) - A(x)}{h} \quad \leq \quad \lim_{h \to 0^+} (x + h)^2$$

$$x^2 \quad \leq \quad \lim_{h \to 0^+} \frac{A(x + h) - A(x)}{h} \quad \leq \quad x^2$$

Therefore, $\lim_{h \to 0^+} \dfrac{A(x+h) - A(x)}{h} = x^2$.

Similarly we can show $\lim_{h \to 0^-} \dfrac{A(x+h) - A(x)}{h} = x^2$

This shows that:

$$\frac{dA}{dx} = x^2$$

Notice the above equation again confirms that the rate of change of the area under a curve and over a variable interval is equal to the height at the right endpoint.

Although we have not found the area $A(x)$ yet, we know that $\dfrac{dA}{dx} = x^2$ and $A(1) = 0$.

What function $A(x)$ satisfies $\dfrac{dA}{dx} = x^2$? Since this equation just says $A(x)$ is a function whose derivative is $f(x) = x^2$, the only possibilities for $A(x)$ are $A(x) = \dfrac{1}{3}x^3 + C$, where C is a constant.

Since $A(1) = 0$, we have $A(1) = \dfrac{1}{3}1^3 + C = \dfrac{1}{3} + C$ or $C = -\dfrac{1}{3}$.

Finally the area over the variable interval $[1, x]$ is $\boxed{A(x) = \dfrac{1}{3}x^3 - \dfrac{1}{3}}$

Let us now generalize these observations.

Derivative of the Area Function

Let $y = f(t)$ be a positive and continuous function. Denote by $A(x)$ the area between the curve and the x-axis over the variable interval $[a, x]$. The picture illustrates that the difference $A(x+h) - A(x)$ corresponds to the area of a thin vertical strip of width h. Our intuition about area suggests that for small h the area of this strip should be approximately equal to the area of a rectangle of base h and height $f(x)$.

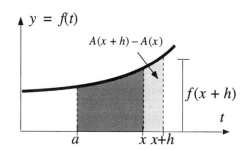

Therefore $A(x+h) - A(x) \sim h\,f(x)$, where "~" means approximately equal.

§4.1 The Area Problem

More precisely, let $h > 0$, then $hm \leq A(x+h) - A(x) \leq hM$, where M is the maximum of $f(x)$ over the interval $[x, x+h]$ and m is the corresponding minimum. By dividing h throughout the inequality we have

$$m \leq \frac{A(x+h) - A(x)}{h} \leq M$$

Since f is continuous, both m and M equal $f(x)$ in the limit as $h \to 0^+$. This is because f is continuous and the interval $[x, x+h]$ shrinks to the point x.

Hence, $\lim_{h \to 0^+} \frac{A(x+h) - A(x)}{h} = f(x)$.

Similarly we can show $\lim_{h \to 0^-} \frac{A(x+h) - A(x)}{h} = f(x)$

Hence we conclude $\quad \frac{dA}{dx} = f(x)$.

Since the variable interval $[a, x]$ shrinks to a point as x approaches a, we have

$$A(a) = 0$$

In summary, we have shown the following theorem:

> **The Area Theorem**
>
> Let $f(x)$ be a positive continous function and let $A(x)$ denote the area under the graph of the curve $y = f(x)$ over the variable interval $[a, x]$.
> Then,
>
> 1) $\quad A(a) = 0 \qquad$ 2) $\quad \frac{dA}{dx} = f(x)$

Notice that with a given $f(x)$ we need to find a function $A(x)$ so that $A'(x) = f(x)$. $A(x)$ is called an "antiderivative" of $f(x)$.
The process of finding an antiderivative of a given function $f(x)$ is the inverse of finding the derivative of $f(x)$.

Traditionally, we use the symbol $\int f(x)dx$ to denote the antiderivative of $f(x)$. We call $\int f(x)dx$ the <u>indefinite integral</u> of $f(x)$. As a rule, the terms indefinite integral and antiderivative are synonyms. Since $\int f(x)dx$ is just the antiderivative of $f(x)$, it is clear that for every differentiation formula, there is a corresponding formula for the indefinite integral.

Thus:
$$\frac{d}{dx}\left(\frac{x^{n+1}}{n+1} + C\right) = x^n, \text{ where } n \neq -1 \text{ corresponds to } \int x^n dx = \frac{x^{n+1}}{n+1} + C.$$

Here the additive constant C does not affect the fact that $\frac{d}{dx}\left(\frac{x^{n+1}}{n+1} + C\right) = x^n$. C is called the "integration constant". This constant can be any arbitrary constant that drops out of the expression when differentiated. From this we can see that any indefinite integral will always have an additive integration constant.

By reversing known differentiation formulas we can easily build up a table of indefinite integrals.

Table 1: Indefinite Integrals

$\frac{d}{dx}\frac{x^{n+1}}{n+1} = x^n$ $\leftrightarrow$ $\int x^n dx = \frac{x^{n+1}}{n+1} + C$

$\frac{d}{dx}\sin x = \cos x$ $\leftrightarrow$ $\int \cos x\, dx = \sin x + C$

$\frac{d}{dx}(-\cos x) = \sin x$ $\leftrightarrow$ $\int \sin x\, dx = -\cos x + C$

$\frac{d}{dx}e^x = e^x$ $\leftrightarrow$ $\int e^x\, dx = e^x + C$

$\frac{d}{dx}\ln|x| = \frac{1}{x}$ $\leftrightarrow$ $\int \frac{1}{x}\, dx = \ln|x| + C$

$\frac{d}{dx}\tan x = \sec^2 x$ $\leftrightarrow$ $\int \sec^2 x\, dx = \tan x + C$

$\frac{d}{dx}(-\cot x) = \csc^2 x$ $\leftrightarrow$ $\int \csc^2 x\, dx = -\cot x + C$

$\frac{d}{dx}\sec x = \sec x \tan x$ $\leftrightarrow$ $\int \sec x \tan x\, dx = \sec x + C$

$\frac{d}{dx}(-\csc x) = \csc x \cot x$ $\leftrightarrow$ $\int \csc x \cot x\, dx = -\csc x + C$

Example 2

Find the area between the curve $y = x^3$ and the x-axis for the interval $[1, 2]$.

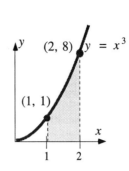

Solution

Let $A(x)$ be the area of the region underneath the curve $y = x^3$ over the variable interval $[1, x]$. Notice that the desired area is $A(2)$. Our strategy will be to find $A(x)$ first. Since $A(x)$ is an antiderivative of x^3, we have

$$A(x) = \int x^3 dx = \frac{1}{4}x^4 + C$$

To determine the constant C we use the fact $A(1) = 0$.
Thus $0 = A(1) = \frac{1}{4} + C$ so that $C = -\frac{1}{4}$.
Hence, $A(x) = \frac{1}{4}x^4 - \frac{1}{4}$. The area under the curve from 1 to 2 corresponds to $A(x)$ when $x = 2$. Thus the answer is:

$$\boxed{A(2) = \frac{1}{4}2^4 - \frac{1}{4} = \frac{15}{4}}$$

Example 3

Find the area between the curve $y = \frac{1}{x}$ and the x-axis for the interval $[1, 2]$.

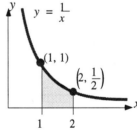

Solution

Let $A(x)$ be the area of the region underneath the curve $y = \frac{1}{x}$ over the variable interval $[1, x]$, where $x > 1$. As in the previous example, the desired area is $A(2)$.
Noting that $A(x)$ is an antiderivative of $\frac{1}{x}$ we obtain:

$$A(x) = \int \frac{1}{x} dx = \ln|x| + C$$

To find the constant C we use $A(1) = 0$ so that $0 = \ln(1) + C$.
Taking $\ln(1) = 0$ into account we see that $C = 0$.
Hence, $A(x) = \ln|x|$ and the desired area is $\boxed{A(2) = \ln 2}$

Example 4 - Sine Arch

Find the area of a sine arch, that is the area of the region under the graph of $y = \sin x$ from $x = 0$ to $x = \pi$.

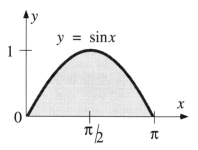

Solution

Let $A(x)$ be the area of the region under the curve $y = \sin x$ over the variable interval $[0, x]$. The area of the sine arch is the value of $A(x)$ at $x = \pi$.
Noting that $A(x)$ is an antiderivative of the function $y = \sin x$, we can write:

$$A(x) = \int \sin x\, dx = -\cos x + C$$

Since $A(0) = 0$, we obtain $0 = -\cos(0) + C$ or $C = 1$. Thus, $A(x) = 1 - \cos x$

The area of the sine arch is $A(\pi) = 1 - \cos \pi = \boxed{2}$

Example 5

Find the area of the region under the graph of the function $f(x) = e^{|x|}$ over the interval $[-1, 1]$.

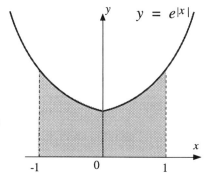

Solution

Since the antiderivative of $e^{|x|}$ is not explicitly known, we proceed slightly differently by sketching the region first. Notice that the region is symmetric with respect to the y-axis. Hence the area is just twice the area of the right half.

On the positive x-axis, the function can be simplified to $e^{|x|} = e^x$ since $x \geq 0$. Hence the area of the right half is the same as the area underneath the curve $y = e^x$ over the interval $[0, 1]$.

Let $A(x)$ be the area of the region underneath the function $f(x) = e^x$ over the variable interval $[0, x]$. We want to find $A(x)$ when $x = 1$ then double it to find the entire shaded area. The desired answer is just $2A(1)$.

$$\text{Now } A(x) = \int e^x dx = e^x + C.$$

Using $A(0) = 0$ we obtain $0 = e^0 + C$ or $C = -1$.
Thus $A(x) = e^x - 1$ so that $A(1) = e^1 - 1$.
Since the area over the interval $[-1, 1]$ is twice this, we find that:

$$\boxed{\text{Area} = 2(e - 1)}$$

§4.1 The Area Problem

Linearity of Indefinite Integrals

We have seen the important role that the antiderivative of a function $f(x)$ plays in the area problem. Two basic properties of indefinite integrals are very useful in finding the area under more complicated functions.

1) $$\int k f(x)\, dx = k \int f(x)\, dx$$

2) $$\int (f(x) + g(x))\, dx = \int f(x)\, dx + \int g(x)\, dx$$

The first property states that the indefinite integral of a constant times a function is equal to the same constant times the indefinite integral of the function. The second property states that the indefinite integral of a sum is the sum of the indefinite integrals.

Recall that derivatives also have the same properties as above. Together, these two properties are what we call "linearity." The proof of the above two properties is just a straight forward consequence of the linearity of derivatives. Using linearity, we can find the indefinite integrals of many more functions than are in Table 1. The next examples illustrate how to find indefinite integrals term by term.

Example 6

Find the indefinite integral $\int \left(5x^8 - 3\sqrt{x}\right) dx$

Solution

Rewriting $3\sqrt{x}$ as $3x^{1/2}$ we have:

$$\int \left(5x^8 - 3x^{1/2}\right) dx = \int 5x^8\, dx - \int 3x^{1/2}\, dx \qquad \text{by linearity (2)}$$

$$= 5 \int x^8\, dx - 3 \int x^{1/2}\, dx \qquad \text{by linearity (1)}$$

$$= \frac{5}{9}x^9 - 3 \cdot \frac{2}{3}x^{3/2} + C$$

$$= \boxed{\frac{5}{9}x^9 - 2x^{3/2} + C}$$

Do not forget the constant of integration C.

Don't Forget

Check

It is a good strategy to check our answer by differentiating:

$$\frac{d}{dx}\left(\frac{5}{9}x^9 - 2x^{3/2} + C\right) = \frac{5}{9}\frac{d}{dx}(x^9) - 2\frac{d}{dx}(x^{3/2}) + \frac{d}{dx}(C)$$

$$= 5x^8 - 3x^{1/2}$$

Example 7

Find the indefinite integral $\int (3\sin x - 5e^x)dx$

Solution

$$\int (3\sin x - 5e^x)dx = 3\int \sin x\, dx - 5\int e^x dx \qquad \text{(by linearity (1) and (2))}$$

$$= -3\cos x - 5e^x + C$$

Check by differentiating:

$$\frac{d}{dx}(-3\cos x - 5e^x + C) = -3\frac{d}{dx}(\cos x) - 5\frac{d}{dx}(e^x) + \frac{d}{dx}C$$

$$= 3\sin x - 5e^x$$

§4.1 The Area Problem   CAFÉ   Page 255

A polygon is called regular if all of its sides have equal length L and all its interior angles are the same. Find the area A for each of the following regular polygons.

1) L

2) L

3) L

4) L

5) What is the area A of a regular polygon having n sides of length L.

6) **Geodesic Greenhouse**

Spike is building a glass greenhouse in which to grow exotic orchids and flowering plants. The greenhouse is to be fabricated in the form of half a geodesic sphere. A complete geodesic sphere consists of 12 pentagons and 20 hexagons. If the length of each side of the pentagons and hexagons is 1 meter, what is the area A of glass needed to build the greenhouse?

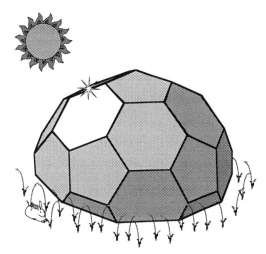

Although the functions are complicated, the following problems are very easy.

7) Let $A(x)$ denote the area under the graph of $f(x) = e^x \sqrt{x^2 + 1}$ over the variable interval $[1, x]$. Find $A(1)$ and $\dfrac{dA}{dx}$.

8) Let $A(x)$ denote the area under the graph of $f(x) = \dfrac{e^x}{e^x + \sqrt{x^2 + \sin x}}$ over the variable interval $[0, x]$. Find $A(0)$ and $\dfrac{dA}{dx}$.

9) Let $A(x)$ denote the area under the graph of $f(x) = \dfrac{\sqrt{3x^2 + 1}}{\sqrt{x + 5} + e^x}$ over the interval variable $[1, x]$. Find $A(1)$ and $\dfrac{dA}{dx}$.

10) Let $A(x)$ denote the area under the graph of the function $y = f(x)$ and over a variable interval of the form $[a, x]$. What is $A(a)$?

11) The function $A(x)$ denotes the area under the graph of the function $y = f(x)$ for a variable interval of the from $[x, b]$. Find $A(b)$ and $\dfrac{dA}{dx}$.

Hint: Select the point a so that $a < x < b$. Then $[a, x] \cup [x, b] = [a, b]$.

Evaluate the following indefinite integrals.
(Use the linearity properties to tackle each integral one term at a time.)

12) $\displaystyle\int \left(4x^3 + 2x^6 - 1\right) dx$ 13) $\displaystyle\int \left(1 + 2x + 3x^2 + 4x^3\right) dx$

14) $\displaystyle\int \left(6e^x + 5x^{1/3} + 4\sin x\right) dx$ 15) $\displaystyle\int \left(\dfrac{3}{x^2} + \dfrac{1}{x}\right) dx$

16) $\displaystyle\int \left(2\sec^2 x - \cos x\right) dx$ 17) $\displaystyle\int \left(3\sin x + 5e^x + \dfrac{2}{x}\right) dx$

Find the area $A(x)$ under the graph of the function $y = f(x)$ over the variable interval $[1, x]$. Then verify that $\dfrac{dA}{dx} = f(x)$.

18) $f(x) = x$ 19) $f(x) = x^2$

20) $f(x) = 3x^2 + 2x$ 21) $f(x) = 1 + x + x^2$

In each of the following exercises, find the area of the region underneath the graph of $f(x)$ over the given interval.

22) $f(x) = 3x^4 + 1$ $[-1, 1]$ 23) $f(x) = 2\sin x + 3$ $[0, 2\pi]$

24) $f(x) = 1 - x^2$ $[-1, 1]$ 25) $f(x) = 2\cos x$ $\left[\dfrac{-\pi}{2}, \dfrac{\pi}{2}\right]$

26) $f(x) = 3e^{-x} + 2$ $[-1, 2\}$ 27) $f(x) = \sqrt{x} - x$ $[0, 1]$

For each of the following exercises, sketch the region under the graph of $f(x)$ for the given interval. Break up the region into appropriate subregions and find its area.

28) $f(x) = |x|$ $[-1, 1]$ 29) $f(x) = |\sin x|$ $[0, 2\pi]$

30) $f(x) = |5x - 1|$ $[-1, 1]$ 31) $f(x) = \left|x - \dfrac{1}{x}\right|$ $\left[\dfrac{1}{2}, 2\right]$

§4.1 The Area Problem CAFÉ Page 257

32) $f(x) = e^{|x|}$ $[-1, 2]$ 33) $f(x) = |1 - x^2|$ $[-2, 2]$

34) $f(x) = |x^3|$ $[-1, 1]$ 35) $f(x) = |2e^x - 1|$ $[0, 2]$

36) $f(x) = |2e^x - 1|$ $[-1, 1]$ 37) $f(x) = |2\sin x - 1|$ $\left[0, \dfrac{\pi}{2}\right]$

ADVANCED EXERCISES

38) Recall the floor function $\lfloor x \rfloor$ denotes the greatest integer $\leq x$. For example $\lfloor 4.5 \rfloor = 4$. Find the area of the region under the graph of $f(x) = \lfloor x \rfloor$ and over the interval $[1, 4.5]$.

39) Find the area of the region bounded above by $f(x) = \lfloor \sin x \rfloor + 2$ over the interval $[0, 2\pi]$. (Hint: the graph of $\lfloor \sin x \rfloor + 2$ is a simple one.)

40) Find the area of the region bounded above by the curve $f(x) = \lfloor x^2 \rfloor$ and over the interval $[0, 2]$. (Hint: sketch the graph of the function first.)

41) **Tile Design**

Consider a ceramic tile decorated with two colors in the geometric pattern shown. For what value of n is the area of the darker region equal to the area of the lighter region? Use Newton's Method to approximate the value of n to an accuracy of four decimal places.
Note that the darker area is eight times the area of the region underneath the curve $y = x^n$ over the interval $\left[0, \dfrac{1}{2}\right]$.

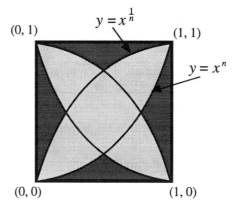

42) Pick an arbitrary point $P(x, x^2)$ on the parabola $y = x^2$ and construct two triangles OPP' and QPP' where PQ is tangent to the curve at P and P' is the projection of P on the x-axis. Let $A(x)$ denote the parabolic region which is bounded above by $y = x^2$ and below the x-axis.

Denote the area of $\triangle OPP'$ by $A(OPP')$ and denote the area of $\triangle QPP'$ by $A(QPP')$. Find the following limits:

a) $\displaystyle\lim_{x \to 0} \frac{A(x)}{A(OPP')}$

b) $\displaystyle\lim_{x \to 0} \frac{A(x)}{A(QPP')}$

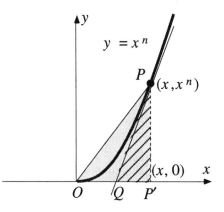

43) This problem is a generalization of the previous problem to the curve $y = x^n$ for $n > 1$. Denote the area of the region under the curve $y = x^n$ over the interval $[0, x]$ by $A(x)$. (see illustration). PQ is tangent to the curve at P. Denote the area of $\triangle OPP'$ by $A(OPP')$ and denote the area of $\triangle QPP'$ by $A(QPP')$. Find the following limits:

a) $\displaystyle\lim_{x \to 0^+} \frac{A(x)}{A(OPP')}$

b) $\displaystyle\lim_{x \to 0^+} \frac{A(x)}{A(QPP')}$

§ 4.2 Area & Definite Integrals

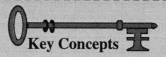
Key Concepts

- Reimann Sums
- Basic Properties of Definite Integrals
- Definition of the Definite Integral

Our focus in this section we will be on the definition of the definite integral in terms of Riemann sums. Before we discuss Riemann sums however, we will introduce a convenient notation for working with summations that will greatly facilitate our work in this section.

Sigma Notation

The symbol $\sum$ is the Greek letter "sigma".
This letter is like our letter <u>s</u> and is used to denote <u>s</u>ummation.
We use the letter as a convenient notation to denote a sum.

For example, $\sum_{i=1}^{3} i = 1 + 2 + 3$

The variable i could be denoted by any other symbol since it is just a <u>dummy variable</u>.

For example $\sum_{j=1}^{3} j = 1 + 2 + 3$, denotes the same summation as above.

Also note that the summation variable i *does* not have to be a specific number like 1 or 3.

For example, we may write: $\sum_{i=1}^{n} i = 1 + 2 + 3 + \cdots + n$ (here n is a variable)

More generally, given any function f, we may write:

$$\sum_{i=1}^{n} f(x_i) = f(x_1) + f(x_2) + f(x_3) + \cdots + f(x_n)$$

We will need the following basic properties of $\sum$.

Properties of Summation

$$\sum_i (a_i \pm b_i) = \sum_i a_i \pm \sum_i b_i$$

$$\sum_i (Ca_i) = C \sum_i a_i, \quad \text{where } C \text{ is a constant}$$

$$\sum_{i=1}^{n} C = nC, \quad \text{where } C \text{ is a constant}$$

Together, the first two properties are called "linearity of summation".
The proof of these properties is simple and is left to the exercises.

Our goal is to introduce a new concept called the <u>definite integral</u>.
Although it may seem a little complicated at first, the definite integral is one of the most important concepts in all of calculus. The definite integral is defined in terms of partitions, norms and limits which we will now introduce.

Norms and Partitions

Consider a function $f(x)$ defined on the interval $[a, b]$.

A <u>partition</u> P of the interval $[a, b]$ is a set of points $x_0, x_1, x_2, \ldots x_{n-1}, x_n$

chosen from $[a, b]$ such that: $a = x_0 < x_1 < x_2 < \cdots < x_{n-1} < x_n = b$.

Such a partition divides the interval $[a, b]$ into n subintervals.

$[x_0, x_1], [x_1, x_2], [x_2, x_3], \ldots, [x_{n-1}, x_n]$

The subintervals are not necessarily of the same length.

We define the <u>norm</u> of a partition to be the length of the <u>largest</u> subinterval.
The norm of the partition P is denoted by the notation $|P|$.

As an illustration, here are two partitions of the interval $[0, 1]$.

Let $P = \{0, 0.1, 0.2, 0.3, 0.4, 0.5, 0.6, 0.7, 0.8, 0.9, 1\}$
In this case, the points of the partition are all equally spaced and the partition is said to be regular. The norm of the partition is just $|P| = 0.1$

Consider a second partition of the interval $Q = \{0, 0.2, 0.5, 0.7, 1\}$
In this case, the subintervals are <u>not</u> equally spaced.
The norm of Q is the length of the largest subinterval: $|Q| = 0.3$

Riemann Sums

Consider a function $f(x)$ defined on the interval $[a, b]$.

Let P denote the partition of the interval $[a, b]$ corresponding to the points:

$$a = x_0 < x_1 < x_2 < \cdots < x_{n-1} < x_n = b$$

Such a partition divides the interval $[a, b]$ into n subintervals.

$$[x_0, x_1], [x_1, x_2], [x_2, x_3], \ldots, [x_{n-1}, x_n]$$

From each subinterval $[x_{i-1}, x_i]$, $1 \leq i \leq n$ we now arbitrarily choose a point and call it x_i^* so that $x_{i-1} \leq x_i^* \leq x_i$. We emphasize that the choice of x_i^* from the interval $[x_{i-1}, x_i]$ is completely arbitrary.

Now form the sum $\sum_{i=1}^{n} f(x_i^*)(x_i - x_{i-1})$

This expression is called a "Riemann Sum" of the function $f(x)$ after the great German mathematician Bernhard Riemann (1826-1866). Don't worry if the above sum seems a little unmotivated at this point - at this stage, we only want to help you understand the notation involved in this sum and how it is calculated given a partition P and the points x_i^*. Expanding the sum out using the definition of the $\sum$ notation, we see that it is a sum of n terms.

$$\sum_{i=1}^{n} f(x_i^*)(x_i - x_{(i-1)}) = f(x_1^*)(x_1 - x_0) + f(x_2^*)(x_2 - x_1) + \ldots + f(x_n^*)(x_n - x_{n-1})$$

Notice that the definition of the Riemann sum depends both on the partition P and the points x_i^* selected from each subinterval.

Interpretation of the Riemann Sum

Now that you are familiar with the notation used in the Riemann sum it is time to see what it means. One interpretation for the Riemann sum is related to the area underneath the graph of $f(x)$ which we studied in the previous section. Notice that the term $f(x_i^*)(x_i - x_{i-1})$ represents the <u>area</u> of the rectangle with height $f(x_i^*)$ and base $(x_i - x_{i-1})$.

Let $\Delta x_i = x_i - x_{i-1}$ denote the length of the ith subinterval.

Then we can see that the Riemann sum $\sum_{i=1}^{n} f(x_i^*) \Delta x_i$ is just the sum of the areas of the rectangles illustrated below.

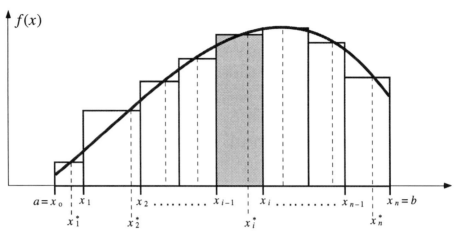

The Riemann Sum corresponds to the sum of the areas of the rectangles.

If enough rectangles are chosen, we can see that the Riemann sum is just an <u>approximation</u> to the <u>area</u> of the region underneath the curve $y = f(x)$ over the interval $[a, b]$.

Of course, a Riemann sum need not stand for area.
For example, let $v(t)$ be the velocity of a particle moving along a straight line.
Choose the interval to be $[0, 1]$. Then a typical Riemann sum of the velocity function $v(t)$ looks like the following:

$$\sum_{i=1}^{n} v(t_i^*) \Delta t_i, \text{ where } \Delta t_i = t_i - t_{i-1},$$

where $0 = t_0 < t_1 < t_2 < ... < t_{n-1} < t_n = 1$ is a <u>partition</u> of $[0, 1]$. Since t_i^* is chosen from the subinterval $[t_{i-1}, t_i]$, $v(t_i^*)$ is the velocity at the time instant t_i^*.
Note that the term $v(t_i^*) \Delta t_i$ is an approximation to the displacement of the particle during the time interval $[t_{i-1}, t_i]$!

Hence, in this case, the Riemann sum $\sum_{i=1}^{n} (t_i^*) \Delta t_i$ can be interpreted as an approximation to the displacement of the particle during the time interval $[0, 1]$.

We now define the definite integral of f from a to b.

Definition of the Definite Integral

$$\int_a^b f(x)\,dx = \lim_{|P| \to 0} \sum_{i=1}^{n} f(x_i^*) \Delta x_i$$

§4.2 Area & Definite Integrals

The above definition will only work provided the limit <u>exists</u> and is <u>independent</u> of the choice of the points x_i^*.

If the definite limit exists, we say f is <u>integrable</u> on the interval $[a, b]$.
Otherwise we say f is <u>not</u> integrable on the interval $[a, b]$.

The next theorem gives a sufficient condition for the existence of the integral $\int_a^b f(x)\,dx$.

> If $f(x)$ is continous on $[a, b]$, then $\int_a^b f(x)\,dx$ exists and is independent of the choice of the points x_i^* used in the construction of the Riemann sums.

Thus, if a function is <u>continuous</u>, its definite integral over a finite interval is guaranteed to exist. The converse need not be true. Many engineering functions are integrable on a given interval even though they are not continuous. However, not all functions are integrable on a given interval. (See exercises.)

Notational Comments

In the notation $\int_a^b f(x)\,dx$, $f(x)$ is called the <u>integrand</u> and the points a and b are called the <u>limits of integration</u>; a is the <u>lower</u> limit and b is the <u>upper</u> limit.
The variable x is called the <u>integration variable</u>. As in the case of the summation variable, it is just a dummy variable and can be replaced by any convenient variable.
The symbol "$\int$" is an elongated "S" signifying that an integral is just a limit of <u>s</u>ums as the number of summands increases to infinity while each summand diminishes to zero.

Although dx has no formal mathematical meaning by itself, dx can be heuristically considered as an infinitesimal version of Δx_i. You are already with the quantity dx appearing in the notation for <u>derivatives</u> such as $\frac{dy}{dx}$. We will see that the term dx appearing in the notation for definite <u>integrals</u> can be manipulated as if it were a very small number much as we did when it appeared in the notation for derivatives. This interpretation for dx is particularly suitable in many geometrical or physical contexts. We'll return to this in the next section.

Example 1

Consider the linear function $y = 1 + 2x$ on the interval $[-2, 2]$.

The partition $P = \{-2, -1.5, -1.1, -0.6, 0.2, 1, 1.3, 2\}$ divides the interval into seven subintervals. One point has been selected from each subinterval as follows:

$$x_1^* = -2,\ x_2^* = -1.4,\ x_3^* = -1,\ x_4^* = 0,\ x_5^* = 0.5,\ x_6^* = 1.3,\ x_7^* = 1.5$$

a) What is the norm of this partition?

a) Find the Riemann sum of the function $y = 1 + 2x$ over the interval $[-2, 2]$ using the above partition and the selected points.

Solution

The partition P divides the interval into the seven subintervals.

$[-2, -1.5]$, $[-1.5, -1.1]$, $[-1.1, -0.6]$, $[-0.6, 0.2]$, $[0.2, 1]$, $[1, 1.3]$ and $[1.3, 2]$

The corresponding lengths are:

$\Delta x_1 = -1.5 - (-2) = 0.5$

$\Delta x_2 = -1.1 - (-1.5) = 0.4$

$\Delta x_3 = -0.6 - (-1.1) = 0.5$

$\Delta x_4 = 0.2 - (-0.6) = 0.8$

$\Delta x_5 = 1 - 0.2 = 0.8$

$\Delta x_6 = 1.3 - 1 = 0.3$

$\Delta x_7 = 2 - 1.3 = 0.7$

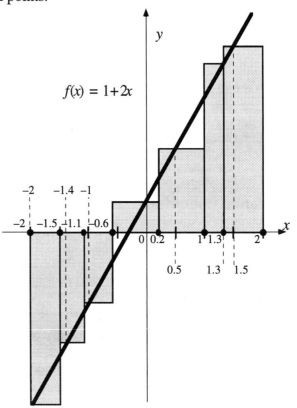

The norm of the partition P is defined as the length of the largest subinterval. Inspecting the above list of lengths we see that the largest number is 0.8 so that the norm of this partition is:

$$|P| = 0.8$$

Using the values of the function at each of the seven points x_i^* selected from the subintervals we find that the value of the Riemann sum is:

$$\sum_{i=1}^{7} f(x_i^*)\Delta x_i = (-3)(0.5) + (-1.8)(0.4) + (-1)(0.5) + (1)(0.8) + (2)(0.8) + (3.6)(0.3) + (4)(0.7)$$

$$= \boxed{3.56}$$

§4.2 Area & Definite Integrals

The Riemann sum can be interpreted as the sum of the areas of rectangles shown provided that those above the x-axis are counted as <u>positive</u> and those below the x-axis are counted as <u>negative</u>!

Geometric Interpretation of the Definite Integral - (Sign Convention)

Recall the fundamental definition of the definite integral:

$$\int_a^b f(x)\,dx = \lim_{\|P\|\to 0} \sum_{i=1}^n f(x_i^*)\Delta x_i$$

If we interpret the Riemann sum in terms of area, we see that the Riemann sum represents an approximation to the area of the region underneath the graph of $y = f(x)$ over the interval $[a, b]$. However, the area of those rectangles <u>below</u> the x-axis must be treated as <u>negative</u>! In the limit as the norm of the partition approaches 0, the approximation gets better and better and ultimately in the limit, the Riemann sum certainly coincides with our intuition of the area of a curved region, provided the area of the region above the x-axis is considered as positive, and that below the x-axis is considered as negative!

Thus the definite integral $\int_a^b f(x)\,dx$ may be considered as the area of the region bounded by the graph of $f(x)$, the x-axis, the vertical line $x = a$, and the vertical line $x = b$, provided we use the sign convention explained above.

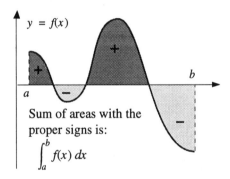

Sum of areas with the proper signs is:
$$\int_a^b f(x)\,dx$$

Two Methods for Finding Areas

In the case of nonnegative functions, we now have <u>two</u> methods by which to calculate the area of the graph under the curve of the function. In the previous section we found the area by first finding an <u>antiderivatiive</u> of the function. The above geometric interpretation of the Riemann sum shows that the area can also be found by evaluating a definite integral.

> **Area as a definite integral**
>
> The definite integral $\int_a^b f(x)\,dx$ of the nonnegative function f, represents the area under the graph of f over the interval $[a, b]$

Example 2

Evaluate the definite integral $\int_0^1 x^n dx$, where $n > 0$.

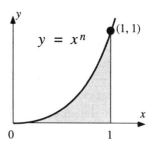

The definite integral corresponds to the shaded area under the curve.

Solution

Note that the integrand x^n is nonnegative on [0, 1]. Thus the definite integral represents the <u>area</u> of the region bounded above by $y = x^n$ over the interval [0, 1]. Let $A(x)$ be the area of the region underneath the curve $y = x^n$ on the <u>variable</u> interval [0, x]. From what we have learned in "The Area Problem", we see that:

$$A(x) = \int x^n dx = \frac{x^{n+1}}{n+1} + C$$

Note that $A(0) = 0$ implies that $C = 0$ and hence that: $A(x) = \frac{x^{n+1}}{n+1}$

Since the area over the original interval [0, 1] is $A(1) = \frac{1}{n+1}$, we have:

$$\int_0^1 x^n dx = \boxed{\frac{1}{n+1}}$$

Rules of Algebra for Definite Integrals

The daily work of engineers and scientists requires the evaluation of some rather complicated definite integrals. A few simple rules of algebra for manipulating definite integrals greatly facilitate this work.

1. **Zero Rule**

$$\int_a^a f(x) dx = 0$$

   The zero rule simply states that when the region degenerates to a vertical line, then the integral (the area under the curve) is zero.

We have already seen how useful the zero rule was in finding the constant of integration for the area function $A(x)$ in the previous section.

2. **Order of Integration**

$$\int_a^b f(x) dx = -\int_b^a f(x) dx$$

   If the endpoints of integration are reversed, the value of the definite integration changes sign!

Together the next two rules of algebra for definite integrals are known as the <u>linearity</u> property.

3. **Sums and differences**

§4.2 Area & Definite Integrals

$$\int_a^b (f(x) \pm g(x))dx = \int_a^b f(x)\,dx \pm \int_a^b g(x)\,dx$$

4. Constant multiples

$$\int_a^b k f(x)\,dx = k \int_a^b f(x)\,dx$$

We will now give a proof for the sum rule.
All the rules in this section can be proved in a very similar manner.
In each case, the definite integrals are written as the limits of Riemann sums.
Known properties about limits and finite sums suffice to complete each proof.

Proof of the Sum Rule:

Using the definition of the definite integral as the limit of a Riemann sum we see that:

$$\int_a^b (f(x) + g(x))dx = \lim_{\|P\| \to 0} \sum_i \left(f(x_i^*) + g(x_i^*) \right) \Delta x_i$$

$$= \lim_{\|P\| \to 0} \left(\sum_i f(x_i^*)\Delta x_i + \sum_i g(x_i^*) \Delta x_i \right)$$

Recall that the limit of a sum is the sum of the limits. Thus:

$$\int_a^b (f(x) + g(x))dx = \lim_{\|P\| \to 0} \sum_i f(x_i^*)\Delta x_i + \lim_{\|P\| \to 0} \sum_i g(x_i^*) \Delta x_i$$

$$= \int_a^b f(x)\,dx + \int_a^b g(x)\,dx$$

End of Proof

---

5. Additivity with respect to the interval of integration

$$\int_a^b f(x)\,dx = \int_a^c f(x)\,dx + \int_c^b f(x)\,dx$$

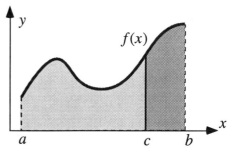

The additivity rule has a simple geometrical interpretation.

The total shaded area is the sum of the areas of the portions to the left and right of the vertical line $x = c$.

For a positive integrand $f(x)$, the definite integrand $\int_a^b f(x)\,dx$ represents the total area under the curve $y = f(x)$ from a to b. A vertical line at $x = c$ cuts this area into a left and right portion whose areas correspond to the definite integrals: $\int_a^c f(x)\,dx$ and $\int_c^b f(x)\,dx$. The additivity rule implies that the total area is the sum of the areas of the left and right portions.

Example 3 - Application of Additivity Rule

Find the definite integral $\int_0^{2\pi} |\sin x|\,dx$

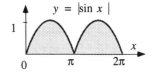

Solution

Notice that the graph of the function consists of <u>two</u> sine arches.
Using the additivity of integration with respect to the interval of integration, we may break the interval up as follows.

$$\int_0^{2\pi} |\sin x|\,dx = \int_0^{\pi} |\sin x|\,dx + \int_{\pi}^{2\pi} |\sin x|\,dx$$

The advantage of breaking up the interval like this is that we can now get rid of the absolute value signs.

Since $\sin x \geq 0$ on $[0, \pi]$, $|\sin x| = \sin x$. Similarly, $|\sin x| = -\sin x$ on $[\pi, 2\pi]$.

Thus the integral becomes: $\int_0^{2\pi} |\sin x|\,dx = \int_0^{\pi} \sin x\,dx + \int_{\pi}^{2\pi} (-\sin x)\,dx$

In the above, each integral represents the area of a <u>sine arch</u>.
By an example in the previous section (The Area Problem), we have

$$\int_0^{2\pi} |\sin x|\,dx = 2 + 2 = \boxed{4}$$

---

§4.2 Area & Definite Integrals   CAFÉ

Here are some further rules of algebra for definite integrals.

6. **Nonnegativeness**

If $f(x) \geq 0$ on $[a, b]$, then $\displaystyle\int_a^b f(x)\,dx \geq 0$

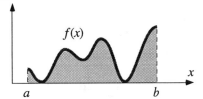

The area is non-negative since the graph lies above the x-axis.

The above rule for nonnegative functions has a useful generalization.
A nonnegative function is a function whose values are never less than zero.
We say that a function f <u>dominates</u> a function g if $\{f(x) \geq g(x)\}$. That is, the values of f are never less than the values of g. Thus, for example, a nonnegative function is a function which dominates the constant function $g(x) = 0$. Using the linearity rule, we can prove:

7. **Domination**

If $f(x) \geq g(x)$ for $a \leq x \leq b$, then $\displaystyle\int_a^b f(x)\,dx \geq \int_a^b g(x)\,dx$

That is, the integral of the larger function is greater!!

An immediate application of the domination rule is the following inequality.
Let $f(x)$ be a function which may assume both positive and negative values.
Since it is clear that $|f(x)| \geq f(x)$ for all x, we can say that $|f|$ dominates the function f.
Using the domination rule we have:

8. **Absolute Value Rule**

$\left|\displaystyle\int_a^b f(x)\,dx\right| \leq \int_a^b |f(x)|\,dx$ for $a \leq b$

The absolute value rule has a simple geometric interpretation as illustrated in the following figures.

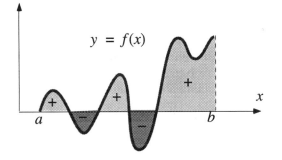
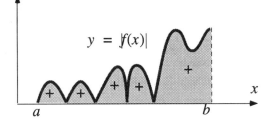

Example 4 - Application of the domination rule.

Show that the definite integral satisfies the inequality $\int_0^1 \frac{x^2}{x^{10}+1} dx \leq \frac{1}{3}$

Solution

Note it would be very difficult to prove this inequality by directly evaluating the integral. Using the <u>domination</u> rule gives a quick and easy proof of the inequality.

Note that the denominator term satisfies $x^{10} + 1 \geq 1$ for x in $[0, 1]$.

Inverting both sides (and the inequality) we obtain:

$$\frac{1}{x^{10}+1} \leq \frac{1}{1} = 1$$

which implies that:

$$\frac{x^2}{x^{10}+1} \leq x^2 \text{ for } x \text{ in } [0, 1].$$

> That is, the original complicated integrand is dominated by a simpler function whose integral is easy to evaluate!

Therefore by the <u>domination</u> rule we have

$$\int_0^1 \frac{x^2}{x^{10}+1} dx \leq \int_0^1 x^2 dx.$$

Using the fact that $\int_0^1 x^2 \, dx$ is the area of the region below the parabola $y = x^2$ over the interval $[0, 1]$ we know $\int_0^1 x^2 \, dx = \frac{1}{3}$. Thus, we have proven the requested inequality:

$$\int_0^1 \frac{x^2}{x^{10}+1} dx \leq \frac{1}{3}.$$

We would like to show an additional application of the Riemann sum. Many interesting equalities involving infinite summations can be easily proven using the interpretation of the Riemann sum as a definite integral.

Example 5

As n goes to infinity, the sum $\sum_{i=1}^{n} \frac{i^3}{n^4}$ approaches a finite limit which is denoted by $\lim_{n \to \infty} \sum_{i=1}^{n} \frac{i^3}{n^4}$.

Find this limit.

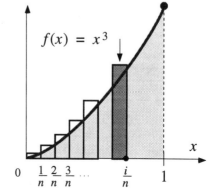

§4.2 Area & Definite Integrals — CAFÉ

Solution

The key to discovering this limit is to interpret it as the limit of a Riemann sum. Each term in a Riemann sum corresponds to the area of a <u>rectangle</u>. Since the area of a rectangle is the product of its base times its height, we must rearrange each term in the summation as a <u>product of two terms</u> in such a way that each term can be <u>interpreted</u> as the area of a rectangle. Although there is no unique way to do this, let us rearrange the summand as follows:

$$\sum_{i=1}^{n} \frac{i^3}{n^4} = \sum_{i=1}^{n} \left(\frac{i}{n}\right)^3 \cdot \frac{1}{n}$$

Note that the product $\left(\left(\frac{i}{n}\right)^3\right)\left(\frac{1}{n}\right)$ corresponds to the area of a rectangle with height $\left(\frac{i}{n}\right)^3$ and base $\frac{1}{n}$. This suggests that the function for the Riemann sum should be x^3 on the interval $[0, 1]$. Using a regular partition with n equal subintervals from $[0, 1]$ the base of each rectangle would be $\frac{1}{n}$ as above. If we select $x_i^* = \frac{i}{n}$ as the right end point of each subinterval, then the height of the rectangle is precisely $\left(\frac{i}{n}\right)^3$.

With this interpretation, the sum $\sum_{i=1}^{n} \frac{i^3}{n^4}$ corresponds to a Riemann sum for the function $y = x^3$ over the interval $[0, 1]$.

Using the interpretation of area as the limit of a Riemann sum we can now find the original limit!

$$\lim_{n \to \infty} \sum_{i=1}^{n} \frac{i^3}{n^4} = \lim_{n \to \infty} \sum_{i=1}^{n} \left(\frac{i}{n}\right)^3 \cdot \frac{1}{n} = \int_0^1 x^3 \, dx = \boxed{\frac{1}{4}}$$

The correspondence of terms in the transition:

from the sum: $\lim_{n \to \infty} \sum_{i=1}^{n} \left(\frac{i}{n}\right)^3 \cdot \frac{1}{n}$

to the integral: $\int_0^1 x^3 \, dx$

is summarized to the right.

Page 272   CAFÉ   §4.2 Area & Definite Integrals

WARMUP EXERCISES

In each of the following exercises, we are given a function f on an interval $[a,b]$.
A partition P of the interval is given and the points x_i^* selected from each of the resulting subintervals are identified. Find the norm of the partition and the corresponding Riemann sum.

1) Consider the function $f(x) = 2 + 3x^2$ on the interval $[0, 5]$.
$P = \{0, 0.8, 1.7, 3, 3.5, 4, 5\}$, x_i^* = left endpoint of the ith subinterval

2) Consider the function $f(x) = \dfrac{x^2}{x^2 + 1}$ on the interval $[-1, 3]$.
$P = \{-1, -0.4, 0.1, 1.3, 1.9, 2.5, 2.7, 3\}$, x_i^* = midpoint of the ith subinterval

3) Consider the function $f(x) = e^x$ on the interval $[-2, 1]$.
P is a regular partition into 10 equal parts,
x_i^* = right endpoint of the ith subinterval

4) Consider the function $f(x) = \dfrac{e^x}{e^x + 1}$ on the interval $[-3, 2]$.
$P = \{-3, -2.5, -1.4, 0, 0.7, 1, 1.4, 2\}$,
$x_1^* = -2.6$, $x_2^* = -1.4$, $x_3^* = -1$, $x_4^* = 0.5$, $x_5^* = 0.7$, $x_6^* = 1.2$, $x_7^* = 1.6$

5) Consider the function $f(x) = \sin x$ on the interval $[0, \pi]$.
P is a regular partition into n equal parts, x_i^* = left endpoint of the ith subinterval

6) Consider the function $f(x) = \ln x$ on the interval $[1, 5]$.
P is a regular partition into n equal parts, x_i^* = right endpoint of the ith subinterval

---

For each of the following exercises, evaluate the given definite integral and state the name of the rules in this section that are most useful in arriving at the answer.

7) Find $\displaystyle\int_1^1 (x^{20} e^x (\cos 5x)) dx$

8) Find $\displaystyle\int_{-1}^{-1} \dfrac{\sin x}{x^2 + e^x} dx$

9) Given that $\displaystyle\int_0^1 x^2 dx = \dfrac{1}{3}$ what is $\displaystyle\int_1^0 x^2 dx$?

10) Evaluate $\displaystyle\int_0^{100} x^{10} dx + \int_{100}^{50} x^{10} dx + \int_{50}^{0} x^{10} dx$

11) Evaluate $\displaystyle\int_1^{10} e^x \ln x \, dx + \int_{10}^{5} e^x \ln x \, dx + \int_5^1 e^x \ln x \, dx$

§4.2 Area & Definite Integrals

12) Evaluate $\int_a^b f(x)\,dx + \int_b^c f(x)\,dx + \int_c^a f(x)\,dx$

13) Given that $\int_0^1 f(x)\,dx = -2$ and $\int_1^3 f(x)\,dx = 3$, find $\int_0^3 f(x)\,dx$.

14) Given that $\int_0^3 f(x)\,dx = 5$ and $\int_0^3 g(x)\,dx = -2$, find $\int_0^3 (2f(x) - 3g(x))\,dx$.

15) Given that $\int_{-1}^1 f(x)\,dx = 2$ and $\int_{-1}^0 f(x)\,dx = 1$, find $\int_0^1 f(x)\,dx$.

16) Given that $\int_{-1}^1 f(x)\,dx = 2$ and $\int_{-1}^1 (3f(x) + 4g(x))\,dx = 5$, find $\int_{-1}^1 g(x)\,dx$

INTERMEDIATE EXERCISES

In the following exercises, first check that the integrand is nonnegative, then use the interpretation of a definite integral as an area to evaluate the integrals.
Consider the area $A(x)$ over a <u>variable</u> interval $[a,x]$.

17) $\int_1^5 \frac{2}{x}\,dx$

18) $\int_1^2 (2x^2 - 1)\,dx$

19) $\int_2^3 (2x^2 + 3x + 4)\,dx$

20) $\int_0^1 (e^x + 2x)\,dx$

21) $\int_2^5 \sqrt{x}\,dx$

22) $\int_{-1}^5 e^{-x}\,dx$

23) $\int_{-1}^1 \cos x\,dx$

24) $\int_1^2 \frac{1}{\sqrt{x}}\,dx$

25) $\int_0^{\pi/4} \sec^2 x\,dx$

In the following exercises, break each interval into subintervals over which the integrand assumes a simpler form. Then evaluate the integral.
(This illustrates an important application of the additivity rule.)

26) $\int_{-1}^2 |x|\,dx$

27) Let $f(x) = \begin{cases} x^2, & 0 \le x \le 1 \\ 2x, & 1 \le x \le 5 \end{cases}$. Evaluate $\int_0^5 f(x)\,dx$.

28) $\int_{\pi/2}^{3\pi/2} |\sin x|\,dx$

29) $\int_{-2}^2 |x^2 - 1|\,dx$

30) $\int_{-1}^2 \sqrt{(x-1)^2}\,dx$

Recall that $\sum_{i=1}^{n} a_i = a_1 + a_2 + \cdots + a_n$.

Prove the following basic properties of summations which are used in the proof of the analogous properties for definite integration. (See the linearity rules.)

31) $\sum_{i=1}^{n} (a_i + b_i) = \sum_{i=1}^{n} a_i + \sum_{i=1}^{n} b_i$

32) $\sum_{i=1}^{n} (k a_i) = k \sum_{i=1}^{n} a_i$, where k is a constant

33) $\sum_{i=1}^{n} k = nk$, where k is a constant

ADVANCED EXERCISES

Evaluate the following definite integrals. The additivity rule will be very useful.

34) $\int_{0}^{4} |e^x - 2| \, dx$

35) $\int_{0}^{\pi} |\sin x - \cos x| \, dx$

36) $\int_{0}^{3} [3 + u(x-1) + 2u(x-2)] \, dx$, where $u(x)$ is the unit step function

37) $\int_{0}^{3} \left[2 + (x^2 - 2) \cdot u(x-1)\right] dx$

38) $\int_{0}^{2} e^x u(x-1) \, dx$

39) Evaluate $\int_{0}^{5} \lfloor x \rfloor \, dx$. Recall that the floor function $\lfloor x \rfloor$ denotes the greatest integer $\leq x$. For example, $\lfloor \pi \rfloor = 3$.
(Hint: Sketch the graph of $\lfloor x \rfloor$ and use the additivity rule.)

40) $\int_{0}^{3} \lfloor 2x \rfloor \, dx$

41) $\int_{-1}^{3} \lfloor |x| \rfloor \, dx$

42) $\int_{0}^{3} \lfloor x^2 \rfloor \, dx$

43) $\int_{0}^{3} \lfloor x^2 + 1 \rfloor \, dx$

§4.2 Area & Definite Integrals

Rules of Algebra for Definite Integrals

44) **a)** Using the definition of the definite integral as a limit of Riemann sums, prove the additivity rule.

$$\int_a^b f(x)\,dx = \int_a^c f(x)\,dx + \int_c^b f(x)\,dx \quad \text{where } a < c < b.$$

b) Use **a)** and rules 1 and 2 to prove

$$\int_a^b f(x)\,dx = \int_a^c f(x)\,dx + \int_c^b f(x)\,dx \text{ where } c \text{ is not necessarily}$$

in the interval $[a, b]$.

45) Using the definition of the definite integral as a limit of Riemann sums, prove the rule for nonnegative functions. (Rule 6)

That is, show that if $f(x) \geq 0$ on $[a, b]$, then $\int_a^b f(x)\,dx \geq 0$.

46) Not every function is integrable. The purpose of this exercise is to show that the limit of a Riemann sum as the norm of the partition approaches zero may not always exist. Below is a well-known pathological case.

$$\text{Let } f(x) = \begin{cases} 0 & \text{if } x \text{ is rational} \\ 1 & \text{if } x \text{ is irrational} \end{cases} \quad \text{where } 0 \leq x \leq 1.$$

Recall that a <u>rational</u> number is a number expressible as a quotient of two integers. For example $\frac{2}{3}$, $-\frac{10}{101}$, etc. A number that is not rational is called <u>irrational</u>, for example $\sqrt{2}$, $\sqrt{3}$, etc. Show that Riemann sums corresponding to the function $f(x)$ can be made equal to 1 or 0 depending on the choice of the points x_i^* no matter how small the norm of partition is!

Thus the definite integral $\int_0^1 f(x)\,dx$ does not exist!!!.

In the following exercises, interpret each sum as a Riemann sum of a function over an appropriate interval. Express the limit as $n \to \infty$ of the summation as an integral, and then evaluate it.

47) $\displaystyle\lim_{n \to \infty} \frac{1}{n}\left(\left(\frac{1}{n}\right)^5 + \left(\frac{2}{n}\right)^5 + \left(\frac{3}{n}\right)^5 + \cdots + \left(\frac{n}{n}\right)^5\right)$

48) $\displaystyle\lim_{n \to \infty} \frac{1}{n}\left(\sin\left(\frac{1}{n}\right) + \sin\left(\frac{2}{n}\right) + \cdots + \sin\left(\frac{n}{n}\right)\right)$

49) $\displaystyle\lim_{n \to \infty} \frac{1}{n}\left(e^{1/n} + e^{2/n} + e^{3/n} + \cdots + e^{n/n}\right)$

50) $\displaystyle\lim_{n\to\infty} \frac{1}{n} \sum_{i=1}^{n} \left(2\left(1+\frac{i}{n}\right)^2 + 1 \right)$

51) $\displaystyle\lim_{n\to\infty} \frac{2}{n} \sum_{i=1}^{n} \sqrt{1+\frac{2i}{n}}$

52) $\displaystyle\lim_{n\to\infty} \frac{2}{n^{3/2}} \left(\sqrt{n+2} + \sqrt{n+4} + \sqrt{n+6} + \cdots + \sqrt{n+2n} \right)$

53) $\displaystyle\lim_{n\to\infty} \frac{2}{n} \sum_{i=1}^{n} \left(1+\frac{2i}{n}\right)^{-1/2}$

54) $\displaystyle\lim_{n\to\infty} \frac{2}{\sqrt{n}} \left(\frac{1}{\sqrt{n+2}} + \frac{1}{\sqrt{n+4}} + \frac{1}{\sqrt{n+6}} + \cdots + \frac{1}{\sqrt{n+2n}} \right)$

55) Let $A_n = \sqrt[n]{\dfrac{n!}{n^n}}$, where $n! = 1\times 2\times 3\times \cdots \times n$ is called n factorial.

For example: $4! = 1\times 2\times 3\times 4 = 24$
Our goal is to find the limit $\displaystyle\lim_{n\to\infty} A_n$.

    a)   First calculate $\ln(A_n)$ and interpret it as a Riemann sum of a function over an appropriate interval. Identify the function and the interval?

    b)   Express $\displaystyle\lim_{n\to\infty} \ln(A_n)$ as an integral (do not evaluate the integral).

    c)   Express $\displaystyle\lim_{n\to\infty} A_n$ in terms of the result in (b).

56) Assume that $f(x)$ is integrable over the interval $[a, b]$.
What is $\displaystyle\lim_{n\to\infty} \frac{b-a}{n} \sum_{i=1}^{n} f\left(a + \frac{b-a}{n} i\right)$?

§ 4.3 The Fundamental Theorems of Calculus

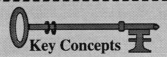
Key Concepts

- The First Fundamental Theorem
- The Second Fundamental Theorem
- The Area Between Curves

Functions defined as integrals

Many important functions of interest to the scientist or engineer are defined as integrals of other functions. For example, given a function f, we often require information about the new function F defined as the integral:

$$F(x) = \int_a^x f(t)\,dt$$

Since functions defined as integrals arise so frequently, it is important to be able to work with these functions just as we would with an ordinary function like a polynomial or trigonometric function. In this section we will illustrate a few of the situations in which functions defined as integrals naturally arise and will develop formulas to manipulate then mathematically just as we would manipulate any other function. In fact, although a function which is defined as an integral may seem a little unfamiliar at first, we will see that in many ways, they are easier to work with than most functions!

❖

Examples of functions defined as integrals

There are many important functions in science and engineering that are defined as integrals of other functions $F(x) = \int_a^x f(t)\,dt$. Some examples are:

1) The <u>error</u> function $\operatorname{erf}(x) = \dfrac{2}{\sqrt{\pi}} \int_0^x e^{-t^2}\,dt$ appears in probability and statistics.

2) The <u>Fresnel</u> functions $\int_0^x \sin(t^2)\,dt$ and $\int_0^x \cos(t^2)\,dt$ arise in optics.

3) The function $s_i(x) = \int_0^x \dfrac{\sin t}{t}\,dt$ is important in signal processing and Fourier analysis.

One may wonder if there are simpler ways to express such functions, but unfortunately, the answer is no. The integral representation of each of these functions remains the simplest and the most elegant way to define these functions.

Let $F(x)$ be the function defined by the integral $F(x) = \int_a^x f(t)\,dt$. Since this is a perfectly good function, we should be able to manipulate it as we would any other function. For example, is the function differentiable, and if so, what is its derivative? The answer to this question is provided by the First Fundamental Theorem of Calculus.

The First Fundamental Theorem of Calculus

> If f is continous on $[a, b]$, then
>
> $$F(x) = \int_a^x f(t)\,dt$$
>
> is differentiable at every point x in (a, b) and
>
> $$F'(x) = \frac{d}{dx}\int_a^x f(t)\,dt = f(x)$$

This Fundamental Theorem assures us that the derivative of $F(x)$ will exist if the integrand f is continuous, and not only that, it tells us precisely what the derivative is:

$$\frac{dF}{dx} = f(x)$$

Thus finding the derivative of a function defined by an integral is actually <u>easier</u> than finding the derivative of more familiar functions. If you reflect on how hard you had to work to find the derivative for some of the functions in the previous chapters, you will see how useful this result is. To help build confidence with the statement of this theorem, we will give a typical example of how it is used before showing the proof.

Example 1

Consider the function $F(x) = \int_0^x \frac{e^t + 1}{\sin t + t^2 + 5}\,dt$ which is defined as an integral.

Find the derivative $\frac{dF}{dx}$.

Solution

Note that the integrand $f(t) = \dfrac{e^t + 1}{\sin t + t^2 + 5}$ is continuous everywhere.

Thus the First Fundamental Theorem applies and we conclude that the derivative is equal to the integrand:

$$\frac{dF}{dx} = f(x) = \boxed{\frac{e^x + 1}{\sin x + x^2 + 5}}$$

Compare how easily we obtained the derivative using the Fundamental Theorem to how hard you would have to work to evaluate the derivative of the rather complicated looking integrand!

We will now show how to prove the fundamental theorem. If you feel a little uncomfortable with proofs, we recommend you move forward and read some more of the example problems. Then return to this proof when you feel comfortable with what the Fundamental Theorem states and how it is used.

Proof of the First Fundamental Theorem

The fundamental theorem is a statement about the value of the derivative $\dfrac{dF}{dx}$ where F is defined by the integral $F(x) = \displaystyle\int_a^x f(t)\,dt$. Recall the definition of the derivative is:

$$\frac{dF}{dx} = \lim_{\Delta x \to 0} \left(\frac{F(x + \Delta x) - F(x)}{\Delta x} \right)$$

Thus we are naturally lead to consider the difference quotient for $F(x)$.

$$\frac{F(x + \Delta x) - F(x)}{\Delta x} = \frac{\displaystyle\int_a^{x+\Delta x} f(t)\,dt - \int_a^x f(t)\,dt}{\Delta x} = \frac{\displaystyle\int_x^{x+\Delta x} f(t)\,dt}{\Delta x}$$

The last step follows from the <u>additivity</u> property of definite integrals over an interval.

Thus, we can prove that $\dfrac{dF}{dx} = f(x)$ if we can prove that:

$$\lim_{\Delta x \to 0} \frac{\displaystyle\int_x^{x+\Delta x} f(t)\,dt}{\Delta x} = f(x)$$

We will show that the above limit holds as $\Delta x \to 0^+$ leaving the case $\Delta x \to 0^-$ as an exercise. Thus assume that $\Delta x > 0$.

Let M and m denote the Maximum and minimum of the function f on $[x, x + \Delta x]$.
By definition, $m \leq f(t) \leq M$ for all t in $[x, x + \Delta x]$.
Integrating this inequality over the interval $[x, x + \Delta x]$ we see that:

$$m\Delta x \leq \int_x^{x+\Delta x} f(t)dt \leq M\Delta x$$

Now $f(x)$, being continuous on $[x, x + \Delta x]$, takes on both the extreme values M and m somewhere in the interval $[x, x + \Delta x]$ by the Extreme Value Theorem.

That is, $M = f(C)$ and $m = f(c)$ for some points C and c in $[x, x + \Delta x]$.

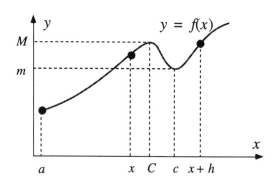

Substituting the equations $M = f(C)$ and $m = f(c)$ into the above inequality yields:

$$f(c)\Delta x \leq \int_x^{x+\Delta x} f(t)dt \leq f(C)\Delta x$$

Since $\Delta x > 0$, dividing through by Δx leads to

$$f(c) \leq \frac{1}{\Delta x}\int_x^{x+\Delta x} f(t)dt \leq f(C)$$

As $\Delta x \to 0^+$, the interval $[x, x + \Delta x]$ shrinks to the single point x.
Since the points c and C lie in the interval $[x, x + \Delta x]$, it is clear that $c \to x$ and $C \to x$.

Evaluating the limit of all three terms of the above inequality and using the continuity of the function f, we obtain:

$$\lim_{\Delta x \to 0^+} f(c) \leq \lim_{\Delta x \to 0^+} \frac{\int_x^{x+\Delta x} f(t)dt}{\Delta x} \leq \lim_{\Delta x \to 0^+} f(C)$$

so that:

$$f(x) \leq \lim_{\Delta x \to 0^+} \frac{\int_x^{x+\Delta x} f(t)dt}{\Delta x} \leq f(x)$$

Noting that both sides of the inequality are the same we conclude:

$$\lim_{\Delta x \to 0^+} \frac{\int_x^{x+\Delta x} f(t)dt}{\Delta x} = f(x) \quad \text{or} \quad \lim_{\Delta x \to 0^+} \frac{F(x+\Delta x) - F(x)}{\Delta x} = f(x)$$

§4.3 Fundamental Theorems

Similarly we can show that $\displaystyle\lim_{\Delta x \to 0^-} \frac{F(x+\Delta x) - F(x)}{\Delta x} = f(x)$.

Thus we have shown that $\dfrac{dF}{dx} = f(x)$ for all x in (a, b)

End of Proof

The First Fundamental Theorem asserts the <u>existence</u> of an antiderivative F of a continuous function f and gives an extremely simple equation for the <u>derivative</u> $\dfrac{dF}{dx} = f(x)$.
However, it says nothing about how to <u>find</u> the antiderivative F !
Indeed, F may not even be expressible in terms of <u>elementary functions</u>. Recall that elementary functions are those functions resulting from combinations of constants, powers of x, $\sin x$, $\cos x$, e^x, $\ln x$, and other trigonometric functions and their inverses, using only elementary algebraic operations i.e. $+, -, \times, \div, \sqrt[k]{}$ and composition.

The First Fundamental Theorem shows that :

> Differentiation and integration are inverse processes.

That is, F is the integral of f and f is the derivative of F.

This is an extremely important observation, and has tremendous impact on the fields of science and mathematics. For instance, we know that velocity v and displacement x are related by

$$\frac{dx}{dt} = v$$

If velocity as a function of t is known, we can discover the displacement $x(t)$ by integrating $v(t)$ with respect to time t. (i.e. $x(t) = \displaystyle\int_a^t v(x)\,dx$.)

Example 2

Find $\dfrac{dy}{dt}$ if $y(t) = \displaystyle\int_0^{t^2} e^{-x^2}\,dx$.

Solution

Let us define the new function $F(u) = \displaystyle\int_0^u e^{-x^2}\,dx$. Then $y = F(u)$, where $u = t^2$.
Notice that y is the composition of the functions F and u, that is $y(t) = F(u(t))$.

By the First Fundamental Theorem we have $\dfrac{dy}{du} = e^{-u^2}$.

To calculate the derivative $\dfrac{dy}{dt}$ we need to use the Chain Rule. $\dfrac{dy}{dt} = \dfrac{dy}{du}\dfrac{du}{dt} = e^{-u^2}(2t)$

Expressing u in terms of the original variable t we obtain: $\quad \dfrac{dy}{dt} = \boxed{2te^{-t^4}}$

Now we come to the remarkable Second Fundamental Theorem. It is a statement that enables us to calculate the value of a definite integral if an antiderivative is known.

The Second Fundamental Theorem of Calculus

> If f is continuous on $[a, b]$ and F is any antiderivative of f on $[a, b]$, then
> $$\int_a^b f(x)\,dx = F(b) - F(a)$$

Using this theorem, you can calculate the definite integral $\int_a^b f(x)\,dx$ in two steps.

1)   Find an antiderivative F of f

2)   Evaluate the antiderivative at the endpoints and subtract: $F(b) - F(a)$

That's it!

However, if we are unable to find an antiderivative F of f <u>explicitly</u>, then the Second Fundamental Theorem is of little use in computing the numerical value of the integral.

When F is a complicated expression, it is convenient to have a short hand notation for the difference $F(b) - F(a)$. Introducing the following notation enables one to write the function F only once (instead of twice)

$$F(x)\Big]_a^b = F(b) - F(a)$$

This allows us to express the Second Fundamental Theorem as:

$$\int_a^b f(x)\,dx = F(x)\Big]_a^b$$

§4.3 Fundamental Theorems

To build confidence using the Second Fundamental Theorem, we first work out an example problem before giving the proof of this theorem.

Example 3

Evaluate the integral $\int_0^1 (e^x - x)\,dx$.

Solution

The function $(e^x - x)$ is continuous on the interval $[0, 1]$.

An antiderivative of $(e^x - x)$ is $F(x) = e^x - \frac{1}{2}x^2$

Thus, the Second Fundamental Theorem implies:

$$\int_0^1 (e^x - x)\,dx = \left(e^x - \frac{1}{2}x^2\right)\Big|_0^1 = \left(e - \frac{1}{2}\right) - \left(e^0 - 0\right) = \boxed{e - \frac{3}{2}}$$

Proof of the Second Fundamental Theorem of Calculus

The proof of the Second Fundamental Theorem requires a statement that although geometrically appealing has not been rigorously justified.

"A differentiable function $f(x)$ that has <u>zero derivative</u> everywhere in (a, b) must be a <u>constant</u> function in (a, b)."

A rigorous proof can be found in the section entitled Mean Value Theorem I.

We must show that $\int_a^b f(x)\,dx = F(b) - F(a)$ where F is an antiderivative of f.

Consider the new function G defined as the integral: $G(x) = \int_a^x f(t)\,dt$

The First Fundamental Theorem implies that $\frac{dG}{dx} = f(x)$ so that $G(x)$ is also an antiderivative of f.

Hence:

$$\frac{d}{dx}(F(x) - G(x)) = \frac{d}{dx}F(x) - \frac{d}{dx}G(x) = f(x) - f(x) = 0 \text{ for } x \text{ in } (a, b).$$

We see that the derivative of the function $(F(x) - G(x))$ is zero everywhere in (a, b).

By the result mentioned at the beginning of the proof we see that for all x in (a, b)

$$F(x) - G(x) = c, \qquad \text{where } c \text{ is a constant.}$$

But $F(x) - G(x)$ is <u>continuous</u> on the closed interval $[a, b]$.
Thus $F(x) - G(x) = c$ at the endpoints of the interval $[a, b]$ as well.

Since is $F(x) - G(x)$ constant, it is certainly true that:

$$F(b) - F(a) = G(b) - G(a)$$

Since $G(a) = \int_a^a f(t)dt = 0$ and $G(b) = \int_a^b f(t)dt$, we conclude that:

$$\int_a^b f(x)dx = F(b) - F(a)$$

𝔈𝔫𝔡 𝔬𝔣 ℜ𝔯𝔬𝔬𝔣

Example 4

Evaluate the definite integral $\int_{1/2}^{2} \left(x - \frac{1}{x}\right) dx$.

Solution

Evaluating the integral term by term we find:

$$\int_{1/2}^{2} \left(x - \frac{1}{x}\right) dx = \int_{1/2}^{2} x\, dx - \int_{1/2}^{2} \frac{1}{x} dx$$

$$= \frac{1}{2}x^2 \bigg]_{1/2}^{2} - \ln x \bigg]_{1/2}^{2}$$

$$= \left(2 - \frac{1}{8}\right) - \left(\ln 2 - \ln \frac{1}{2}\right)$$

$$= \boxed{\frac{15}{8} - 2\ln 2}$$

Areas Between Curves

Suppose that $y = f(x)$ and $y = g(x)$ are continuous functions such that $f(x) \geq g(x)$ over the interval $[a, b]$. Then the graph of $f(x)$ lies above the graph of g over the interval $[a, b]$. How do we find the area A of the region between these two curves?

§4.3 Fundamental Theorems

Let us approximate the area using Riemann sums. First we divide the interval $a \leq x \leq b$ into n subintervals of equal length $\dfrac{b-a}{n}$ so that the ith division point $x_i = a + \dfrac{b-a}{n}(i-1),\ 1 \leq i \leq n$.

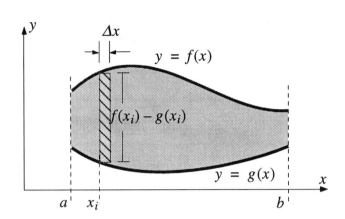

The ith rectangle in the shaded region has height $f(x_i) - g(x_i)$ and width Δx, where $\Delta x = \dfrac{b-a}{n}$. Therefore, $(f(x_i) - g(x_i))\Delta x$ represents its area.

Thus the Riemann sum $\sum\limits_{i=1}^{n} (f(x_i) - g(x_i))\Delta x$ is an approximation of the true area A of the region. In the limit as $n \to \infty$, the Riemann sum gives the exact area.

$$A = \lim_{n \to \infty} \left(\sum_{i=1}^{n} (f(x_i) - g(x_i))\, \Delta x \right) = \int_a^b (f(x) - g(x))\,dx$$

Thus the area A of the region between the two curves is $A = \displaystyle\int_a^b (f(x) - g(x))\,dx$.

The Area Between Two Curves

If $f(x) \geq g(x)$ for $a \leq x \leq b$, then the area between the graphs of f and g from a to b is:

$$\text{Area} = \int_a^b (f(x) - g(x))\,dx.$$

In other words, the area is the integral of (the upper function − the lower function).

Geometric Interpretation of the Area Formula

The area formula has a simple geometric interpretation. Denote the term after the integral sign by
$dA = (f(x) - g(x))dx$

The term dA can be interpreted as the area of an infinitesimal rectangle of width dx and height $(f(x) - g(x))$.

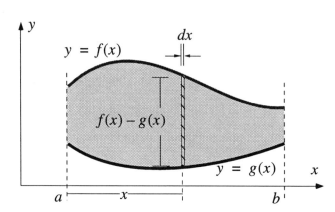

Area of infinitesimal rectangle is
$dA = (f(x) - g(x))dx$.

Example 5

Find the area of the region bounded by the curves $y = x^2$ and $y = x^{1/2}$.

Solution

We sketch the graph for $y = x^2$ and $y = x^{1/2}$ to see which curve lies above the other:

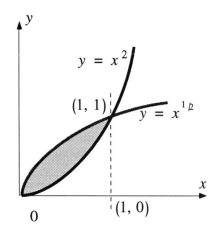

Since $x^2 \leq x^{1/2}$ for $0 \leq x \leq 1$, we see $y = x^{1/2}$ is the upper curve. The desired expression for the area is:

$$\int_0^1 (x^{1/2} - x^2)dx = \int_0^1 x^{1/2}dx - \int_0^1 x^2 dx$$
(by linearity)

$$= \frac{2}{3}x^{3/2}\bigg]_0^1 - \frac{1}{3}x^3\bigg]_0^1$$

$$= \frac{2}{3} - \frac{1}{3} = \boxed{\frac{1}{3}}$$

Example 6

A river bends around a meadow, following a curve given by the parabola:

$$y = 3(x-x^2)$$

A highway cuts across the meadow along the line:

$$y = \frac{1}{2}x$$

What is the area of meadow between the highway and the river? (Assume x and y are measured in miles.)

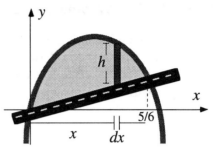

Meadow is bounded above by the river and below by the highway.

Solution

Consider an infinitesimal rectangle of width dx located a distance x from the y-axis. The <u>height</u> of this rectangle is:

$$h = 3(x-x^2) - \frac{1}{2}x = -3x^2 + \frac{5}{2}x$$

Thus the <u>area</u> of the infinitesimal rectangle is $dA = \left(-3x^2 + \frac{5}{2}x\right)dx$

By solving $\begin{cases} y = 3(x-x^2) \\ y = \frac{1}{2}x \end{cases}$ simultaneously, we find that the river intersects the road at the points $x = 0$ and $x = \frac{5}{6}$.

The area A of the meadow is the sum of the areas of all such infinitesimal rectangles as x ranges from 0 to 5/6.

Thus:

$$A = \int_0^{5/6} \left(-3x^2 + \frac{5}{2}x\right)dx$$

To evaluate the definite integral, note that $-x^3 + \frac{5}{4}x^2$ is an antiderivative of $-3x^2 + \frac{5}{2}x$.

By the Second Fundamental Theorem, the area of the meadow is:

$$A = \int_0^{5/6} \left(-3x^2 + \frac{5}{2}x\right)dx = -x^3 + \frac{5}{4}x^2 \Big|_0^{5/6} = \boxed{0.2893 \text{ square miles}}$$

Example 7 – Archimedes' Formula for Parabolic Arches

Even before the invention of calculus, Archimedes using an ingenious method, discovered that the area A under a parabolic arch is two-thirds the base b times the height h:

$$A = \frac{2}{3} bh$$

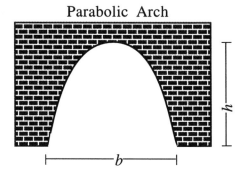

Parabolic Arch

Solution

We can rediscover his beautiful result using calculus.

The area of the arch corresponds to the area under the parabola $f(x) = h\left(1 - \left(\frac{2x}{b}\right)^2\right)$ over the interval $\left[-\frac{b}{2}, \frac{b}{2}\right]$. Thus the area is given as the definite integral

$$\int_{-b/2}^{+b/2} h\left(1 - 4\frac{x^2}{b^2}\right) dx = h\left(x - 4\frac{x^3}{3b^2}\right)\Bigg|_{-b/2}^{b/2} = h\left(b - \frac{b}{3}\right) = \boxed{\frac{2}{3} bh}$$

§4.3 Fundamental Theorems

In the following exercises, find the derivative of the given function with respect to *x*.

1) $\displaystyle\int_0^x \frac{\sin t}{t^2 + 1}\, dt$

2) $\displaystyle\int_{-1}^x \sqrt{t^4 + 1}\, dt$

3) $\displaystyle\int_x^2 \frac{\sin t}{\cos^2 t + 1}\, dt$

4) $\displaystyle\int_x^{10} \sin(t^2)\, dt$

5) $\displaystyle\int_x^2 \frac{e^t}{t^4 + 5}\, dt$

6) $\displaystyle\int_0^x e^{-t^2}\, dt$

❖

Evaluate the following integrals using the Fundamental Theorems.

7) $\displaystyle\int_1^2 (x+3)\, dx$

8) $\displaystyle\int_{-\pi/2}^{\pi/2} \cos x\, dx$

9) $\displaystyle\int_0^1 (3e^x - 2)\, dx$

10) $\displaystyle\int_1^2 \frac{1 - \sqrt{t}}{\sqrt{t}}\, dt$

11) $\displaystyle\int_0^1 (x^2 + 1)^2\, dx$

12) $\displaystyle\int_0^{\pi/2} (\sin 2t)\, dt$

13) $\displaystyle\int_1^2 \left(t^2 + \frac{1}{t^2}\right) dt$

14) $\displaystyle\int_0^1 (3x + 5)\, dx$

15) $\displaystyle\int_0^1 (1 - x^2)\, dx$

16) $\displaystyle\int_1^4 \sqrt{x}\, dx$

17) $\displaystyle\int_0^{2\pi} \sin x\, dx$

18) $\displaystyle\int_1^2 \left(t - \frac{1}{t}\right) dt$

INTERMEDIATE EXERCISES

Find the area of the planar region bounded by the given curves.

19) $y = x^3 \quad\quad y = x$

20) $y = \sqrt{x} \quad\quad y = x^4$

21) $y = x^2 - 1 \quad\quad y = 1 - x^2$

22) $y = 3 - x^2 \quad\quad y = -2x$

23) $y = 2 \quad\quad y = \cosh x$

24) $x = y^3 \quad\quad x = y^4$

Find the derivative $\dfrac{dy}{dx}$ of the following functions.

25) $y = \displaystyle\int_0^{x^2} \dfrac{\sin t}{t}\, dt$

26) $y = \displaystyle\int_{x^3}^{0} \dfrac{e^t - 1}{t}\, dt$

27) $y = \displaystyle\int_0^{\sqrt{x}} \sin(t^2)\, dt$

28) $y = \displaystyle\int_0^{g(x)} \dfrac{dt}{1 + \sqrt{1 + t}}$

29) $y = \displaystyle\int_{2x+1}^{x^2} e^{t^2}\, dt$

30) $y = \displaystyle\int_{h(x)}^{x^2} e^{t^2}\, dt$

31) $y = \displaystyle\int_{h(x)}^{g(x)} e^{t^2}\, dt$

32) $y = \displaystyle\int_{x+1}^{x} t^t\, dt$

33) Find $f(1)$ if $\displaystyle\int_0^x f(t)\, dt = \dfrac{\sin x}{x}$.

34) Find $f(2)$ if $\displaystyle\int_0^{x^2} f(t)\, dt = e^x$.

§4.3 Fundamental Theorems — CAFÉ

ADVANCED EXERCISES

35) Find the area of the shaded region between the curves $y = \sin x$ and $y = \cos x$ (see illustration)

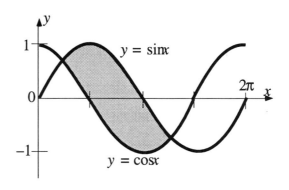

36) **The Quarrel**

  a) Two brothers quarrel over how to divide a piece of land evenly with the shape as shown. The younger suggests to divide the land using a horizontal line. What must be the height of this horizontal line?

  b) The elder brother suggests to cut the land evenly by using a straight line passing through $(1, 1)$. What must be the slope of the line? Use Newton's method to calculate the slope to an accuracy of four decimal places.

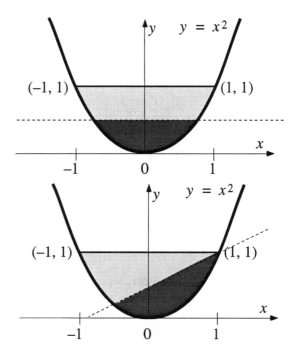

---

37) If f is continuous and g and h are differentiable, find $\dfrac{d}{dx}\displaystyle\int_{h(x)}^{g(x)} f(t)\,dt$.

38) Let $x \geq 0$. Find $\displaystyle\int_0^x \lfloor \sin t \rfloor \, dt$. Is it true that $\dfrac{d}{dx}\displaystyle\int_0^x \lfloor \sin t \rfloor \, dt = \lfloor \sin x \rfloor$ for $x \geq 0$?
(Hint: Sketch the graph of $\lfloor \sin t \rfloor$.)

39) Consider the piecewise function $f(t) = \begin{cases} 0 & \text{if } 0 \le t \le 1 \\ t & \text{if } 1 < t \end{cases}$

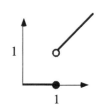

Find $\int_0^x f(t)\,dt$. Is it true that $\dfrac{d}{dx}\int_0^x f(t)\,dt = f(x)$ for all $x \ge 0$?

40) Evaluate $\int_0^3 \sin(\lfloor t \rfloor)\,dt$. (Hint: Sketch the graph of $\sin(\lfloor t \rfloor)$.)

41) Evaluate $\int_{-2}^2 \lfloor |x| \rfloor\,dx$.

42) Evaluate $\int_{-2}^2 \lfloor\!\lfloor x \rfloor\!\rfloor\,dx$.

---

43) The current in a river has speed $v = v_0 + ky(a - y)$, where a is the width of the river. (v_0 and k are constants.) The rowboat is pointed directly towards the opposite shore and rowed in such a way that its y coordinate is $y(t) = 4t$. The width of the river is 2 miles and the speed of the current is 2 miles/hr near the shore and 4 miles/hr mid-stream.

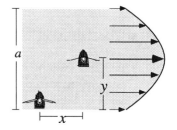

a) Find a, v_0, and k.

b) Find $x(t)$, the x coordinate of the boat as a function of time.

c) When the boat is 1 mile down stream, how far away is it from the opposite bank? Use Newton's Method to reach an accuracy of 4 decimal places.

44) **Fuel Consumption**

The gas consumption rate of a jet liner is shown. Let v be the speed of the plane. The gas consumption $g(v)$ is defined as the amount of fuel consumed per hour when traveling at the speed v. The jet liner accelerates at a constant rate from a speed of 300 miles/hour to 700 miles/hour in 1/2 hour for a cross-Atlantic flight.

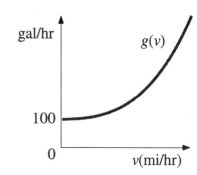

a) Find the total gas consumed in this time interval. Assume the consumption curve to be $g(v) = 100 + \dfrac{1}{1000}v^2$.

Hint: The gas consumption (in gallons) during an infinitesimal time interval of length dt at time t is $g(v(t))\,dt$.

b) How far will the jet liner fly for the first 100 gallons of fuel?

§ 4.4 Substitution

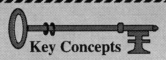

Key Concepts

❖ Substitution for Indefinite Integrals   ❖ Substitution for Definite Integrals

The Fundamental Theorem of Calculus enables us to solve many integration problems by antidifferentiation. However, the Fundamental Theorem does not tell us explicitly how to actually find the antiderivative for a given function. Therefore, there remains the important practical problem of developing techniques to find antiderivatives. This problem is very distinct from the problem of finding a derivative. In previous chapters we have discovered rules which enabled us to differentiate any function obtained by combining constants, powers of x, and functions like $\sin x$, $\cos x$, e^x, and their inverses, using only elementary algebraic operations $\left(\text{i.e. } +, -, \times, \div, \sqrt[k]{}, \text{ and composition}\right)$. Such functions are called <u>elementary</u> functions. For example, despite its complex appearance, the function $f(x) = \dfrac{\sqrt{\sin x + x^2}}{xe^{x^2} - 1} + 3 \ln x$
is an elementary function. Although the derivative of any elementary function is also an elementary function, the same does not hold true for integrals. In fact, most elementary functions do <u>not</u> have antiderivatives expressible in terms of other elementary functions. So in a certain sense, we are very lucky when we find the antiderivative of a given function. Examples of functions whose integrals <u>cannot</u> be expressed in terms of elementary functions include:

$$\int e^{x^2} dx \qquad \int \frac{1}{\ln x} dx \qquad \int \frac{e^x}{x} dx$$

Despite this basic difficulty associated with finding antiderivatives, we shall develop some simple integration techniques that will enable you to find antiderivatives for most of the functions you will encounter in your engineering career. Most of the fundamental integration techniques are just integral versions of the differentiation techniques introduced earlier.

> To every rule for differentiating functions there is a corresponding integration rule.

For example, just as the derivative of a sum of functions is the sum of their derivatives, the integral of a sum of functions is the sum of their integrals.

One of the most important differentiation rules is the chain rule. The corresponding integration rule is called <u>substitution</u> or <u>change of variable</u> and is the central focus of this section.

❖

Let $f(x)$ and $g(x)$ be functions. Let $F(x)$ be an antiderivative of the function $f(x)$, and consider the composition: $F(g(x))$
Using the chain rule, we find:

$$\frac{d}{dx}(F(g(x))) = f(g(x))g'(x)$$

To convert this differentiation rule into an integration rule, we just integrate both sides of the equation.

$$\int \frac{d}{dx}(F(g(x)))\,dx = \int f(g(x))g'(x)\,dx$$

By the Fundamental Theorem, $\int \frac{d}{dx}(F(g(x)))\,dx = F(g(x)) + C$.

Thus we may write.

> **Method of Substitution (Long Form)**
>
> $$\int f(g(x))g'(x)\,dx = F(g(x)) + C$$

This integration rule can be easily remembered using a shorthand notation.
Introduce the substitution $u = g(x)$. Then $\frac{du}{dx} = g'(x)$ or $du = g'(x)dx$.
Substituting these expressions into the integral yields:

> **Method of Substitution (Short Form)**
>
> $$\int f(u)\,du = F(u) + C$$
>
> Where F is an antiderivative of f.

The antiderivative is easily expressed in terms of the original variable x using the substitution $u = g(x)$.

The same method can be extended to handle definite integrals, that is integrals having limits of integration. However, it is important to note that the limits of integration must be changed. Using the Fundamental Theorem:

$$\int_a^b f(g(x))g'(x)\,dx = F(g(x))\Big]_a^b = F(g(b)) - F(g(a))$$

§4.4 Substitution   CAFÉ   Page 295

Thus we may write:

$$\int_a^b f(g(x))g'(x)dx = \int_{g(a)}^{g(b)} f(u)du = F(g(b)) - F(g(a))$$

Notice that the limits of integration with respect to $u = g(x)$ are from $g(a)$ to $g(b)$ instead of from a to b.

The difficult part in applying substitution to find an antiderivative lies in choosing the function u. There is no general rule that tells us which substitution to make. Further, the method of substitution does not always work. However, there are a few good strategies that can help one make the correct substitution in the case that the method does work. These strategies are illustrated in the following examples.

Example 1

Evaluate the indefinite integral $\int 2x(1+x^2)^{10}\,dx$ using the method of substitution.

Solution

Inside the tenth power is the term $u = 1+x^2$ and the matching derivative term is $du = 2x\,dx$. Rewriting the original integrand in terms of the new variable u we obtain:

$$\int 2x(1+x^2)^{10}\,dx = \int u^{10}\,du$$

After making this substitution, the original integrand is now in a form which is easy to integrate:

$$\int u^{10}\,du = \frac{u^{11}}{11} + C$$

Converting back to the original variable x one obtains:

$$\int 2x(1+x^2)^{10}\,dx = \boxed{\frac{(1+x^2)^{11}}{11} + C}$$

Check

As is always the case with integration, we can check the result by differentiation.

$$\frac{d}{dx}\left(\frac{(1+x^2)^{11}}{11} + C\right) = (1+x^2)^{10}\,2x$$

Example 2

Find the indefinite integral $\int \dfrac{x+2}{x^2+4x+7}\,dx$ using the method of substitution.

Solution

This time the choice of the u function is not so clear. However, notice that the numerator is half the derivative of the denominator. Thus if we introduce the substitution, $u = x^2+4x+7$ the corresponding derivative is $du = (2x+4)dx = 2(x+2)dx$.

Rewriting the original integrand in terms of the new variable u we obtain:

$$\int \dfrac{x+2}{x^2+4x+7}\,dx = \dfrac{1}{2}\int \dfrac{2(x+2)}{x^2+4x+7}\,dx = \dfrac{1}{2}\int \dfrac{du}{u} = \dfrac{1}{2}\ln|u| + C$$

Note that the $\dfrac{1}{2}$ was inserted into the expression because the numerator is half of the substituted value of du. Remember that constants can be pulled out of the integrand. Converting back to the original variable x one obtains:

$$\int \dfrac{x+2}{x^2+4x+7}\,dx = \dfrac{1}{2}\ln|u| + C = \dfrac{1}{2}\ln(x^2+4x+7) + C$$

Notice that the absolute value sign was dropped since the function $x^2+4x+7 = (x+2)^2 + 3$ is always positive. Using properties of the logarithm the answer could also be expressed in the form:

$$\int \dfrac{x+2}{x^2+4x+7}\,dx = \boxed{\ln\sqrt{x^2+4x+7} + C}$$

Check   We can check the result by differentiation.

$$\dfrac{d}{dx}\left(\dfrac{1}{2}\ln(x^2+4x+7) + C\right) = \dfrac{1}{2}\dfrac{2x+4}{x^2+4x+7} = \dfrac{x+2}{x^2+4x+7}$$

Examples like this, where the numerator is equal to the derivative of the denominator (up to a constant) arise quite frequently as the next two examples show.

§4.4 Substitution

Example 3

Use the method of substitution to find the integral of the tangent function.
That is find $\int \tan x \, dx$.

Solution

By the definition of the tangent function: $\quad \int \tan x \, dx = \int \frac{\sin x}{\cos x} \, dx$

After the substitution $\boxed{u = \cos x} \quad \boxed{du = -\sin x \, dx}$ the integral becomes:

$$\int \tan x \, dx = -\int \frac{1}{u} \, du = -\ln|u| + C = -\ln(|\cos x|) + C$$

Using the logarithmic identity $\ln\left|\frac{1}{z}\right| = -\ln|z|$ and the definition $\sec x = \frac{1}{\cos x}$ gives:

$$\int \tan x \, dx = \boxed{\ln(|\sec x|) + C}$$

Example 4

Use the method of substitution to find the integral of the secant function.
That is find $\int \sec x \, dx$.

Solution

The key to discovering the integral of the secant function is to multiply it by the curious combination $\frac{\sec x + \tan x}{\sec x + \tan x} = 1$. This is just one of those little tricks that works.

$$\int \sec x \, dx = \int \sec x \left(\frac{\sec x + \tan x}{\sec x + \tan x}\right) dx$$

$$= \int \left(\frac{\sec^2 x + \sec x \tan x}{\sec x + \tan x}\right) dx$$

We can now see that the numerator is the derivative of the denominator by adding the identities.

$$\frac{d}{dx} \tan x = \sec^2 x \qquad \text{and} \qquad \frac{d}{dx} \sec x = \sec x \tan x$$

This suggests the substitution

$$\boxed{u = \sec x + \tan x} \qquad \boxed{du = (\sec x \tan x + \sec^2 x) \, dx}$$

after which the integral becomes:

$$\int \sec x\, dx \quad = \quad \int \frac{1}{u} du = \ln(|u|) + C = \ln(|\sec x + \tan x|) + C$$

Thus the indefinite integral of the secant function is:

$$\int \sec x\, dx = \boxed{\ln(|\sec x + \tan x|) + C}$$

Example 5

Evaluate the definite integral $\int_0^{\pi/2} \cos^3 x\, dx$ using the method of substitution.

Solution

Method 1 - (In terms of the original variable x).
Although we may be tempted to make the substitution $u = \cos x$, this turns out to be of no help because the derivative of the cosine does not appear in the integrand. However, using the trigonometric identity $\sin^2 x + \cos^2 x = 1$, we can rewrite the integrand as $\cos^3 x = (1 - \sin^2 x)\cos x$. Notice that the remaining $\cos x$ term will make the substitution $u = \sin x$ work. Letting $\boxed{u = \sin x}$ and $\boxed{du = \sin x\, dx}$ we obtain:

$$\int \cos^3 x\, dx = \int (1 - \sin^2 x)\cos x\, dx = \int (1 - u^2)\, du = \int 1\, du - \int u^2\, du$$

$$= u - \frac{1}{3}u^3 + C$$

Since $u = \sin x$,

$$\int \cos^3 x\, dx = \sin x - \frac{1}{3}\sin^3 x + C$$

Using the Fundamental Theorem of Calculus, the original definite integral can now be found by evaluating the antiderivative at the endpoints and subtracting.

$$\int_0^{\pi/2} \cos^3 x\, dx = \left(\sin x - \frac{1}{3}\sin^3 x\right)\Big]_0^{\pi/2} = \left(1 - \frac{1}{3}\right) - (0 - 0) = \boxed{\frac{2}{3}}$$

Method 2 - (In terms of the new variable u.)
Another way to look at the same problem is to express the limits of integration in terms of the new variable u. Since $u = \sin x$, when $x = 0$ we have $u(0) = \sin(0) = 0$ and when $x = \frac{\pi}{2}$ we have $u\left(\frac{\pi}{2}\right) = \sin\left(\frac{\pi}{2}\right) = 1$.

$$\int_{x=0}^{x=\pi/2} \cos^3 x\, dx = \int_{x=0}^{x=\pi/2} (1 - \sin^2 x)\cos x\, dx = \int_{u=0}^{u=1} (1 - u^2)\, du$$

$$= \left(u - \frac{1}{3}u^3\right)\Big]_0^1 = \boxed{\frac{2}{3}}$$

Both methods give the same answer. Notice that in the second approach, the antiderivative is expressed in terms of the new variable u, not in terms of the original variable x.

Example 6 - Tsunami (Giant Ocean Wave)

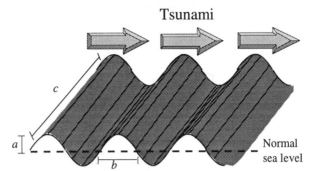

Tsunami

Occasionally, earthquakes occur under the ocean floor causing giant waves on the surface called tsunamis. These waves may have crests over one hundred feet high.

At a given instant, the height y of an ocean wave above the normal sea level is given by the function $y = a \sin \frac{\pi x}{b}$ where a and b are constants. Thus, in the interval from $x = 0$ to $x = b$ the ocean surface is raised above normal sea level and forms a giant wall of water. The entire force of this wall of water may strike a coastal community with catastrophic consequences. Note that the constant a corresponds to the maximum height of the ocean wave. Find the volume V of water raised above the normal sea level in a single ocean wave as it enters a harbor of width c.

Solution

The volume of water contained in a single ocean wave as it crashes into the harbor is $V = Ac$ where c is the width of the harbor and A is the area under the sine curve $y = a \sin \frac{\pi x}{b}$ over the interval from 0 to b.

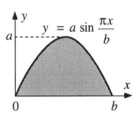

The area under the sine curve is $A = \int_0^b a \sin \frac{\pi x}{b} \, dx$.

This integral is easier to evaluate after the substitution: $\boxed{u = \frac{\pi x}{b}}$ $\boxed{du = \frac{\pi}{b} dx}$

Thus $dx = \frac{b}{\pi} du$ and the integral becomes:

$$A = \int_0^b a \sin \frac{\pi x}{b} \, dx = \frac{ab}{\pi} \int_0^\pi \sin u \, du = \frac{ab}{\pi}(-\cos u)\Big]_0^\pi = \frac{2ab}{\pi}$$

Thus the volume of water crashing into the harbor in each wave is $V = Ac = \boxed{\dfrac{2abc}{\pi}}$

Example 7   Evaluate the integral $\int \sin 5x \cos 3x \, dx$

Solution

Notice the integrand is a product of a sine and a cosine term. To evaluate such integrals it is always helpful to use the product-to-sum formulas first, and then to perform the integral termwise. Using the product-to-sum formula

$$\sin A \cos B = \frac{1}{2}\{\sin(A-B) + \sin(A+B)\}$$

we obtain:

$$\int \sin 5x \cos 3x \, dx = \int \frac{1}{2}(\sin(5x-3x) + \sin(5x+3x))dx$$

$$= \frac{1}{2}\int (\sin 2x + \sin 8x)dx$$

$$= \frac{1}{2}\int \sin 2x \, dx + \frac{1}{2}\int \sin 8x \, dx$$

We then use the substitution $u = 2x$ for the first integral and $u = 8x$ for the second to get:

$$\int \sin 5x \cos 3x \, dx = \boxed{-\frac{1}{4}\cos 2x - \frac{1}{16}\cos 8x + C}$$

Using the product-to-sum formulas, any product of two sines or cosines can be integrated as in the last example. The product-to-sum formulas listed below will help with many of the exercises in this section.

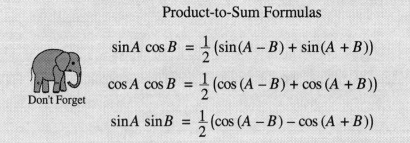

Product-to-Sum Formulas

$$\sin A \cos B = \frac{1}{2}(\sin(A-B) + \sin(A+B))$$

$$\cos A \cos B = \frac{1}{2}(\cos(A-B) + \cos(A+B))$$

$$\sin A \sin B = \frac{1}{2}(\cos(A-B) - \cos(A+B))$$

Don't Forget

§4.4 Substitution   CAFÉ   Page 301

WARMUP EXERCISES

1) Find the derivatives of the functions $\sqrt{1+x^2}$, $\sin(1+x^2)$ and $\tan(1+x^2)$.

Evaluate the following indefinite integrals with the help of your results from above.

a) $\displaystyle\int x\cos(1+x^2)\,dx$ b) $\displaystyle\int \frac{x}{\sqrt{1+x^2}}\,dx$ c) $\displaystyle\int \frac{x}{\cos^2(1+x^2)}\,dx$

Evaluate the following integrals with the help of the suggested substitutions.

2) $\displaystyle\int_0^{\pi/2} \cos(2x)\,dx$ Let $u = 2x$

3) $\displaystyle\int_0^1 (3+2x)^{2000}\,dx$ Let $u = 3+2x$

4) $\displaystyle\int \cos^{10}t \,\sin t\,dt$ Let $u = \cos t$

5) $\displaystyle\int \cos(\sin x)\cos x\,dx$ Let $u = \sin x$

6) $\displaystyle\int_0^1 \frac{e^x}{e^x+1}\,dx$ Let $u = e^x+1$

7) $\displaystyle\int e^{x^2}x\,dx$ Let $u = x^2$

8) $\displaystyle\int \frac{dx}{x\ln x}$ Let $u = \ln x$

9) $\displaystyle\int x\sqrt{x^2+1}\,dx$ Let $u = x^2+1$

10) $\displaystyle\int \cos^3 x\,\sin x\,dx$ Let $u = \cos x$

11) $\displaystyle\int \frac{dx}{x\sqrt{\ln x}}$ Let $u = \ln x$

12) $\displaystyle\int (x^2+1)^{100}x\,dx$ Let $u = x^2+1$

13) $\displaystyle\int_0^3 x\sqrt{x+1}\,dx$ Let $u = x+1$

INTERMEDIATE EXERCISES

Evaluate the following integrals using the method of substitution.

14) $\int \sin(7x)\, dx$

15) $\int \sin\left(\dfrac{60t}{2\pi}\right) dt$

16) $\int \cos(7x+11)\, dx$

17) $\int \cos(\omega t + \phi)\, dt$

18) $\int 2x \sin(1+x^2)\, dx$

19) $\int x \tan(1+x^2)\, dx$

20) $\int x e^{x^2}\, dx$

21) $\int \dfrac{e^{\sqrt{x}}}{\sqrt{x}}\, dx$

22) $\int \sqrt{2x+1}\, dx$

23) $\int \dfrac{1}{2x+1}\, dx$

24) $\int \dfrac{x}{1+x^2}\, dx$

25) $\int \dfrac{x^2}{1+x^3}\, dx$

26) $\int \dfrac{\sin x}{\cos x}\, dx$

27) $\int \dfrac{1+2x}{x+x^2}\, dx$

28) $\int \dfrac{x^3}{1+x^4}\, dx$

29) $\int \sqrt{x^3+1}\, x^2\, dx$

30) $\int \dfrac{(1+\sqrt{x})^5}{\sqrt{x}}\, dx$

31) $\int \dfrac{\sin\sqrt{x}}{\sqrt{x}}\, dx$

32) $\int x\sqrt{x^2+4}\, dx$

33) $\int x\sqrt{x^2-4}\, dx$

34) $\int x\sqrt{1+x}\, dx$

35) $\int x^2 \sqrt{1+x}\, dx$

36) $\int x\sqrt{4-x^2}\, dx$

37) $\int \dfrac{x^2\, dx}{\sqrt{x^3+1}}$

38) $\int \cos^3 x \sin x\, dx$

39) $\int \sec^2(4x)\, dx$

40) $\int \sec(2x)\tan(2x)\, dx$

The following integrals require some basic trigonometric identities.

41) $\int \sin^2 x\, dx$ Hint: Use $\sin^2 x = \dfrac{1-\cos 2x}{2}$

42) $\int_0^\pi \cos^2 x\, dx$ Hint: Use $\cos^2 x = \dfrac{1+\cos 2x}{2}$

43) $\int \tan^3 x\, dx$ Hint: Use $\tan^2 x = \sec^2 x - 1$

44) $\int \sin^3 x\, dx$

45) $\int \cos^2(2x)\, dx$

46) $\int \sin^2(3x)\, dx$

§4.4 Substitution

Use product-to-sum formulas to evaluate the following integrals.

47) $\int \cos 5x \cos 7x \, dx$

48) $\int_0^\pi \sin 5x \sin 8x \, dx$

49) $\int_0^{2\pi} \sin 5x \sin 6x \, dx$

50) $\int_0^{2\pi} \sin 4x \cos 3x \, dx$

51) $\int_0^{2\pi} \sin mx \cos nx \, dx \quad m \neq n, m \text{ and } n \text{ are integers}$

52) $\int_0^{2\pi} \sin mx \sin nx \, dx \quad m \neq n, m \text{ and } n \text{ are integers}$

53) $\int_0^{2\pi} \cos mx \cos nx \, dx \quad m \neq n, m \text{ and } n \text{ are integers}$

Evaluate each of the following indefinite integrals in two ways. One approach is to multiply out all the terms and then integrate. A second approach is to use a substitution. In each case note which method involves less work.

54) $\int 2(3+2x)^3 \, dx$

55) $\int 2x(3+x^2)^3 \, dx$

Evaluate each of the following indefinite integrals either by using a substitution or by multiplying out all the terms. Choose the method that involves the least effort.

56) $\int (2+5x)^{100} \, dx$

57) $\int (1+x^3)^3 x^2 \, dx$

58) $\int (1+x^3)^3 \, dx$

Suppose that $F(x)$ is known to be an antiderivative of the function $f(x)$.
Evaluate each of the following integrals in terms of the function $F(x)$.

59) $\int f(2x) \, dx$

60) $\int 2x f(x^2) \, dx$

61) $\int f(\sin x) \cos x \, dx$

62) $\int f(e^x) e^x \, dx$

63) $\int f(1+3x) \, dx$

64) $\int f(\cos x) \sin x \, dx$

65) Hydroelectric Power Station

The power P (in kilowatts) produced by a hydroelectric generator due to water flowing through its turbines can be reasonably modeled by the function

$$P(f) = 1000\left(1 - \frac{500}{f}\right)$$

where f is the flow rate in cubic meters per second. No power is generated if the flow rate falls below 500 m³/sec. Note that the maximum power output is 1000 kilowatts.

The average flow rate through the turbines is 1000 m³/sec. However, due to a brief storm, the flow increases for one hour as shown in the illustration. The total energy produced is the integral of the power:

$E = \int P\,dt.$ Find the energy E (in kilowatt-hours) generated during the hour.

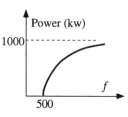

Power output as a function of flow rate.

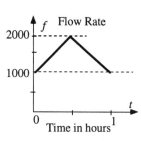

Flow rate during one hour storm

§ 4.5 Mean Value Theorem
Part 1

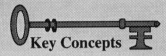
Key Concepts

- Rolle's Theorem
- Counting Zeroes of a Function
- Mean Value Theorem
- First Acceleration Theorem

In this section we introduce the Mean Value Theorem and touch on some of its many consequences and applications. It is one of the most fundamental results in calculus. However, we first explore <u>Rolle's Theorem</u>, which is a special case of the Mean Value Theorem and is a little easier to visualize.

When a juggler tosses a ball up into the air and catches it, the ball first rises and then falls. If he holds his hands at the same height, then the height of the ball when thrown is the same as the height of the ball when caught. At the peak of the trajectory, the ball stops rising and begins to fall. Clearly that at the highest point, the vertical component of the ball's velocity must be zero.
This is an example of Rolle's Theorem.

<u>Rolle's Theorem</u>: The height of the ball is the same when thrown and when caught. At some point during the interval of flight, the derivative of the height function is zero.

Rolle's Theorem

Consider a function f which has the same value at two points a and b.

That is $f(a) = f(b)$.

If the function f is differentiable on (a, b) and continuous on $[a, b]$ then there is at least one point c in (a, b) such that $f'(c) = 0$.

Illustration of Rolle's Theorem

The two functions illustrated to the right both satisfy the hypotheses of Rolle's Theorem. In each case, there is at least one point c where the tangent to the graph is horizontal and $f'(c) = 0$.

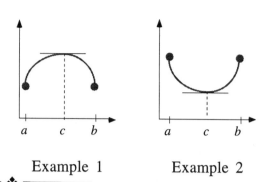

Example 1   Example 2

Proof of Rolle's Theorem

We must show that there is some point c in the interval (a, b) such that $f'(c) = 0$. Either the function $f(x)$ is constant on $[a, b]$ or it is not. If it is constant (the graph is a straight, horizontal line) then we are done because for any point c in (a, b) the derivative is zero so $f'(c) = 0$. If the function is not constant, then there is some point x_0 in the interval (a, b) where the function has a <u>different</u> value than the value at the endpoints. That is:

$$f(x_0) \neq f(a)$$

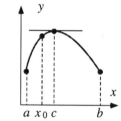

Thus either $f(x_0) > f(a)$ or $f(x_0) < f(a)$.
We shall consider the first case $f(x_0) > f(a)$ and leave the second one as an exercise. Since the function is continuous on $[a, b]$ it must assume a **maximum** value at some point c in the closed interval $[a, b]$ by the **Extreme Value Theorem**. The point c where the maximum occurs cannot be one of the endpoints because we have assumed that the value at x_0 is larger than the value at the endpoints: $f(x_0) > f(a)$. Thus c lies in the open interval (a, b). Since the function is differentiable, then by **Fermat's Theorem** we have $f'(c) = 0$.

End of Proof

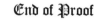

Let us try a few sample problems to strengthen our understanding of Rolle's theorem.

Example 1

Verify that the cubic polynomial $f(x) = x^3 - 4x + 2$ satisfies the <u>hypotheses</u> of Rolle's Theorem on the interval $[0, 2]$. Then find all numbers c that satisfy the <u>conclusion</u> of Rolle's Theorem.

Solution

First we will verify that the polynomial satisfies the hypotheses of Rolle's Theorem.

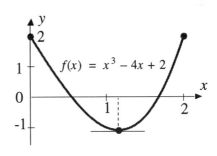

1) The values of the polynomial are equal at the endpoints of the interval $[0, 2]$ since

$$f(0) = f(2) = 2$$

2) Since all polynomials are continuous and differentiable everywhere, it follows as a special case that this cubic polynomial is continuous on $[0, 2]$ and differentiable on $(0, 2)$.

Thus the hypotheses of Rolle's Theorem apply and we conclude that there is some point c in $(0, 2)$ such that $f'(c) = 0$.

Next we will find all points c in the interval $(0, 2)$ where the derivative is zero.

Since the derivative of $f(x) = x^3 - 4x + 2$ is $f'(x) = 3x^2 - 4$, the point c must satisfy

$$3c^2 - 4 = 0$$

Solving for the unknown c gives two roots: $c = \pm \dfrac{2}{\sqrt{3}}$.

Only the positive solution $\boxed{c = +\dfrac{2}{\sqrt{3}}}$ lies in the interval $(0, 2)$ and this is the point where $f'(c) = 0$.

The next example shows that Rolle's Theorem when combined with the Intermediate Value Theorem can sometimes be used to find the number of zeroes of a function in a given interval.

Example 2 - Counting Zeroes or Roots

Show that the polynomial $f(x) = x^5 + 5x + 3$ has exactly one real root.

Solution

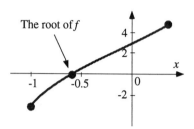

Recall that a point x is a <u>root</u> of a function if $f(x) = 0$. A polynomial of degree five is called a <u>quintic</u> polynomial. It is known that there is, in general, no elementary expression in terms of radicals for the roots of a polynomial of degree five or more. Thus we cannot expect to find the root(s) explicitly. We will show that the polynomial has exactly one root in two steps.
First we use the Intermediate Value Theorem to approximately locate one root, and then we use Rolle's Theorem to show that there cannot be another. Thus there must be exactly one root.

1)   There is at least one root.
By testing the values of the function we find that $f(0) = +3$ and $f(-1) = -3$. Thus the function changes sign and by the Intermediate Value Theorem, there is at least one root x_1 in the interval $(-1, 0)$.

2)   There are no other roots.
To show there are no other roots, we <u>tentatively</u> assume that there is a second root x_2 and arrive at a contradiction. This method of arguing is called <u>Proof by Contradiction</u>.

Since x_1 and x_2 are both supposed to be roots, we have $f(x_1) = f(x_2) = 0$.
On the interval with endpoints x_1 and x_2, the hypotheses of Rolle's Theorem are satisfied.
Thus there exists a point c between x_1 and x_2 such that $f'(c) = 0$.
The derivative of $f(x) = x^5 + 5x + 3$ is $f'(x) = 5x^4 + 5$.
Thus the derivative is always greater than 5 and so cannot possibly be zero.
We have a <u>contradiction</u>!!! Thus our tentative assumption that there is a second root is wrong. We conclude that the function $f(x) = x^5 + 5x + 3$ has precisely one root somewhere between -1 and 0.

We will now use Rolle's theorem to prove the more general Mean Value Theorem.

> ### Mean Value Theorem (MVT)
> Let the function f be continuous on the closed interval $[a, b]$ and differentiable on the open interval (a, b).
> Let $m = \dfrac{f(b) - f(a)}{b - a}$ denote the slope of the secant line through the endpoints of the graph of f over the interval $[a, b]$.
> Then there is some point c in (a, b) such that $f'(c) = m$

Geometric Interpretation

Although the formal statement of the Mean Value Theorem sounds a little complicated, the geometric interpretation is really quite simple. At the point c, the slope of the tangent line is equal to the slope of the secant line through the endpoints.

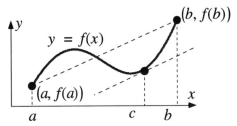

The tangent line at $x = c$ is parallel to the secant line through the end points.

We will now present a proof of the MVT. If you feel a little uncomfortable with proofs, you may want to read Example 3 first and then return to tackle this proof.

Proof of the MVT (Derivative Version)

We must show that there is some point c in (a, b) such that $f'(c) = m$.
The proof begins by finding the equation for the secant line joining the endpoints $(a, f(a))$ and $(b, f(b))$. The slope m of this secant line is:

$$m = \frac{\text{rise}}{\text{run}} = \frac{f(b)-f(a)}{b-a}$$

Thus the equation of the secant line is:

$$y = f(a) + m(x-a)$$

Next we find an expression for the height $h(x)$ of the graph of f above the secant line.

$$h(x) = f(x) - [f(a) + m(x-a)]$$

Note that $h(x)$ is zero at the endpoints a and b.

$$h(a) = f(a) - [f(a) + m(a-a)] = 0$$

$$h(b) = f(b) - [f(a) + m(b-a)] = 0 \quad \text{since } m = \frac{f(b)-f(a)}{b-a}.$$

Further, the height $h(x)$ is continuous on the closed interval $[a,b]$ and differentiable on the open interval (a, b) because $f(x)$ is. So $h(x)$ satisfies the hypotheses of Rolle's Theorem.

Thus by Rolle's Theorem, there is a point c in (a, b) such that $h'(c) = 0$.
Differentiating the function $h(x) = f(x) - [f(a) + m(x-a)]$ gives:

$$h'(x) = f'(x) - m \qquad \text{so that} \qquad h'(c) = f'(c) - m.$$

Thus at the point c in (a, b) such that $h'(c) = 0$, we also have $f'(c) = m$.
    This completes the proof of the Mean Value Theorem.

𝔈𝔫𝔡 𝔬𝔣 𝔓𝔯𝔬𝔬𝔣

Example 3

Consider the quadratic polynomial $f(x) = x^2$. Show that this function satisfies the hypotheses of the Mean Value Theorem on the interval $[0, 1]$ and find all values c that satisfy the conclusion of the theorem.

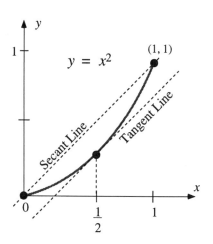

Solution

First we show that the function satisfies the hypotheses of the Mean Value Theorem. Recall that all polynomials are continuous and differentiable everywhere. Thus as a special case, the quadratic polynomial $f(x) = x^2$ is continuous on $[0, 1]$ and differentiable on $(0, 1)$ and hence satisfies the hypotheses of the Mean Value Theorem.

The conclusion of the MVT is that there is at least one point c in $(0, 1)$ such that:

$$f'(c) = m = \frac{f(1)-f(0)}{1-0} = \frac{1^2-0^2}{1-0} = 1$$

Since the derivative is $f'(x) = 2x$, we conclude that there is a point c such that $2c = 1$.

Clearly $\boxed{c = 1/2}$. Geometrically, the tangent line at $c = 1/2$ is parallel to the secant line through the endpoints. (Both lines have the same slope of 1.)

In proving the Second Fundamental Theorem of Calculus in an earlier section, we make use of the following geometrically intuitive result. The MVT can be used to rigorously justify this result.

> **Corollary of the MVT**
>
> If the derivative of a function is zero at all points in an interval (a, b), then the function is constant on (a, b).
> That is, if $f'(x) = 0$ for all x in (a, b), then f is constant

§4.5 Mean Value Theorem I   CAFÉ   Page 311

𝔓roof

Select two points x_1 and x_2 in (a, b) satisfying $x_1 < x_2$.

We will show that $f(x_1) = f(x_2)$ for all possible values of x_1 and x_2 in (a, b). This will establish that the function f is constant in (a, b).

First we note that the hypotheses of the Mean Value Theorem are satisfied on any such interval $[x_1, x_2]$. Indeed, since f is differentiable on (a, b), it must be differentiable on the subinterval (x_1, x_2). Since a function must be continuous wherever it is differentiable, we also see that f is continuous on $[x_1, x_2]$. Applying the Mean Value Theorem to the interval $[x_1, x_2]$ shows that there exists a number c in (x_1, x_2) such that:

$$\frac{f(x_2) - f(x_1)}{x_2 - x_1} = f'(c)$$

Since by assumption the derivative of f is zero at all points, we have $f'(c) = 0$. Thus $f(x_2) - f(x_1) = 0$ so that $f(x_2) = f(x_1)$.

𝔈nd of 𝔓roof

Example 4 - Squirt Classic

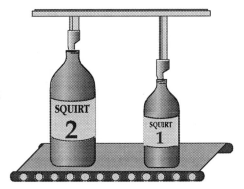

A soft drink distributor markets both one liter and two liter bottles. The bottling machinery adjusts its flow rate so that it takes the <u>same</u> amount of time (10 seconds) to fill one liter bottles as it does to fill 2 liter bottles.

Assume a one liter and a two liter bottle begin filling at the same time. Show that there is a time instant when Squirt Classic is flowing into the two liter bottle at twice the rate of the one liter bottle.

Solution

Let $V_1(t)$ denote the volume (in liters) of Squirt Classic in the one liter bottle at time t (in seconds). Since there is no Squirt in the bottle at $t = 0$, $V_1(0) = 0$. Since the one liter bottle is full after 10 seconds, $V_1(10) = 1$ liter. Let $V_2(t)$ be similarly defined for the two liter bottle, and notice that $V_2(0) = 0$ and $V_2(10) = 2$ liters. Now let us define a new function:

$$g(t) = V_2(t) - 2V_1(t)$$

Notice that $\quad\quad g(0) = V_2(0) - 2V_1(0) = 0 - 2 \times 0 = 0 \quad\quad$ and

$$g(10) = V_2(10) - 2V_1(10) = 2 - 2 \times 1 = 0$$

Thus $g(t)$ satisfies the assumptions of Rolle's Theorem. We conclude that there exists a time instant t_0 such that,

$$g'(t_0) = 0 \quad\quad \text{where } 0 < t_0 < 10$$

Since $g'(t_0) = V_2'(t_0) - 2V_1'(t_0) = 0$ we have,

$$V_2'(t_0) = 2V_1'(t_0)$$

Thus at the time t_0, Squirt Classic is flowing into the 2 liter bottle at twice the rate of the one liter bottle.

If a particle with initial velocity v_i, travels along a straight line with <u>constant acceleration</u> a, then in a time interval Δt the particle travels a distance:

$$\Delta x = v_i \Delta t + \frac{1}{2} a (\Delta t)^2$$

The First Acceleration Theorem is an elegant generalization this equation to the case where the acceleration a is not constant.

The First Acceleration Theorem

The location $x(t)$ of a particle moving along the x-axis is monitored during the time interval $[t_i, t_f]$. Here t_i denotes the initial time and t_f denotes the final time.
Let $\Delta t = t_f - t_i$ and let $\Delta x = x(t_f) - x(t_i)$.
Here we only assume the velocity function satisfies the assumption of the MVT on the interval $[t_i, t_f]$. The acceleration is <u>not</u> assumed to be constant!

Then there exists a moment of time $t = c$ in the interval $[t_i, t_f]$ at which the acceleration $a(t)$ of the particle satisfies:

$$\Delta x = v_i \Delta t + \frac{1}{2} a(c) (\Delta t)^2$$

In problems this can be solved for the acceleration: $\quad a(c) = \dfrac{2(\Delta x - v_i \Delta t)}{(\Delta t)^2}$

In the special case where the initial velocity is zero, the First Acceleration Theorem reduces to:

$$\Delta x = \frac{1}{2} a(c) (\Delta t)^2$$

To gain confidence with the statement of this theorem, we will work out an example problem before presenting the proof.

§4.5 Mean Value Theorem I   CAFÉ   Page 313

Example 5 - Hundred Meter Race Event

A sprinter finishes the 100 meter event in 10 seconds.

a) Show that his speed at some point must be exactly 10 m/s. (Use the MVT)

b) Show that there must be some instant when his acceleration is exactly 2 m/s². (Use the First Acceleration Theorem)

Solution

a) The average speed of the sprinter during the race is:

$$v_{ave} = \frac{100 \text{ meters}}{10 \text{ seconds}} = 10 \, \frac{m}{sec}$$

By the Mean Value Theorem, there is an instant of time t during the race where his instantaneous speed equals his average speed, that is $v(t) = 10 \, \frac{m}{sec}$.

b) Of course the sprinter begins the race starting from rest.
By the First Acceleration Theorem, there is an instant of time $t = c$ such that:

$$\Delta x = \frac{1}{2} a(c) (\Delta t)^2 \quad \text{or} \quad a(c) = \frac{2\Delta x}{(\Delta t)^2} = \frac{2 \cdot (100 \text{ meters})}{(10 \text{ sec})^2} = 2 \, \frac{m}{sec^2}.$$

Proof of the Acceleration Theorem

We only give the proof for the case where $v_i = 0$ (i.e., the case where the initial velocity of the particle is zero). The general case will be left as an exercise. Consider the new function:

$$L(t) = m(t - t_i), \text{ where}$$

m is the constant $\frac{2\Delta x}{(\Delta t)^2}$.

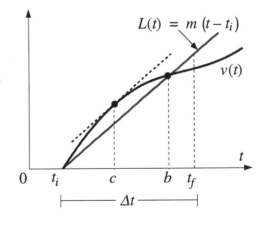

Note that the graph of $L(t)$ is a straight line with slope m and the region underneath $L(t)$ over the interval $[\,t_i, t_f\,]$ is a right triangle with base Δt and height $(\Delta t)m$ or $2\dfrac{\Delta x}{\Delta t}$. Hence the area of this right triangle is:

$$\frac{1}{2} \times \text{base} \times \text{height} \;=\; \frac{1}{2}(\Delta t)\left(2\frac{\Delta x}{\Delta t}\right) \;=\; \Delta x$$

Recall that Δx, the distance traveled by the particle during the time interval $[\,t_i, t_f\,]$, is also the area underneath $v(t)$. Obviously the graph of the particle's velocity function $v(t)$ can not stay <u>entirely</u> above the line $L(t)$ because this would imply that the area underneath $v(t)$ over $[\,t_i, t_f\,]$ is bigger than that underneath the line $L(t)$. This would imply the contradiction that $\Delta x > \Delta x$ Similarly, the argument works for the case where the graph of $v(t)$ lies <u>entirely</u> underneath $L(t)$. Hence the graph of $v(t)$ must intersect $L(t)$ somewhere in $[\,t_i, t_f\,]$, say at the instant $t = b$. Applying the MVT to the function $v(t)$ on $[\,t_i, b\,]$ we conclude that there exists a moment c in $(\,t_i, b\,)$ such that

$$v'(c) \;=\; m$$

Notice that $v'(c) = a(c)$, the acceleration at c.

Hence we have shown that there is a moment $t = c$ when $\Delta x \;=\; \dfrac{1}{2}\,a(c)\,(\Delta t)^2$.

End of Proof

§4.5 Mean Value Theorem I   CAFÉ   Page 315

WARMUP EXERCISES

1) The mayor of Metropolis proudly announced that his new 'get tough on crime' policy resulted in the arrest of 14 drug dealers in the last 7 days. Thus an average of two dealers were arrested each day. Can the Mean Value Theorem be used to show that there must have been one day on which precisely two dealers were removed from the city's mean streets? (Hint: If you are thinking the answer might be yes, consider what would happen if there had been 15 arrests instead of 14.)

2) The function illustrated in the figure satisfies $f(a) = f(b) = 0$ on each of the indicated intervals $[a, b]$.

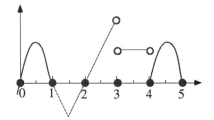

i) [0, 1]   **ii)** [1, 2]   **iii)** [2, 3]   **iv)** [3, 4]   **v)** [4, 5]

On which of these intervals does the function satisfy the <u>hypotheses</u> of Rolle's theorem? For these intervals, locate any points c such that $f'(c) = 0$ with $a < c < b$.

For each of the following exercises, verify that the <u>hypotheses</u> of Rolle's Theorem are satisfied on the given interval. Then find all values of c that satisfy the <u>conclusion</u> of Rolle's Theorem.

3) $f(x) = x(x-2)$ [0, 2]   4) $f(x) = x^2 - 4$ [-2, 2]

5) $f(x) = x^2 - 4x + 3$ [1, 3]   6) $f(x) = \sin(x)$ [0, 2π]

7) $f(x) = x^3 - x$ [-1, 1]   8) $f(x) = x^4 - x^2$ [-1, 1]

9) $f(x) = x - 10\sqrt{x}$ [0, 100]   10) $f(x) = \dfrac{x^2 - 9}{x - 5}$ [-3, 3]

11) Consider the polynomial $f(x) = (x-1)(x-2)^2(x-3)$

    a) Verify the conclusion of Rolle's Theorem in the interval [1,3].
    b) Verify the conclusion of Rolle's Theorem in the interval [1,2].

12) Consider the rational function $f(x) = \dfrac{x^2 - a^2}{x - b}$ where a and b are positive constants.

   a) For what values of the constant b are the hypotheses of Rolle's Theorem satisfied on the interval $[-a, a]$? (Check for discontinuities!)

   b) For those values of b that do satisfy the hypotheses, find all values of c that satisfy the conclusion of Rolle's Theorem.

13) Find a quadratic polynomial that satisfies the hypotheses of Rolle's Theorem on the interval $[0, 100]$. Then find all values of c that satisfy the conclusion of Rolle's Theorem.

For each of the following exercises, verify that the hypotheses of the Mean Value Theorem are satisfied on the given interval. Then find all values of c that satisfy the conclusion of this theorem.

14) $f(x) = x^2$ $[0, 10]$ 15) $f(x) = x^2 + 2x$ $[-1, 9]$

16) $f(x) = 2x^2 + x$ $[-1, 9]$ 17) $f(x) = x^2 + 2$ $[-1, 9]$

18) In each of the above problems, you may have noted that the graph of the function was a parabola and that the point c was always the midpoint of the given interval. We will show that the point c satisfying the conclusion of the Mean Value Theorem is the midpoint of the given interval whenever the function is quadratic. Consider a graph of the typical quadratic function
$f(x) = Ax^2 + Bx + C$ on the interval $[a, b]$.
(Assume $A \neq 0$).

   a) Show that the slope m of the secant line through the endpoints of the graph is $m = A(a+b) + B$.

   b) Show that the point c satisfying $f'(c) = m$ is the midpoint of the interval. That is, show that $c = \dfrac{a+b}{2}$.

For each of the following exercises, verify that the hypotheses of the Mean Value Theorem are satisfied on the given interval. Then find all values of c that satisfy the conclusion of this theorem.

19) $f(x) = x^3 - 2x$ $[-1, 2]$ 20) $f(x) = \dfrac{1}{x}$ $[1, 5]$

21) $f(x) = e^x$ $[-1, 3]$ 22) $f(x) = x \ln x$ $[1, 2]$

23) $f(x) = \sin x$ $[0, \pi/2]$ 24) $f(x) = \sin x + 2\cos x$ $[0, \pi]$

25) $f(x) = x^{1/3}$ $[0, 5]$ 26) $f(x) = e^x - 5x + 1$ $[-1, 3]$

§4.5 Mean Value Theorem I   CAFÉ   Page 317

INTERMEDIATE EXERCISES

27) Consider the function $f(x) = \sqrt{1-x^2}$.

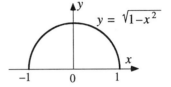

a) Verify that although the function is not differentiable at the points $x = \pm 1$, the hypotheses of the Mean Value Theorem are satisfied on each of the following intervals.

i) $[0, 1]$ ii) $[-1, 1]$

b) For each of the above intervals, find that value of c satisfying the conclusion of the MVT.

28) Consider the function $f(x) = \frac{1}{x}$.

a) Verify that the hypotheses of the Mean Value Theorem are satisfied on the interval $[1, 4]$ and find all values of c that satisfy the conclusion of this theorem.

b) Show that there is no value of c satisfying the conclusion of the MVT on the interval $[-1, 1]$. Why does this not violate the MVT?

29) Let $f(x) = |x|$. Is there a number c in $[-1, 5]$ so that:

$$f(5) - f(-1) = f'(c)(5 - (-1))?$$

Why?

30) Prove that if $f'(x) = g'(x)$ for all x in an interval (a, b), then $f(x) = g(x) + c$, where c is a constant.
(Hint: consider $F(x) = f(x) - g(x)$, then apply the corollary of the MVT in the text to $F(x)$)

31) Show that $x^7 + 4x^3 + x - 6 = 0$ has exactly one real root.

32) Show that $2x^5 + 3x + 1 = 0$ has exactly one real root.

33) After his karate class, Spike likes to relax in the sauna. He notices it takes five minutes for the temperature in the sauna to rise from 25°C to 45°C. Show that there is an instant of time when the temperature is rising at the rate of $4\,\frac{°C}{min}$. State any assumptions that you have made about the temperature $T(t)$ as a function of time t.

34) A perfect cubic crystal is being grown in a materials lab. The dimensions x of each edge of the crystal grew from 1 mm to 5 mm in two days.

a) Show that there exists a moment in time t_1 when the edges of the crystal are growing at the rate of $2 \frac{\text{mm}}{\text{day}}$.

b) Show that there exists a moment in time t_2 when the cross-sectional area $A = x^2$ of the crystal is growing at the rate of $12 \frac{\text{mm}^2}{\text{day}}$.

c) Show that there exists a moment in time t_3 when the volume $V = x^3$ of the crystal is growing at the rate of $62 \frac{\text{mm}^3}{\text{day}}$.

d) Are the three time instants t_1, t_2 and t_3 necessarily the same?

35) This exercise is a continuation of the previous exercise. Assume that the dimensions x of each edge of the crystal are given by the <u>linear</u> function $x(t) = 1 + 2t$ where t is in days. Find all possible values for the moments in time t_1, t_2 and t_3 from the previous problem.

36) Speeding Ticket

A motorcyclist purchased a fare card at the entrance to a tollway at 8:00 PM. At 9:30 PM she arrived at the exit booth, 120 miles further down the highway. Spike, who works as a highway patrol officer on the weekends, issued her a speeding ticket after being summoned by the booth operator.

a) Prove that the motorcyclist exceeded the speed limit of 55 mph.

b) The fine for speeding is 50 dollars plus 10 dollars for each mph above the 55 mph speed limit. What would be an appropriate fine?

§4.5 Mean Value Theorem I   CAFÉ   Page 319

37) **Filling Up Tankers** - Two <u>identical</u> oil tankers are being filled with crude oil. Although the rate of pumping is <u>not</u> constant, each tanker crew begins and finishes the pumping at the same moments by coincidence. The tankers hold 1,000,000 barrels of oil and each is filled in a total of 10 hours.

a) Show that there must exist a time instant when both tankers are being filled at the same rate.

(Hint: Consider the function $D(t) = V_1(t) - V_2(t)$ where $V_i(t)$ denotes the volume of oil pumped into the ith tanker as a function of time where $i = 1, 2$).

b) Due to the weight of the oil, each tanker sits 2 meters deeper in the water when full.
Show that there exists an instant when the water line on both tankers is rising at the same rate.

c) Show for each tanker that there exists an instant when the water line is rising at the rate of $v = 20 \frac{cm}{hr}$. Must this instant be the same for each tanker?

38) **Marathon** - A marathon runner finishes a race of length L in time T. At time $\frac{T}{2}$, he had finished the first half of the course. Show that there are two time instants, differing by $\frac{T}{2}$, when his running speeds are equal.
(Hint: Let $D(t)$ be the distance covered by the runner at time t so that $D(0) = 0$ and $D(T) = L$. Apply MVT to $F(t) := D\left(t + \frac{T}{2}\right) - D(t)$)

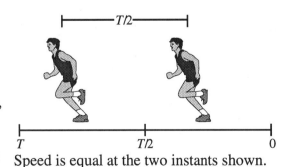

Speed is equal at the two instants shown.

39) **Drag Race** - In drag racing, competitors race down a 1/4 mile track, starting from rest. If a driver finishes the race in 6 seconds, show that there is an instant of time when the acceleration of the car is $\frac{1}{72}$ mile/s².
[Hint: use the First Acceleration Theorem]

40) In 30 minutes the capacitor has stored a charge of 10 mC. Assume that in the first 15 minutes the capacitor has stored 5 mC. Show that there are two time instants separated by exactly 15 minutes when the readings of the ammeter are equal.
[Hint: Apply the MVT to
$g(t) = Q(t+15) - Q(t)$, where $Q(t)$ is the amount of charge stored in the capacitor at time t.]

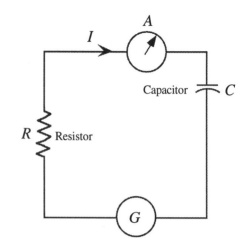

41) **Rodeo** - At the moment when a cowboy starts from rest to chase a feisty bull running along a straight track, the bull has speed a of 8 m/s and is 10 m ahead. Ten seconds later, the cowboy catches up with the bull. Show that there is a time instant when the acceleration of the cowboy exceeds the acceleration of the bull by exactly $\frac{9}{5} \frac{\text{m}}{\text{s}^2}$.

[Hint: use the First Acceleration theorem]

42) **Space Shuttle** - Twenty seconds after launch, the space shuttle has reached a height of 4 miles above the earth's surface. Show that there is an instant of time when an 80 kg astronaut has an apparent weight of 3358.4 N. The apparent weight is the weight that the astronauts would feel. (1 mile = 1609 meters)
(Hint: Apply the First Acceleration Theorem)

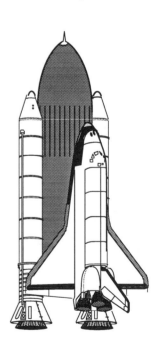

43) Prove the First Acceleration Theorem for the case $v_i \neq 0$

44) **Magic of Two**

Assume that the function f is continuous on the closed interval $[0,1]$ and differentiable on the open interval $(0,1)$. Suppose f satisfies:

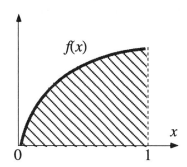

1) $f(0) = 0$ &   2) $\int_0^1 f(x)dx = 1$

Show that there exists a point β in $(0,1)$ such that $f'(\beta) = 2$.
[Hint: If the line $y = 2x$ intersects the graph of $f(x)$ then we are done. Why? (use MVT) Otherwise the graph of $f(x)$ either lies entirely above or entirely below the line $y = 2x$. This would violate part **(b)** above. Why?]

NOTES

§ 4.6 Mean Value Theorem
Part 2

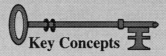
Key Concepts

- Discrete Averages
- Average Value of a Function
- Mean Value Theorem for Integrals
- Second Acceleration Theorem

In this section, we will explore many aspects of the notion of <u>average</u> value. We begin by reviewing the notion of discrete average and then extend this to the notion of the average value of a function over an interval. The Mean Value Theorem for Integrals identifies under what conditions a function is guaranteed to assume its average value and has many other important applications to the foundations of calculus. Finally, we discover a second acceleration theorem which has many applications in kinematic problems.

Discrete Averages

The <u>average</u> of n numbers $x_1, x_2, \cdots x_n$ is defined to be:

$$\text{Ave} = \frac{x_1 + x_2 + \cdots + x_n}{n}$$

Thus the average of the three numbers 1, 2 and 3 is:

$$\text{Ave} = \frac{1 + 2 + 3}{3} = 2$$

In this case, the average value 2 is one of the original three numbers but this is rarely the case. For example: The average of the three numbers 1, 2 and 6 is:

$$\text{Ave} = \frac{1 + 2 + 6}{3} = 3$$

Notice that in this case, the average value 3 is <u>not</u> one of the numbers being averaged.

Average Value of a Function

Just as we can calculate the average value of a set of numbers, we can find the average value of a function on a given interval. Of course even on a finite interval, a typical function such as a polynomial assumes an <u>infinite</u> number of values so we will have to think carefully about what we mean by the average value of a function.

Test Drive - Spike's brother, a professional football player, was away in Dallas. Having nothing better to do, Spike decided to take his brother's silver Ferrari 456 GT out for a test spin on a straight track of road. With its powerful quad-cam V12 engine, the Ferrari easily covered 300 miles in two hours.

This Ferrari covered 300 miles in two hours.

Thus its average velocity v_{ave} during the test drive was:

$$v_{ave} = \frac{\Delta x}{\Delta t} = \frac{300 \text{ miles}}{2 \text{ hours}} = 150 \frac{\text{mi}}{\text{hr}}$$

This simple calculation is actually our first example of finding the average value of a function. In this case the function is the Ferrari's velocity $v(t)$. By thinking about the method used to arrive at the average velocity, we will be able to find the average value of any function. During the test drive, the actual velocity $v(t)$ was not constant, instead varied with time.

We can find the total distance Δx traveled by the car over the time interval from $t = a$ to $t = b$ by integrating:

$$\Delta x = \int_a^b v(t)\, dt$$

Thus, the average velocity can be written in terms of the instantaneous velocity function as:

$$v_{ave} = \frac{\Delta x}{\Delta t} = \frac{1}{b-a} \int_a^b v(t)\, dt$$

§4.6 Mean Value Theorem II

Example 1 - Average Velocity using Integrals

A car accelerates from rest so that its velocity satisfies $v(t) = 10t$ meters per second. Find the average velocity v_{ave} of the car over the time interval $[1, 3]$.

Solution

$$v_{ave} = \frac{1}{b-a}\int_a^b v(t)\,dt = \frac{1}{3-1}\int_1^3 10t\,dt = \frac{1}{2}\left[5t^2\right]_1^3 = \boxed{20\,\frac{m}{sec}}$$

The integral expression we found above for the average velocity can be generalized to find the average of many other important functions. What is the average current flowing in an electric circuit? What is the average pressure in a reaction vessel? What is the average temperature during the year?

> **Definition**
> Let the function f be integrable on the closed interval $[a, b]$. The <u>average value</u> of the <u>function f</u> over this interval is defined to be:
> $$f_{ave} = \frac{1}{b-a}\int_a^b f(x)\,dx$$

Notice that this definition reduces to our formula for the average velocity when f is the velocity function and the independent variable is time.

Example 2

Find the average value of the quadratic function $f(x) = x^2$ on the interval $[0, 1]$.

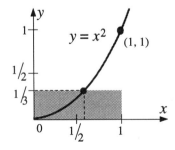

Solution

Using the definition of average value we find:

$$f_{ave} = \frac{1}{b-a}\int_a^b f(x)\,dx = \frac{1}{1-0}\int_0^1 x^2\,dx = \boxed{\frac{1}{3}}$$

Thus the average value is 1/3.

Geometric Interpretation

The average value of a function has a very elegant geometric interpretation. In the illustration, a rectangle is drawn with a height $h = f_{\text{ave}}$ equal to the average value of the function f over the interval $[a,b]$.

The area A of the rectangle is the same as the area under the graph of the function $f(x)$!

(Here f is ≥ 0 over $[a, b]$.)

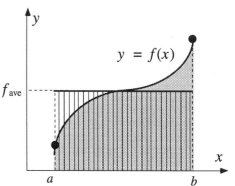

If the height of the rectangle is f_{ave} then its area is the same as the area between the curve and the x-axis over the interval $[a, b]$.

To see this note that the area of the rectangle is equal to the product of its base times its height.

$$A = \text{base} \times \text{height} = (b-a) f_{\text{ave}}$$

By definition, the average value of the function f over the interval $[a, b]$ is:

$$f_{\text{ave}} = \frac{1}{b-a} \int_a^b f(x)\, dx$$

Thus the area of the rectangle is $A = (b-a) f_{\text{ave}} = \int_a^b f(x)\, dx$

which is precisely the area between the graph of the function $f(x)$ and the x-axis over the interval $[a, b]$.

Example 3 - Harbor - Civil Engineering

The depth of water $h(t)$ in a harbor fluctuates up and down due to daily tides. A low water depth is particularly hazardous to ships due to the risk of running aground. Assume the depth of water (in feet) satisfies the equation:

$$h(t) = 30 + 5 \sin(\pi t/6)$$

where the time t is measured in hours. Note there are two high tides per day.

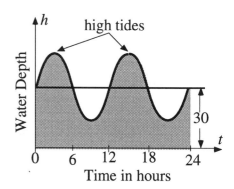

§4.6 Mean Value Theorem II   CAFÉ   Page 327

a) Find the average water depth in the harbor over the interval $[0, 6]$.

b) Find the average water depth in the harbor over the entire day.

Solution

a) The average depth of the water over the interval $[0, 6]$ hours is found from:

$$h_{ave} = \frac{1}{6-0}\int_0^6 (30 + 5\sin(\pi t/6))\,dt = \frac{1}{6}\left(30t - \frac{5\cos(\pi t/6)}{\pi/6}\right)\bigg|_0^6 = \boxed{30 + \frac{10}{\pi}}$$

b) The average depth of the water over the interval $[0, 24]$ hours is found from:

$$h_{ave} = \frac{1}{24-0}\int_0^{24} (30 + 5\sin(\pi t/6))\,dt = \frac{1}{6}\left(30t - \frac{5\cos(\pi t/6)}{\pi/6}\right)\bigg|_0^{24} = \boxed{30}$$

ALERT

A function does not necessarily assume its average value.

The figure shows the graph of the discontinuous function defined by:

$$f(t) = \begin{cases} 1 & \text{if } t < 1 \\ 3 & \text{if } t \geq 1 \end{cases}$$

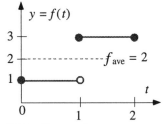

This discontinuous function does not assume its average value on the interval $[0, 2]$.

Over the interval from 0 to 2, the function has average $f_{ave} = 2$. Notice however, that the function never assumes this average value! This example shows that only some functions will assume their average value on an interval. The fact that the average value of discontinuous functions need not be assumed is very similar to the situation with discrete averages. The average value of the numbers 1, 2 and 6 is 3 yet the value 3 is not one of the numbers being averaged. However, for many functions of interest, the average value of the function on an interval will be assumed by the function somewhere in the interval! The next theorem gives conditions under which this will always be the case.

Mean Value Theorem for Integrals

Consider a function f which is continuous on the closed interval $[a, b]$ and let f_{ave} denote the average value of f over this interval. Then there exists a number c in the interval $[a, b]$ such that
$$f(c) = f_{ave}$$

That is, the function assumes its average,

$$f_{ave} = \frac{\int_a^b f(x)\, dx}{b-a}$$

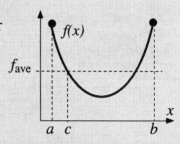

This theorem is called the <u>mean</u> value theorem because another word for average is mean. Before giving the proof of the MVT, we will work through an example to help clarify the statement of the theorem.

Example 4

Let f_{ave} denote the average value (or mean value) of the function $f(x) = 1 + 2x$ over the interval $[0, 2]$. Since the function is continuous, the MVT for integrals implies that there is a point c in the interval $[0, 2]$ such that $f(c) = f_{ave}$.
Find all values of c which satisfy the conclusion of the MVT for integrals.

Solution

Let us first find the average value f_{ave}.
With $a = 0$ and $b = 2$ we have:

$$f_{ave} = \frac{1}{b-a}\int_a^b f(x)\, dx = \frac{1}{2-0}\int_0^2 (1+2x)\, dx = 3$$

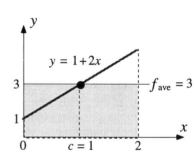

Next we find that point c where f equals its average value.
The condition $f(c) = f_{ave} = 3$ implies that

$$1 + 2c = 3 \text{ so } \boxed{c = 1}$$

The illustration above clearly shows that the area of the trapezoid equals the area of the rectangle whose height is $h = f_{ave}$.

Comment: Any linear function will assume its average value over an interval at the <u>midpoint</u> of the interval. This is not true for other functions in general.

---

§4.6 Mean Value Theorem II

Proof of the MVT (for integrals)
We must show that there exists a point c in the interval $[a, b]$ such that $f(c) = f_{\text{ave}}$.
The proof of the MVT for integrals will actually use the Mean Value Theorem for derivatives which we encountered earlier. To be able to apply the MVT for derivatives we will need to introduce a new function $F(x)$ which is defined as the integral:

$$F(x) = \int_a^x f(t)\,dt, \qquad \text{for } x \text{ in the interval } [a, b].$$

Since $F(x)$ is an antiderivative of $f(x)$, we have $\dfrac{dF}{dx} = f(x)$.

The Second Fundamental Theorem of Calculus states:

$$F(b) - F(a) = \int_a^b f(t)\,dt$$

Dividing the last equation by $(b - a)$ gives an expression for the average value f_{ave} in terms of the new function $F(x)$.

$$\frac{F(b) - F(a)}{b - a} = \frac{1}{b-a}\int_a^b f(t)\,dt = f_{\text{ave}}$$

Our assumption that $f(x)$ is continuous on $[a, b]$, implies that $F(x)$ satisfies the hypotheses for the Mean Value Theorem for derivatives.
That is, the new function $F(x)$ is continuous on $[a, b]$ and differentiable on (a, b).
By the MVT for derivatives there exists a point c in the interval (a, b) such that

$$F'(c) = \frac{F(b) - F(a)}{b - a}$$

Noting that $F'(c) = f(c)$ and $f_{\text{ave}} = \dfrac{F(b) - F(a)}{b - a}$, this last equation can be written:

$$f(c) = f_{\text{ave}}$$

End of Proof

Example 5

Greenhouse Effect
The concentration of carbon dioxide in the atmosphere has been increasing steadily since the industrial revolution. A reasonable model of the carbon dioxide concentration $C(t)$ is given by the equation:

$$C(t) = 338 + 1.6t + 3\cos(2\pi t)$$

where t is the number of years since 1980 and the concentration is given in parts per million. Thus, $t = 0$ at the start of 1980 and $t = 10$ at the end of 1990.

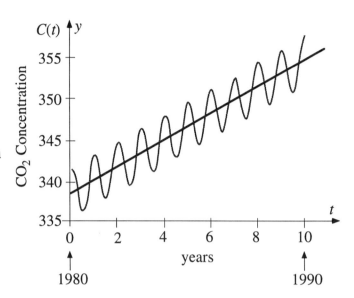

The seasonal fluctuation seen in the graph of $C(t)$ reveals a high concentration of CO_2 in winter and a low concentration in summer. This is due to the increased photosynthesis by plants in the summer. These seasonal fluctuations make it difficult to see the trend in the data and to predict the level of carbon dioxide in the future. However, we can <u>average</u> out the seasonal fluctuations by introducing the function:

$$C_{\text{ave}}(t) = \int_{t-1/2}^{t+1/2} C(\tau)\, d\tau$$

which is just the average of the concentration over a year.

$$C_{\text{ave}}(t) = \int_{t-1/2}^{t+1/2} (338 + 1.6\tau + 3\cos(2\pi\tau))\, d\tau = \boxed{338 + 1.6t}$$

With the seasonal fluctuations averaged out, it is much easier to see the increasing trend. The averaged function is just the line $C_{\text{ave}}(t) = 338 + 1.6t$ and is plotted in the above graph. This clearly reveals that the concentration is increasing at the rate of about 1.6 parts per million per year!!! This example hints at the many applications of averaging in the analysis of data, analysis of noisy signals and image processing.

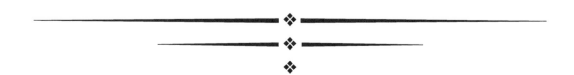

§4.6 Mean Value Theorem II

1) Find the average value of each of the following sets of numbers.
   a) 1,3,5   b) 5,3,5,1,1   c) −1,5,6,2

2) Three mineral samples weighing 2, 4 and 8 pounds have been collected by a prospector.
   a) What is the average weight of the samples?
   b) In handling, one of the samples breaks in half. What is the average weight of the resulting four samples? Does the answer depend on which of the three samples broke?

In the following exercises, find the average value f_{ave} of the function f on the given interval.

3) $f(x) = x$ $[-1, 1]$ **4)** $f(x) = x^2$ $[-1, 1]$

5) $f(x) = \sin(x)$ $[0, \pi]$ **6)** $f(x) = \sqrt{1-x^2}$ $[-1, 1]$

7) $f(x) = a + b\sin(x)$ $[0, 2\pi]$ **8)** $f(x) = 1 + 2x$ $[a, b]$

9) $f(x) = k + mx$ on $[a, b]$ **10)** $f(x) = \sqrt{x}$ $[0, 9]$

Find the average velocity of the velocity function $v(t)$ of a particle moving along a straight line on the given time interval.

11) $v(t) = 2 + 9.8t$ $[1, 10]$ **12)** $v(t) = 10 - 3.5e^{-2t}$ $[1, 5]$

13) $v(t) = \sin 2t$ $[0, 1]$ **14)** $v(t) = 2 + 1.5e^{-t}$ $[2, 4]$

A geometric approach to the average of a function

Let A denote the <u>area</u> between the curve $y = f(x)$ and the x-axis over the interval $[a, b]$ and let $L = b-a$ denote the <u>length</u> of the interval. Then the <u>average value</u> of f can be written in the form:

$$f_{ave} = \frac{1}{b-a}\int_a^b f(x)\, dx = \frac{A}{L}$$

In each of the following exercises, find the area A and the length L by <u>geometric</u> arguments and then calculate the average of f using the equation $f_{\text{ave}} = \dfrac{A}{L}$.

15)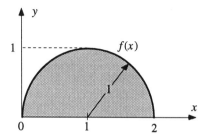

16)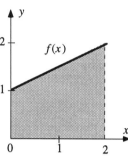

17)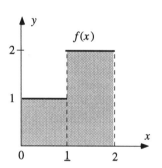

18)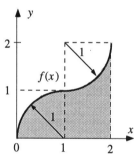

19)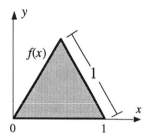

An equilateral triangle of side 1

20)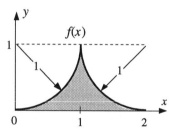

§4.6 Mean Value Theorem II   CAFÉ

A statistical interpretation of the average of a function

21) Let $f(x) = x^2$ be defined on the interval $[1, 2]$. In statistics the term "arithmetic mean" is used for the average of a set of numbers. Thus the arithmetic mean of $f(1)$ and $f(2)$ is $\frac{1}{2}(f(1) + f(2)) = \frac{1}{2}(1^2 + 2^2) = \frac{5}{2}$. This notion coincides with the notion of discrete average introduced in this section.

   a) Let's take 6 sample values of the function.

$$f(1), f\left(1 + \tfrac{1}{5}\right), f\left(1 + \tfrac{2}{5}\right), f\left(1 + \tfrac{3}{5}\right), f\left(1 + \tfrac{4}{5}\right), \text{ and } f\left(1 + \tfrac{5}{5}\right)$$

Calculate the arithmetic mean of these six values.

   b) Now consider the following $n + 1$ values of the function.

$$f(1), f\left(1 + \tfrac{1}{n}\right), f\left(1 + \tfrac{2}{n}\right), \ldots, \text{ and } f\left(1 + \tfrac{n}{n}\right)$$

Find an expression representing the arithmetic mean of these $n + 1$ values.

   c) Calculate the limit of the arithmetic mean in (b) as $n \to \infty$
(Hint: Consider the arithmetic mean as a Riemann sum.)

   d) Show that the limit in (c) is equal to the average of the function $f(x) = x^2$ over $[1, 2]$ in the sense defined in the text.

22) Consider the function $f(x) = e^x$ over the interval $[0, 1]$.

   a) Calculate the limit as $n \to \infty$ of the arithmetic mean of the $n + 1$ function values which are equally spaced and are chosen from $[0, 1]$.

   b) Show that the limit in (a) is just the average value of the function over the interval $[0, 1]$.

23) Let $f(x)$ be an integrable function defined on the interval $[a, b]$.

   a) Calculate the limit as $n \to \infty$ of the arithmetic mean of $n + 1$ function values which are equally spaced and are chosen from $[a, b]$.

   b) Show that the limit in (a) is just the average value of the function over the interval $[a, b]$.

Mean Value Theorem for Integrals

In the following exercises, find the average value of the function f on the given interval. Then find all values c on the interval where the function assumes its average.

24) $f(x) = 1 + x^3$ $[-1, 1]$ 25) $f(x) = e^x + 1$ $[-2, 2]$

26) $f(x) = 2 + \sin x$ $[0, 2\pi]$ 27) $f(x) = 2 + \cos x$ $[0, \pi]$

28) $f(x) = 1 - x^2$ $[-1, 1]$ 29) $f(x) = 1 - x^2$ $[0, 1]$

30) Each of the three functions $f(x) = \sqrt{x}$, $g(x) = x$, and $h(x) = x^2$ increases from 0 to 1 on the interval $[0, 1]$. Which function has the greatest average value? Which has the smallest average value?
Let c_1, c_2 and c_3 respectively denote the points where the functions f, g and h assume their average values. Which of the numbers c_1, c_2 and c_3 is the smallest (largest)?

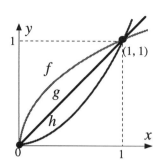

31) During the day, the temperature $T(t)$ at a certain location varies according to the formula $T(t) = 70 + 10 \sin\left(\dfrac{\pi t}{12}\right)$ where t is the time in hours after ten in the morning. That is $t = 0$ at 10 AM and $t = 2$ at noon. The temperature is given in Fahrenheit.

   a) Find the average temperature between 10 AM and 4 PM.

   b) Find the average temperature over an entire day.

32) One end of a one meter metal rod is immersed in ice water while the other is immersed in boiling water. As a result, the temperature along the rod rises from 0°C to 100°C as we move from one end to the other. The temperature at the point x along the rod is given by the function $T(x) = 100x$.
Find the average temperature T_{ave} along the length of the rod using integration and compare your answer with the temperature at the rod's midpoint.

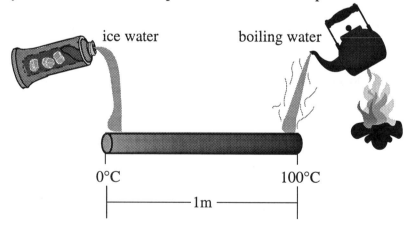

33) The mass density $\rho(x)$ of a non-uniform rod of length L is a function of the location x along the rod. Show that the total mass M of the rod is given by $M = \rho_{ave} L$.

Calculus in Action

Second Acceleration Theorem

An interesting application of the Mean Value Theorems that we have studied is the Second Acceleration Theorem. This theorem gives information about the acceleration experienced by moving particles. Such information is important in the study of the g-forces experienced by test pilots and astronauts and also in the design of trains, cars and machine parts that undergo significant changes in both velocity and acceleration.

Average Acceleration

A particle moves along a straight line with initial velocity v_i and final velocity v_f during an interval of time Δt. Its average acceleration during this time interval is:

$$a_{\text{ave}} = \frac{\Delta v}{\Delta t} = \frac{v_f - v_i}{\Delta t}$$

According to the MVT for integrals, there is a time instant t_0 such that $a(t_0) = a_{\text{ave}}$. That is, the average acceleration is assumed. However, this fact does not tell us by how much the instantaneous acceleration might deviate from the average acceleration over an interval. The answer to this question is provided by our Second Acceleration Theorem.

The Second Acceleration Theorem

A particle moves along a straight line with initial velocity v_i and final velocity v_f during an interval of time Δt. Let $\bar{v} = \dfrac{v_i + v_f}{2}$ be the discrete average of the initial and final velocities and assume that the velocity function $v(t)$ satisfies the assumptions of the MVT for <u>derivatives</u>. Then there exists a time t_0 such that:

$$|a(t_0) - a_{\text{ave}}| = \frac{4}{\Delta t}|v_{\text{ave}} - \bar{v}|$$

That is, the instantaneous acceleration will differ from the average acceleration by an amount equal to $\dfrac{4}{\Delta t}|v_{\text{ave}} - \bar{v}|$ at some instant.

Note that the acceleration of the particle is not assumed to be constant! It is only assumed that the velocity $v(t)$ satisfies the assumptions of the MVT for derivatives during the time interval in question. Of course, if the acceleration is constant, the theorem is just a triviality since both sides of the equation reduce to zero!

Proof of the Second Acceleration Theorem

If the average velocity satisfies $v_{\text{ave}} = \bar{v}$, then the right side is zero and the acceleration theorem just says that there is an instant of time t_0 when $a(t_0) - a_{\text{ave}} = 0$. (This is just what we saw from the MVT for Integrals.)

However, if $v_{\text{ave}} \neq \bar{v}$ we define the new function:

$$u(t) = v(t) - (v_i + a_{\text{ave}}(t - t_i))$$

The function $u(t)$ is zero at both endpoints of the time interval $[t_i, t_f]$ with $\Delta t = t_f - t_i$. That is:

$$u(t_i) = u(t_f) = 0$$

Also note that $u'(t) = a(t) - a_{\text{ave}}$.

This introduction of the new function $u(t)$ is very similar to the way we proved the MVT for derivatives using Rolle's Theorem.

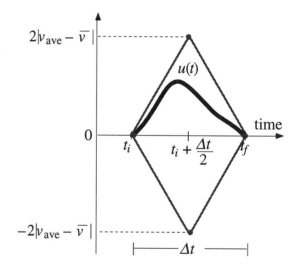

Now let $m = \dfrac{4}{\Delta t} |v_{\text{ave}} - \bar{v}|$.

Consider the top left line segment shown in the figure of slope m. We add three additional line segments to form a perfect <u>diamond</u> as shown. The acceleration theorem readily follows form the fact that during the time interval $[t_i, t_f]$, the graph of the function u must touch at least one side of the diamond!

For example, if $u(t)$ touches the top left line segment of the diamond, then by the MVT for derivatives there is a time t_o such that:

$$u'(t_0) = m$$

so

$$a(t_0) - a_{\text{ave}} = m$$

For some time in the interval $[t_i, t_f]$, the function $u(t)$ must intersect the diamond so that the situation shown here is not possible.

Taking the absolute value of both sides we have:

$$|a(t_0) - a_{\text{ave}}| = \frac{4}{\Delta t}|v_{\text{ave}} - \bar{v}|$$

which is what we want to prove. A similar argument works if any of the other sides of the diamond are touched. Thus we only need show that u must touch the diamond. To show this, first note that the area A of the top half of the diamond is:

$$A = (\Delta t)|v_{\text{ave}} - \bar{v}|, \text{ where } \Delta t = t_f - t_i$$

§4.6 Mean Value Theorem II — CAFÉ

If u does not touch the diamond, then it lies entirely within the diamond, and the area between the graph of u and the t-axis must clearly be <u>strictly</u> less than the area A of the top half of the diamond. But the magnitude of the area between the graph of u and the time axis is:

$$\left| \int_{t_i}^{t_f} u(t)dt \right| = \left| \int_{t_i}^{t_f} v(t)dt - \int_{t_i}^{t_f} (v_i + a_{\text{ave}}(t - t_i))dt \right|$$

$$= \left| (\Delta t)\, v_{\text{ave}} - \left(v_i(\Delta t) + \frac{1}{2}(v_f - v_i)(\Delta t) \right) \right|$$

$$= (\Delta t)\, |v_{\text{ave}} - \overline{v}|$$

But this shows that $A < A$ which is clearly a contradiction. Thus the function u must touch the diamond showing that $|a(t_0) - a_{\text{ave}}| = \dfrac{4}{\Delta t}|v_{\text{ave}} - \overline{v}|$ for some time t_0 in the interval.

End of Proof

Some typical applications of the Second Acceleration Theorem are highlighted in the following examples.

Bullet Train

A bullet train starts from rest along a straight track from station A and moves toward station B which is 200 miles down the track. The train comes to a stop at station B one hour later.

Show that there must be an instant when the acceleration of the train has magnitude $800\,\dfrac{\text{miles}}{\text{hr}^2}$.

Solution

The Second Acceleration Theorem states: there exists a time instant t_o such that,

$$|a(t_0) - a_{\text{ave}}| = \frac{4}{\Delta t}|v_{\text{ave}} - \overline{v}|$$

Now $\Delta t = 1$ (hr) and since $v_i = v_f = 0$ so that,

$$a_{\text{ave}} = \frac{v_f - v_i}{\Delta t} = 0$$

and $\quad \overline{v} = \dfrac{v_i + v_f}{2} = 0$

Notice that $(\Delta t) \times v_{\text{ave}} =$ distance traveled $= 200$ miles

or $v_{ave} = 200$ miles/hr

Plugging these qualities into the theorem gives,

$$|a(t_0)| = \frac{4}{1}|200 - 0| = 800 \text{ miles/hr}^2$$

Exercises A, B, and C illustrate the diverse applications of the Second Acceleration Theorem.

A) Monitoring a Bullet Train

To monitor the movement of this high speed bullet train, an observation stop is established at city C and city D, which are 100 miles apart along the straight track. In a test run it was recorded that the speed of the train was 196 mi/hr at C and 1/2 hours later when it appeared at city D, its speed was 200 mi/hr. Show that during this half hour period, it is impossible that the acceleration $a(t)$ of the train satisfies:

$|a(t) - 8| < 16$ for all t $a(t)$ is in mi/hr^2

(Hint: Use the Second Acceleration Theorem)

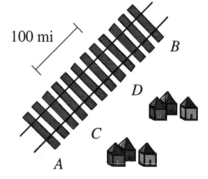

B) Elevator Ride

An elevator takes 30 seconds to reach the observation deck of a high tower that is 150 meters up. The elevator starts from rest and comes to a complete stop at the top. As an experiment, Spike has brought along a bathroom scale calibrated in kilograms. Assuming that Spike's body has a mass of 60 kg, show that there is a time instant when the reading in kg of the scale is either $60 + \frac{40}{g}$ or $60 - \frac{40}{g}$, where $g = 9.8 \text{m/s}^2$ is the acceleration of gravity.

(Hint: Use the Second Acceleration Theorem)

Going up !

§4.6 Mean Value Theorem II — CAFÉ

C) Magic of Four

A function $f(x)$ defined on the interval $[0, 1]$ has the following properties:

a) $f(x)$ is continuous on $[0, 1]$

b) $f'(x)$ exists in $(0, 1)$

c) $f(0) = f(1) = 0$

d) $\displaystyle\int_0^1 f(t)\,dt = 1$

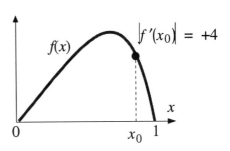

Show that there exists a point x_0 in the interval $(0, 1)$ such that $|f'(x_0)| = 4$.

[Hint: Show the graph of $f(x)$ must be below the graph of

$$g(x) := \begin{cases} 4x, & 0 \le x \le \dfrac{1}{2} \\ 4 - 4x, & \dfrac{1}{2} \le x \le 1 \end{cases}$$

Notes

§ 4.7 Integration by Parts

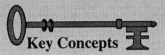
Key Concepts

- ❖ Integration by Parts (Indefinite Version)   ❖ Repeated Use of Parts
- ❖ Integration by Parts (Definite Version)   ❖ Reduction Formulas

To every rule for differentiating functions there is a corresponding integration rule. For example, the corresponding integration rule for chain rule is the substitution rule as we have discussed in the previous section. One of the most important differentiation rules is the Product Rule. The corresponding integration rule is called "Integration by Parts" and is the subject of this section. The product rule for differentiation can be written as

$$\frac{d}{dx}(f(x)\,g(x)) = f'(x)\,g(x) + f(x)\,g'(x).$$

If we integrate both sides and apply the Fundamental Theorem we find:

$$f(x)\,g(x) = \int f'(x)\,g(x)\,dx + \int f(x)\,g'(x)\,dx$$

The power of this formula is that it relates two integrals. If one of the integrals is easier to evaluate than the other, the formula will be useful for evaluating the harder integral. In this spirit, let us recast the formula in a form which expresses one of the integrals in terms of the other.

Integration by Parts (Long Form)

$$\int f(x)\,g'(x)\,dx = f(x)\,g(x) - \int f'(x)\,g(x)\,dx$$

This last equation is called the formula for integration by parts. It also appears in another equivalent form. If we let $u = f(x)$ and $v = g(x)$ the corresponding differentials are $du = f'(x)\,dx$ and $dv = g'(x)\,dx$. Substituting these expressions into the integration by parts formula gives the shorter but equivalent form:

Integration by Parts (Short Form)

$$\int u\,dv = uv - \int v\,du$$

The following examples illustrate how integration by parts can be used to evaluate integrals.

Example 1

Find the indefinite integral $\int x e^x \, dx$ using integration by parts.

Solution

The required integral $\int x e^x \, dx$ does not have an obvious antiderivative. Could applying the integration by parts formula convert it to a simpler integral?

$$\text{Let} \quad u = \boxed{x} \quad dv = \boxed{e^x \, dx}$$

$$\text{Then} \quad du = \boxed{dx} \quad v = \boxed{e^x}$$

<u>Remark:</u> Given u and dv we then have to calculate du and v where du is the differential of u and $v = \int dv$. In this case, the differential of u is dx and since $\int dv = \int e^x \, dx = e^x + C$, we can choose $v = e^x$. Note we will always omit the constant when calculating v from dv. However, when evaluating indefinite integrals by this method, an integration constant must be added at the end.

Substituting into the integration by parts formula $\int u \, dv = uv - \int v \, du$ gives:

$$\int x e^x \, dx = x e^x - \int e^x \, dx = x e^x - e^x + C = \boxed{(x-1) e^x + C}$$

The great virtue of this method can be seen by noticing how it has enabled us to recast the original integral $\int x e^x \, dx$ in terms of the simple integral $\int e^x \, dx$.

Check   As always with integration, we should check the result by differentiation.

$$\frac{d}{dx}\left((x-1) e^x + C\right) = e^x + (x-1) e^x = x e^x$$

§4.7 Integration by Parts

Notice the pattern of substitutions which we have used in the last problem.

$$\text{Let} \quad u = \boxed{} \quad dv = \boxed{}$$

$$\text{Then} \quad du = \boxed{} \quad v = \boxed{}$$

This pattern provides a convenient way to organize your integration by parts calculations.

Example 2 Find the indefinite integral $\int x \cos x \, dx$.

Solution

Let $\quad u = \boxed{x} \quad dv = \boxed{\cos x \, dx}$

Then $\quad du = \boxed{dx} \quad v = \boxed{\sin x}$

The integration by parts formula gives:

$$\int x \cos x \, dx = x \sin x - \int \sin x \, dx = \boxed{x \sin x + \cos x + C}$$

Check Always check an integration result by differentiation.

$$\frac{d}{dx}(x \sin x + \cos x + C) = (\sin x + x \cos x) + (-\sin x) = x \cos x$$

Sometimes an integrand may not look like a product of two terms and yet integration by parts still works. Knowing how to best factor an integrand into a product of two terms so that integration by parts may be applied is really more a matter of intuition and creativity than a deductive process.

Example 3 Find the indefinite integral $\int \ln x \, dx$.

Solution

Although the integrand is only a single term, we may view it as a product of two terms by writing $\ln x = \ln x \cdot (1)$

Let $\quad u = \boxed{\ln x} \quad dv = \boxed{1 \, dx}$

Then $\quad du = \boxed{\dfrac{1}{x} dx} \quad v = \boxed{x}$

The integration by parts formula gives:

$$\int \ln x \, dx = x \ln x - \int \left(x \cdot \frac{1}{x}\right) dx = \boxed{x \ln x - x + C}$$

Check Always check an integration result by differentiation.

$$\frac{d}{dx}(x \ln x - x + C) = \left(\ln x + x \cdot \frac{1}{x}\right) - 1 = \ln x$$

The technique used in the last example, $f(x) = f(x) \cdot (1)$, is useful for integrating many of the inverse functions we have encountered. Here is another example of the same technique applied to a different inverse function.

Don't Forget

In the next example we need the derivative of the inverse tangent function. You will need to recall that:

$$\frac{d}{dx} \tan^{-1} x = \frac{1}{1 + x^2}$$

Example 4 Find the indefinite integral $\int \tan^{-1} x \, dx$.

Solution

In this case, let us view the integrand as the product $\tan^{-1} x = \tan^{-1} x \cdot (1)$

$$\text{Let} \qquad u = \boxed{\tan^{-1} x} \qquad dv = \boxed{1 \cdot dx}$$

$$\text{Then} \qquad du = \boxed{\frac{1}{1+x^2} dx} \qquad v = \boxed{x}$$

The integration by parts formula gives:

$$\int \tan^{-1} x \, dx = x \tan^{-1} x - \int \frac{x}{1+x^2} dx = \boxed{x \tan^{-1} x - \ln \sqrt{1+x^2} + C}$$

In the last result we used the identity $\ln \sqrt{z} = \frac{1}{2} \ln z$ and we also noted that since the argument $1 + x^2$ is always positive, we need not put an absolute value function inside the logarithm.

Check Always check an integration result by differentiation.

$$\frac{d}{dx}\left(x \tan^{-1} x - \ln \sqrt{1+x^2} + C\right) = \left(\tan^{-1} x + \frac{x}{1+x^2}\right) - \frac{1}{2}\frac{2x}{1+x^2} = \tan^{-1} x$$

§4.7 Integration by Parts

Repeated Use of Integration by Parts.

Some integrals require more than one application of integration by parts as illustrated in the next two examples.

Example 5 - Repeated Use

Find the indefinite integral $\int x^2 e^{2x}\, dx$.

Solution

$$\text{Let} \quad u = \boxed{x^2} \quad dv = \boxed{e^{2x}\, dx}$$

$$\text{Then} \quad du = \boxed{2x\, dx} \quad v = \boxed{\dfrac{e^{2x}}{2}}$$

The integration by parts formula gives:

$$\int x^2 e^{2x}\, dx = x^2 \frac{e^{2x}}{2} - \int \frac{e^{2x}}{2} \cdot 2x\, dx = x^2 \frac{e^{2x}}{2} - \int x e^{2x}\, dx \qquad (1)$$

The integral $\int x^2 e^{2x}\, dx$ has been replaced by the simpler integral $\int x e^{2x}\, dx$ but we are still not done. We will apply integration by parts a <u>second</u> time to evaluate the integral $\int x e^{2x}\, dx$.

$$\text{Let} \quad u = \boxed{x} \quad dv = \boxed{e^{2x}\, dx}$$

$$\text{Then} \quad du = \boxed{dx} \quad v = \boxed{\dfrac{e^{2x}}{2}}$$

Integration by parts yields:

$$\int x e^{2x}\, dx = x \frac{e^{2x}}{2} - \int \frac{e^{2x}}{2} \cdot 1\, dx = x \frac{e^{2x}}{2} - \frac{e^{2x}}{4} + \hat{C}$$

Inserting this result back into equation 1 and letting $C = -\hat{C}$ gives:

$$\int x^2 e^{2x}\, dx = x^2 \frac{e^{2x}}{2} - x \frac{e^{2x}}{2} + \frac{e^{2x}}{4} + C = \boxed{\frac{e^{2x}}{4}(2x^2 - 2x + 1) + C}$$

Check As always, we should check an integration result by differentiation.

$$\frac{d}{dx}\left(\frac{e^{2x}}{4}(2x^2 - 2x + 1) + C\right) = \frac{e^{2x}}{2}(2x^2 - 2x + 1) + \frac{e^{2x}}{4}(4x - 2) = x^2 e^{2x}$$

Example 6 - Repeated Use

Find the indefinite integral $\int e^x \sin x \, dx$.

Solution

$$\text{Let} \quad u = \boxed{e^x} \quad dv = \boxed{\sin x \, dx}$$

$$\text{Then} \quad du = \boxed{e^x \, dx} \quad v = \boxed{-\cos x}$$

The integration by parts formula leads to:

$$\int e^x \sin x \, dx = -e^x \cos x + \int e^x \cos x \, dx \tag{1}$$

At first it may appear that the new integral $\int e^x \cos x \, dx$ is no better than the original integral we started with. However, let us apply integration by parts a <u>second</u> time to the new integral. This time we choose the following factorization.

$$\text{Let} \quad u = \boxed{e^x} \quad dv = \boxed{\cos x \, dx}$$

$$\text{Then} \quad du = \boxed{e^x \, dx} \quad v = \boxed{\sin x}$$

Integration by parts yields:

$$\int e^x \cos x \, dx = e^x \sin x - \int e^x \sin x \, dx \tag{2}$$

Inserting this result back into equation 1 we find:

$$\int e^x \sin x \, dx = -e^x \cos x + \left(e^x \sin x - \int e^x \sin x \, dx \right)$$

The unknown integral occurs on both sides. Isolating the unknown we find:

$$2 \int e^x \sin x \, dx = -e^x \cos x + e^x \sin x$$

Dividing by 2 and adding the constant of integration we finally obtain :

$$\int e^x \sin x \, dx = \boxed{\frac{e^x}{2}(\sin x - \cos x) + C}$$

Check As always, we should check an integration result by differentiation.

$$\frac{d}{dx}\left(\frac{e^x}{2}(\sin x - \cos x) + C\right) = \frac{e^x}{2}(\sin x - \cos x) + \frac{e^x}{2}(\cos x + \sin x) = e^x \sin x$$

Integration by Parts for Definite Integrals

By combining the Fundamental Theorem of Calculus with Integration by Parts we can also evaluate <u>definite</u> integrals by this method. We simply evaluate both sides of the integration by parts formula between the limits a and b.

$$\int_a^b f(x)g'(x)\,dx \;=\; f(x)g(x)\Big]_a^b - \int_a^b f'(x)g(x)\,dx$$

The above equation can also be expressed using the shorter notation. Recall $u = f(x)$ and $v = g(x)$. Let u_1 and v_1 denote the values of u and v when $x = a$. Let u_2 and v_2 denote the values of u and v when $x = b$. Then:

$$\int_{v_1}^{v_2} u\,dv \;=\; (u_2 v_2 - u_1 v_1) - \int_{u_1}^{u_2} v\,du$$

Geometric Interpretation

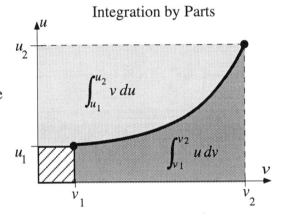

Integration by Parts

The four terms in this Integration by Parts formula have a simple geometric interpretation. The term $u_2 v_2$ is just the area of the large rectangle in the illustration and $u_1 v_1$ is the area of the smaller rectangle in the lower left. The darkly shaded area is the integral $\int_{v_1}^{v_2} u\,dv$ and the lightly shaded area is the integral $\int_{u_1}^{u_2} v\,du$.

Since the area of the larger rectangle is the sum of the three smaller areas we have:

$$u_2 v_2 \;=\; u_1 v_1 + \int_{v_1}^{v_2} u\,dv + \int_{u_1}^{u_2} v\,du$$

Rearranging we obtain the integration by parts formula!

> **Integration by Parts for Definite Integrals**
>
> $$\int_{v_1}^{v_2} u\,dv \;=\; (u_2 v_2 - u_1 v_1) - \int_{u_1}^{u_2} v\,du$$

Example 7 - Integration by Parts for Definite Integrals

Find the definite integral $\int_1^e x^3 \ln x \, dx$.

Solution

$$\text{Let} \quad u = \boxed{\ln x} \quad dv = \boxed{x^3 \, dx}$$

$$\text{Then} \quad du = \boxed{\frac{1}{x} \, dx} \quad v = \boxed{\frac{x^4}{4}}$$

The integration by parts formula (definite version) gives:

$$\int_1^e x^3 \ln x \, dx = \left(\frac{x^4}{4} \ln x\right)\Big]_1^e - \int_1^e \frac{x^4}{4} \cdot \frac{1}{x} \, dx$$

$$= \left(\frac{x^4}{4} \ln x - \frac{x^4}{16}\right)\Big]_1^e = \left(\frac{e^4}{4} \ln e - \frac{e^4}{16}\right) - \left(\frac{1}{4} \ln 1 - \frac{1}{16}\right)$$

Since $\ln e = 1$ and $\ln 1 = 0$, the integral simplifies to:

$$\int_1^e x^3 \ln x \, dx = \frac{e^4}{4} - \frac{e^4}{16} + \frac{1}{16} = \boxed{\frac{3e^4 + 1}{16}}$$

§4.7 Integration by Parts   CAFÉ   Page 349

WARMUP EXERCISES

Evaluate the following indefinite integrals using integration by parts. Use the suggested factorization of the integrand.

1) $\int x\, e^{4x}\, dx$ Let $u = x$, $dv = e^{4x}\, dx$

2) $\int x \ln x\, dx$ Let $u = \ln x$, $dv = x\, dx$

3) $\int x \cos 3x\, dx$ Let $u = x$, $dv = \cos 3x\, dx$

Evaluate the following indefinite integrals using integration by parts.

Check every answer by differentiating your result!

4) $\int t e^{-t}\, dt$ 5) $\int x \cos 5x\, dx$

6) $\int x \sin 3x\, dx$ 7) $\int x \ln 2x\, dx$

Evaluate the following definite integrals using integration by parts.

8) $\int_0^1 x\, e^x\, dx$ 9) $\int_1^e x \ln x\, dx$

10) $\int_0^\pi x \sin x\, dx$ 11) $\int_0^{\pi/2} x \cos x\, dx$

INTERMEDIATE EXERCISES

Evaluate the following integrals. You may need to review the derivatives of some of the functions in this group.

12) $\int x\, 2^x\, dx$ 13) $\int_0^1 \tan^{-1} 2x\, dx$

14) $\int x \sinh x\, dx$ 15) $\int (2x+3) \cosh x\, dx$

16) $\displaystyle\int_0^1 x \tan^{-1} x \, dx$

17) $\displaystyle\int_0^1 \sin^{-1} x \, dx$

Evaluate the following indefinite integrals using integration by parts.
You may have to use parts <u>repeatedly</u>. Check each result by differentiating your answer.

18) $\displaystyle\int x^2 \ln x \, dx$

19) $\displaystyle\int (x^2 + 3x + 1) \ln x \, dx$

20) $\displaystyle\int x^2 e^x \, dx$

21) $\displaystyle\int x^2 \sin x \, dx$

22) $\displaystyle\int x^2 \cos ax \, dx$

23) $\displaystyle\int x^2 \sinh x \, dx$

24) $\displaystyle\int x^2 \cosh x \, dx$

25) $\displaystyle\int x^3 e^{x^2} \, dx$

26) $\displaystyle\int x^5 e^{-x^2} \, dx$

27) $\displaystyle\int x^5 e^{x^3} \, dx$

28) $\displaystyle\int e^x \sin ax \, dx$

29) $\displaystyle\int e^{-x} \cos 2x \, dx$

Evaluate the following indefinite integrals using integration by parts.
You may have to use parts <u>repeatedly</u>.

30) $\displaystyle\int_1^9 \sqrt{x} \ln x \, dx$

31) $\displaystyle\int_1^5 x^2 \ln x^2 \, dx$

32) $\displaystyle\int_0^1 x^2 \sin^{-1} x \, dx$

33) $\displaystyle\int_1^2 x^3 e^{-x} \, dx$

34) $\displaystyle\int_0^\pi x^5 \sin x^2 \, dx$

35) $\displaystyle\int_0^1 x^3 \cosh x \, dx$

Integration by parts is often useful for integrands of the form $\displaystyle\int f(\sqrt{x}) \, dx$ or $\displaystyle\int f(\ln x) \, dx$.
Even though neither integrand is a product, a 'factorization' of the integrand is suggested in each case by considering the chain rule. If F is an antiderivative of f we may write:

$$\int f(\sqrt{x}) \, dx = \int 2\sqrt{x} \left(\frac{f(\sqrt{x})}{2\sqrt{x}}\right) dx = 2\sqrt{x}\, F(\sqrt{x}) - \int \frac{F(\sqrt{x})}{\sqrt{x}} \, dx$$

$$\int f(\ln x) \, dx = \int x \left(\frac{f(\ln x)}{x}\right) dx = x F(\ln x) - \int F(\ln x) \, dx$$

§4.7 Integration by Parts

Evaluate the following integrals using the above suggestions.

36) $\int e^{\sqrt{x}}\, dx$

37) $\int_0^{\pi^2} \cos\sqrt{x}\, dx$

38) $\int \cosh\sqrt{x}\, dx$

39) $\int \sin(1+\sqrt{x})\, dx$

40) $\int \sin(\ln x)\, dx$

41) $\int_1^e \cos(\ln x)\, dx$

---

42) a) Use integration by parts to prove the reduction formula
$$\int_0^{\pi/2} \sin^n x\, dx = \frac{n-1}{n}\int_0^{\pi/2} \sin^{n-2} x\, dx \text{ where } n \geq 2 \text{ is an integer.}$$

b) Use the reduction formula to evaluate $\int_0^{\pi/2} \sin^3 x\, dx$ and $\int_0^{\pi/2} \sin^5 x\, dx$.

c) Show by induction that if n is odd ($n \geq 3$) then:
$$\int_0^{\pi/2} \sin^n x\, dx = \frac{2\cdot 4\cdot 6\cdot\ \ldots\ (n-1)}{3\cdot 5\cdot 7\cdot\ \ldots\ n}.$$

d) Obtain a similar formula if n is even. (The answer will involve π).

43) Find a reduction formula for:

a) $\int_0^1 (\ln x)^n\, dx$

b) then evaluate $\int_0^1 (\ln x)^3\, dx$

44) Find a reduction formula for:

a) $\int_0^1 x^n e^x\, dx$

b) then evaluate $\int_0^1 x^4 e^x\, dx$

45) Find a reduction formula for:

a) $\int_0^1 x^n(1-x)^m\, dx$

b) then evaluate $\int_0^1 x^3(1-x)^{10}\, dx$

46) Find a reduction formula for:

a) $\int_0^{\pi/4} \tan^n x\, dx$

b) then evaluate $\int_0^{\pi/4} \tan^4 x\, dx$

Notice that the next two problems involve nearly identical integrals even though the physical interpretation is completely different.

47) Due to gravitational forces, the density of a planet often increases towards its center. Assume that a small planet is a sphere of radius R and that its density ρ is reasonably modeled by the equation

$$\rho(r) = \rho_0 e^{-kr} \; \frac{\text{kg}}{\text{m}^3}$$

where r is the distance from the center, k is a positive constant and ρ_0 is the density at its center.

Select an infinitesimal spherical shell of radius r and thickness dr (see illustration). Recall that the surface area of such a spherical shell is $4\pi r^2$.

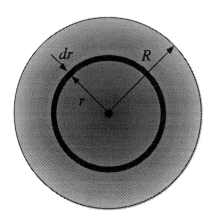

Cross-section of Planet

The volume of the infinitesimal spherical shell is $dV = 4\pi r^2 \, dr$ and its mass is $dm = \rho \, dV = 4\pi \rho_0 e^{-kr} r^2 \, dr$.
Therefore, the total mass of the planet is given by the integral:

$$M = \int_0^R dm = \int_0^R 4\pi \rho_0 e^{-kr} r^2 \, dr$$

Use integration by parts <u>twice</u> to calculate the total mass M of the planet.

48) Friction between spinning circular surfaces is important in the design of pivot bearings, clutch plates and disk brakes. In the clutch mechanism shown, two flat circular disks of radius R are mounted so that they can be brought into contact causing a net frictional force. The contact pressure in general depends on the distance from the center and in one specific case may be modeled by the equation: $P(r) = P_0 e^{-kr}$ where k is a positive constant and r is the distance from the center of the discs.

Clutch Mechanism

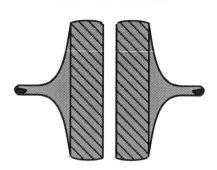

a) Evaluate the integral $F = 2\pi \int_0^R P(r) r \, dr$ which gives the total contact force between the discs.

b) Evaluate the integral: $T = 2\pi\mu \int_0^R P(r) r^2 \, dr$ which gives the total torque between the discs. The constant μ is the coefficient of friction.

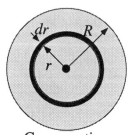

Cross-section

§ 4.8 Approximate Integration

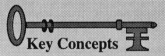

Key Concepts

- ❖ Left & Right Rectangle Rules
- ❖ Midpoint & Trapezoid Rules
- ❖ Simpson's Rule

Why do we need to approximate integrals?

The Fundamental Theorem of Calculus is usually the best way to evaluate a definite integral such as $\int_a^b f(x)\,dx$. If we can find an antiderivative F of the integrand f, then the value of the definite integral is easily found using

$$\int_a^b f(x)\,dx = F(b) - F(a)$$

The fly in the ointment is that it may not be so easy or even possible to find an antiderivative F for the integrand f. It has been proven that for functions such as $\sqrt{1+x^4}$ and e^{x^2} it is not possible to express the antiderivative in terms of any of the elementary functions we have discussed so far. In a situation like this, where one cannot find a functional expression for F, the only choice is to try to approximate the integral.

The situation when actual scientific or engineering data must be 'integrated' underscores the need for approximate integration techniques in an even more compelling fashion. Experimental data consists of a <u>finite</u> set of measurements of some quantity Q of interest. How can we integrate Q knowing only a few of its values?

The fuel for a jet is stored in its wings. Since we do not know a mathematical function describing the actual contours of the tank, we can only estimate the amount of jet fuel stored by taking measurements of the tank.

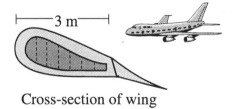

Cross-section of wing

In summary, when either an antiderivative cannot be found or when the value of the integrand is only known at a few data points, approximate integration schemes will have to be used. In this section, we shall discuss several different numerical procedures for approximating a definite integral.

How to approximate an integral.

There are many different procedures for approximating an integral. Some of these procedures follow directly from the definition of an integral in terms of Riemann sums.

Consider the definite integral: $I = \int_a^b f(x)\,dx$

This integral is <u>defined</u> by the following process. First the interval $[a, b]$ is divided into n subintervals. To keep the discussion simple, we shall assume the subintervals are equally spaced so that each interval has width $\Delta x = \dfrac{b-a}{n}$.
The endpoints of the intervals are denoted by: $a = x_0, x_1, \cdots, x_n = b$.
These points are illustrated below for the case $n = 5$.

Next a point x_i^* is arbitrarily selected from the ith subinterval and the integral is defined as the limit:

$$\int_a^b f(x)\,dx = \lim_{n \to \infty} \left(\dfrac{b-a}{n}\right) \cdot \sum_{i=1}^{n} f(x_i^*)$$

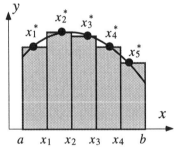

This summation corresponds to approximating the area under the curve by the area of the illustrated rectangles. The width of each rectangle is $\dfrac{b-a}{n}$ and the height is $f(x_i^*)$.
Therefore, the rectangles would have the area: base × height or $\dfrac{b-a}{n} \times f(x_i^*)$.

By choosing any finite number n we may approximate the integral by the formula:

Approximation of the Integral

$$\int_a^b f(x)\,dx \approx \Delta x \sum_{i=1}^{n} f(x_i^*)$$

Note that the symbol $\approx$ means that the quantities are approximately equal.

Of course, there is always the possibility that the approximation will be grossly inaccurate if the number of subintervals n is not large enough.

§4.8 Approximate Integration

Left and Right Rectangles

We can select the points x_i^* in different ways obtaining different approximation formulas.

Estimating the area using left rectangles corresponds to selecting each point x_i^* as the left endpoint from each subinterval. Thus $x_i^* = x_{i-1}$ and the approximation becomes:

$$L(n) = \Delta x \sum_{i=0}^{n-1} f(x_i)$$

or

$$L(n) = \Delta x \{f(x_0) + f(x_1) + f(x_2) + \cdots + f(x_{n-1})\}$$

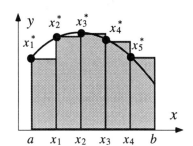

Left endpoint approximation ($n = 5$)

This approximation is called the left rectangle rule.

Estimating the area using right rectangles corresponds to selecting each point x_i^* as the right endpoint from each subinterval. Thus $x_i^* = x_i$ and the approximation becomes:

$$R(n) = \Delta x \sum_{i=1}^{n} f(x_i)$$

or

$$R(n) = \Delta x \{f(x_1) + f(x_2) + f(x_3) + \cdots + f(x_n)\}$$

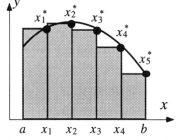

Right endpoint approximation ($n = 5$)

This approximation is called the right rectangle rule.

Although the left and right rectangle rules are not as accurate as some of the other numerical methods we shall consider in this section, we shall start with the left and right-rectangle methods because they are the simplest, plus their obvious short comings will enable us to discover more powerful methods.

Consider the integral $\int_0^1 \frac{1}{1+x} dx = \ln(1+x) \Big|_0^1 = (\ln 2 - \ln 1) = \ln 2$

The value of this integral was easy to find from the antiderivative $F(x) = \ln(1+x)$ using the Fundamental Theorem of Calculus. Our goal in this example is to see how well various approximations perform in estimating this integral. We can then apply the best approximations to cases where we do not know the actual value of the integral.

Example 1 - Left & Right Rectangles - Test Driving the New Formula

Approximate the integral $\int_0^1 \frac{1}{1+x}\, dx$ using just two left-sided rectangles and then using two right-sided rectangles.

Solution

Dividing the interval [0, 1] into two subintervals leads to the vertices

$$\left\{x_0 = 0, x_1 = \tfrac{1}{2}, x_2 = 1\right\}$$

The values of the integrand $f(x) = \frac{1}{1+x}$ at these points are
$\left\{f(0) = 1,\ f(0.5) = \tfrac{2}{3},\ f(1) = \tfrac{1}{2}\right\}$.

The base of each rectangle has width $\Delta x = \tfrac{1}{2}$.

The area of the two left-sided rectangles shown in the adjacent figure is somewhat larger than the true area under the curve $f(x) = \frac{1}{1+x}$. Thus if we approximate the true area by area of the two rectangles, our estimate will be close but somewhat large.

$$L(2) = \left[f(0) + f\!\left(\tfrac{1}{2}\right)\right]\Delta x = \left(1 + \tfrac{2}{3}\right)\tfrac{1}{2} = \tfrac{5}{6} \approx \boxed{0.8333}$$

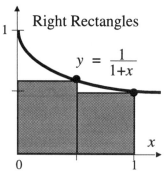

The area of these two left-sided rectangles overestimates the true area.

The approximation using two right rectangles is

$$R(2) = \left[f\!\left(\tfrac{1}{2}\right) + f(1)\right]\Delta x = \left(\tfrac{2}{3} + \tfrac{1}{2}\right)\tfrac{1}{2} = \tfrac{7}{12} \approx \boxed{0.5833}$$

Notice that compared to the actual value of $\ln 2 \approx 0.6931$, both these approximations are really quite bad. The reason is fairly obvious to see.

Since the function $f(x) = \frac{1}{1+x}$ is a decreasing function, each left-sided rectangle overestimates the true area. Similarly, each right-sided rectangle underestimates the true area. This will be true no matter how many subintervals n we divide [0, 1] into.

The area of these two right-sided rectangles underestimates the true area.

We repeated these calculations using a greater number n of rectangles (both left and right).

§4.8 Approximate Integration

The following table summarizes the resulting approximations obtained for our integral
$\int_0^1 \frac{1}{1+x} dx = \ln 2 \approx 0.6931$ for various values of n.

Table 1

n	Left	Right
2	0.833	0.583
4	0.760	0.635
8	0.725	0.663
16	0.709	0.678
32	0.701	0.685
64	0.697	0.689
128	0.695	0.691

Looking at the data in the table we see that the approximation using left-sided rectangles is always too large, whereas the approximation using right-sided rectangles is too small.

This is a simple consequence of the fact that the function $f(x) = \frac{1}{1+x}$ decreases over the interval [0, 1].

We will now generalize our observations to any function $f(x)$ which decreases over an interval $[a, b]$. We will need the following definition.

> **Definition:** A function is said to be <u>monotonic</u> on an interval if it either increases throughout the interval or decreases throughout the interval.

Let $f(x)$ be a <u>decreasing</u> function on the interval $[a, b]$.

Then it is obvious geometrically that, for any number n of subintervals the right rectangle rule $R(n)$ underestimates the true value of the integral and the left rectangle rule $L(n)$ overestimates the true value. That is,

$$R(n) \le \int_a^b f(x)\, dx \le L(n)$$

On the other hand, if $f(x)$ is an <u>increasing</u> function on the interval $[a,b]$ then

$$L(n) \le \int_a^b f(x)\, dx \le R(n)$$

In fact, we can show the following stronger theorem.

Theorem

Let $f(x)$ be an monotonic function on the interval $[a, b]$. Let $A(n)$ denote either of the approximations $L(n)$ or $R(n)$.

Then the error of approximation $\left| A(n) - \int_a^b f(x)\,dx \right|$ satisfies

$$\left| A(n) - \int_a^b f(x)\,dx \right| \le (\Delta x)|f(a) - f(b)|$$

Proof

We shall give the proof in the case that $f(x)$ is decreasing on the interval $[a, b]$.

Each rectangle has base $\Delta x = \dfrac{b-a}{n}$. The fact that $L(n)$ is bigger than $\int_a^b f(x)\,dx$ is obvious since the function is decreasing and each left rectangle lies above the graph of $f(x)$. We shall actually prove the inequality:

$$L(n) - \int_a^b f(x)\,dx \le \Delta x\,(f(a) - f(b))$$

The difference between the approximation $L(n)$ and the actual area $\int_a^b f(x)\,dx$ is equal to the shaded area above the curve consisting of incomplete rectangles. Now augment each partial rectangle into a complete rectangle and translate the resulting rectangles to the very left column. Because $f(x)$ decreases, these rectangles stack up <u>without</u> overlapping to form a larger rectangle with height $f(a) - f(b)$ and base Δx.

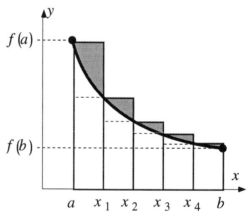

Since the area of the partial rectangles is less that the total area of the complete rectangles, we have the following inequality.

$$L(n) - \int_a^b f(x)\,dx \le \Delta x\,(f(a) - f(b))$$

A similar proof works for the other inequality.

End of Proof

Consider the data in Table 1 again. Notice that the errors using the left-sided rectangles are almost equal and opposite to the errors using the right-sided rectangles. Although the errors get increasingly small as the number n of rectangles increases, we have to work harder for each additional correct digit.

Midpoint and Trapezoidal Rule

We have seen that the approximation of the integral $\int_0^1 \frac{1}{1+x}\,dx = \ln 2$ using left-sided rectangles was too large and the approximation using right-sided rectangles was too small by about the same amount. The situation is similar to the children's story about Goldilocks and the three bears. The porridge in the bowl was too hot at one end of the table, too cold at the other end, and just right in the middle. Could we obtain better approximations to the exact value $\ln 2$ by selecting the heights of the rectangles at the midpoint instead of the left or right point? The answer is yes, and this approximation is called the Midpoint Rule. (See illustration). Notice how each rectangle in the midpoint rule is both above and below the actual curve. This affords an opportunity for the errors to cancel each other out.

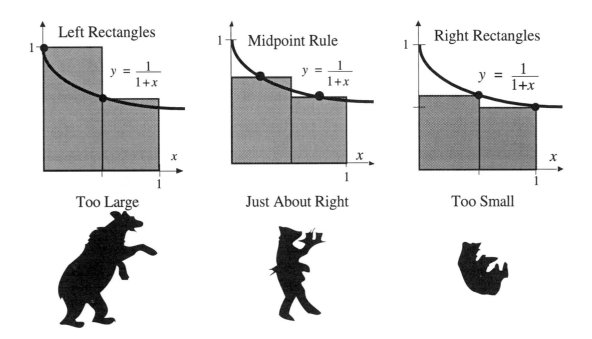

There is yet another way for Goldilocks to prepare her porridge at 'just the right temperature'. She could mix together the bowl that was too hot and the bowl that was too cold to make a new bowl that was just about right. In our example, the approximation $L(n)$ was too large and the approximation $R(n)$ was too small. If we average these two approximations we should get a better answer. This new method is called the Trapezoid Rule.

$$\text{Trap}(n) = \frac{L(n) + R(n)}{2}$$

Geometric Interpretation of the Trapezoid Rule

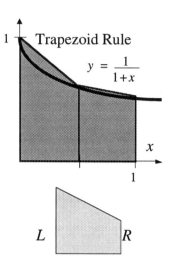

Although it is not obvious from our derivation, using the Trapezoid Rule is the same as approximating the area over each subinterval by a trapezoid as in the adjacent illustration.

Recall the area A of a trapezoid with base Δx, left edge of height L and right edge of height R is:

$$A = \frac{L+R}{2} \Delta x$$

Notice that this formula for the area of the trapezoid is analogous to the rule

$$\text{Trap}(n) = \frac{L(n)+R(n)}{2}.$$

To give an explicit formula for Trap(n), recall that:

$$L(n) = (\Delta x)(f(x_0) + f(x_1) + f(x_2) + \ldots + f(x_{n-1}))$$

and

$$R(n) = (\Delta x)(f(x_1) + f(x_2) + f(x_3) + \ldots + f(x_n))$$

Since the Trapezoid Rule is defined by $\text{Trap}(n) = \dfrac{L(n)+R(n)}{2}$, we see that:

$$\text{Trap}(n) = \Delta x \left\{ \frac{f(x_0)}{2} + f(x_1) + f(x_2) + \cdots + f(x_{n-1}) + \frac{f(x_n)}{2} \right\}$$

§4.8 Approximate Integration

Example 2 - Midpoint & Trapezoid Rules

Approximate the integral $\int_0^1 \frac{1}{1+x} dx = \ln 2 \approx 0.6931$ using both the Midpoint Rule and the Trapezoid Rule with $n = 2$. Compare the results with the values obtained previously using the left and right-sided rectangles.

Solution

The midpoint of the interval [0, 0.5] is 1/4 and the midpoint of the interval [0.5, 1] is 3/4.

Thus the approximation using midpoints is:

$$\text{Mid}(2) = [f(1/4) + f(3/4)]\, \Delta x = \left(\frac{4}{5} + \frac{4}{7}\right)\frac{1}{2} = \frac{24}{35} = \boxed{0.6857}$$

The Trapezoidal Rule is defined as the average of the left and right values. Thus:

$$\text{Trap}(2) = \frac{L(2) + R(2)}{2} = \frac{0.8333 + 0.5833}{2} = \boxed{0.7083}$$

Both these values are closer to the exact value $\ln 2 \approx 0.6931$ than the previous estimates using two left rectangles or two right rectangles.

Notice that the error using the trapezoid rule $(0.7083 - 0.6931 = +0.0152)$ is about double and opposite the error using the midpoint rule $(0.6857 - 0.6931 = -.0074)$. This will turn out to be important later.

The following table summarizes the approximations obtained to our integral

$\int_0^1 \frac{1}{1+x} dx = \ln 2 \approx 0.693147$ using both the Trapezoid and Midpoint Rules with larger values of n.

n	Midpoint	Trapezoid
2	0.685714	0.708333
4	0.691220	0.697024
8	0.692661	0.694122
16	0.693025	0.693391
32	0.693117	0.693208
64	0.693140	0.693162
128	0.693145	0.693151

For each value of n, the errors using the trapezoid rule are almost twice the errors using the midpoint rule. Can we design a better numerical routine that will cancel these errors out? Before attempting to discover such an enhanced routine, let us stop to consider how the errors behave when using Midpoint and Trapezoidal rules.

When does the Trapezoid Rule give values that are too large or too small?

If the function is concave up on an interval, then each trapezoid will lie above the graph of the function resulting in an overestimate. If the function is concave down, then each trapezoid will lie below the graph of the function resulting in an underestimate.

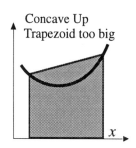
Concave Up
Trapezoid too big

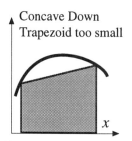
Concave Down
Trapezoid too small

When does the Midpoint Rule give values that are too large or too small?

The answer is the exact opposite of what happens for the Trapezoid Rule.

To see this, consider a rectangle through some midpoint and draw a tangent to the curve at the midpoint forming a new trapezoid. (See illustration.) Note the new trapezoid has the same area as the rectangle because the two triangles shown are congruent.

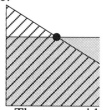
The trapezoid has the same area as the rectangle.

* The new trapezoid is not the same as the trapezoid in the Trapezoid Rule. It does not intersect the function at the endpoints of the interval!

Replacing the midpoint rectangles with these new trapezoids having the same area we see that:

If the function is concave up, then the midpoint rule gives an underestimate.
If the function is concave down, then the midpoint rule gives an overestimate.

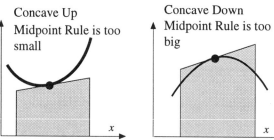

Midpoint rectangles have been replaced by equivalent trapezoids of equal area tangent to the curve.

Let us summarize all these observations.

How concavity affects the midpoint and trapezoid approximations

If $f(x)$ is concave up on the interval $[a, b]$ then

$$\text{Mid}(n) \leq \int_a^b f(x)\, dx \leq \text{Trap}(n)$$

If $f(x)$ is concave down on the interval $[a, b]$ then

$$\text{Trap}(n) \leq \int_a^b f(x)\, dx \leq \text{Mid}(n)$$

§4.8 Approximate Integration — CAFÉ — Page 363

Simpson's Rule and Archimedes' Formula

Archimedes, the greatest mathematician and engineer of antiquity, discovered without calculus that the area under a parabolic arch of height h and base b is given by

$$A = \frac{2}{3}bh$$

Using his exact result we can compare the errors made in estimating the area under a parabolic arch using both the Trapezoid Rule and the Midpoint Rule. We will divide the base of the arch into just two subintervals so that $n = 2$.

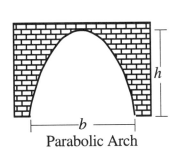
Parabolic Arch

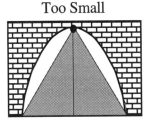
Too Small
Area approximated by trapezoid rule ($n = 2$)

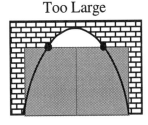
Too Large
Area approximated by midpoint rule ($n = 2$)

The area estimate using the Trapezoid Rule corresponds to the area of the two triangles shown in the middle illustration (the triangles are degenerate trapezoids). Thus $A_{Trap} = \frac{1}{2}bh$ which is too small by an amount $\frac{2}{3}bh - \frac{1}{2}bh = \frac{1}{6}bh$. On the other hand, the area estimate using the Midpoint Rule corresponds to the two rectangles shown in the right illustration. The height of each rectangle is $\frac{3}{4}h$.

To see this, note that the equation of the parabolic arch is
$y = h\left(1 - \frac{4}{b^2}x^2\right)$ with the origin placed as in the illustration.
At the midpoints, $x = \pm\frac{b}{4}$ and $y = \frac{3}{4}h$.

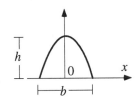

Thus the area using the midpoint rule is $A_{Mid} = \frac{3}{4}bh$ which is $\frac{1}{12}bh$ too large. Note that the magnitude of the error using the Trapezoid Rule $\left(\frac{1}{6}bh\right)$ is twice as large as the error using the Midpoint Rule $\left(\frac{1}{12}bh\right)$. However, while one estimate is too large, the other is too small. By combining the estimates in the ratio of two to one, the errors magically cancel out!

$$\frac{2 A_{Mid} + A_{Trap}}{3} = \frac{2}{3}bh \quad \longleftarrow$$

The new rule we have discovered is called Simpson's Rule.

$$\text{Simp} = \frac{2\,\text{Mid} + \text{Trap}}{3}$$

Together, the points $x_0, x_1, \cdots, x_n$ used in the trapezoidal rule and the midpoints $\overline{x}_1, \cdots, \overline{x}_n$ divide the interval $[a, b]$ into $2n$ subintervals. Thus we define

$$\text{Simp}(2n) = \frac{2\,\text{Mid}(n) + \text{Trap}(n)}{3}$$

Example 3 - Simpson's Rule

Approximate the integral $\int_0^1 \frac{1}{1+x} dx = \ln 2 \approx 0.6931$ using Simpson's Rule.
Compare the approximation with the values obtained previously.

Solution

We have already worked out the following approximations in a previous example.

$$\text{Mid}(2) = 0.6857 \qquad \text{Trap}(2) = 0.7083$$

Simpson's Rule combines these values in such a way as to cancel most of their individual errors:

$$\text{Simp}(4) = \frac{2\,\text{Mid}(2) + \text{Trap}(2)}{3} = \frac{2\,(0.6857) + 0.7083}{3} = \boxed{0.6932}$$

This result differs from the calculator value of 0.6931 only in the last decimal! Thus, Simpson's Rule gives a very precise answer in this case.

In certain situations, we may want to apply Simpson's Rule without first calculating the Midpoint Rule and Trapezoid Rule and then applying the equation $\text{Simp} = \frac{2\,\text{Mid} + \text{Trap}}{3}$. We will now give an explicit formula for Simpson's Rule. Using the formulas for the Midpoint and Trapezoid rules, one can show that Simpson's Rule corresponds to the formula:

Simpson's Rule

$$\text{Simp}(n) = \frac{\Delta x}{3} \left(f(x_0) + 4f(x_1) + 2f(x_2) + 4f(x_3) + \ldots + 2f(x_{n-2}) + 4f(x_{n-1}) + f(x_n) \right)$$

where $\Delta x = \frac{b-a}{n}$.

It is important to note that Simpson's Rule can only be used when n is even.

---

§4.8 Approximate Integration

Example 4 - Chemical Spill Cleanup

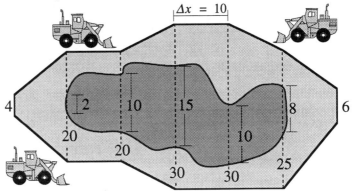

The darkly shaded region in the illustration reveals the actual extent of a chemical spill. Fortunately, the spill is confined in the topsoil by an impermeable clay layer at a depth of two feet. After determining the extent of the spill, environmental engineers plan to clean up the site by removing the top two feet of soil using powerful excavators. To be on the safe side, all the soil within the larger polygonal area is to be removed.
Width measurements for both the spill and the larger excavation area are given in 10 foot intervals.

- a) Estimate the volume V of soil that is to be removed.
- b) Estimate the area A of the actual spill.
  In each part, choose the method (trapezoid rule or Simpson's rule) that will give the most accurate answer.

Solution

a) *Estimate the volume of soil removed.*
Since the larger excavation area consists of six trapezoids, we clearly should use the Trapezoidal Rule. In this particular case, the Trapezoidal Rule gives the exact answer!

$$\text{Trap}(6) = \Delta x \left[\frac{f(x_0)}{2} + f(x_1) + f(x_2) + f(x_3) + f(x_4) + f(x_5) + \frac{f(x_6)}{2} \right]$$

Using $\Delta x = 10$ feet and the dimensions from the illustration gives:

$$\text{Trap}(6) = 10 \left[\frac{4}{2} + 20 + 20 + 30 + 30 + 25 + \frac{6}{2} \right] = 1300 \text{ ft}^2$$

Since the topsoil is to be removed down to a depth of two feet, the total volume V of soil that has to be excavated is:

$$\boxed{V = 2600 \text{ ft}^3}$$

b) *Estimate the area of the actual spill.*
Since the region occupied by the actual spill (darker area) has gently curved boundaries, Simpson's Rule is more likely to give an accurate answer than the Trapezoid Rule. Notice that the Trapezoid Rule would underestimate the actual area in this particular case. Also note that on each subinterval, the boundaries can be reasonably approximated as quadratic or cubic functions. This is a good reason to use Simpson's Rule. (From the illustration, we see that $n = 4$.)

$$\text{Simp}(4) = \frac{\Delta x}{3} \left[f(x_0) + 4f(x_1) + 2f(x_2) + 4f(x_3) + f(x_4) \right]$$

Using $\Delta x = 10$ feet, and the dimensions from the illustration gives:

$$\text{Simp} = \frac{10}{3}\left[2+4\cdot(10)+2\cdot(15)+4\cdot(10)+8\right] = 400 \text{ ft}^2$$

Thus the actual spill covers an area of about 400 square feet.

Summary of Approximation Techniques

In this section we have learned five methods for approximating definite integrals of the form $\int_a^b f(x)\,dx$. In each method, the interval $[a, b]$ is first divided into n subintervals by the sequence of equally spaced points $\{x_0=a, x_1, x_2, \cdots, x_{n-1}, x_n=b\}$. The width of each subinterval is $\Delta x = \frac{b-a}{n}$ so the coordinates of the points are $x_i = a + i\Delta x$ for $i = 0$ to n. With this notation, the left-rectangle, right-rectangle and trapezoid rules are given by the formulas:

$$L(n) = \Delta x \left[f(x_0)+f(x_1)+\cdots+f(x_{n-1})\right]$$

$$R(n) = \Delta x \left[f(x_1)+f(x_2)+\cdots+f(x_n)\right]$$

$$\text{Trap}(n) = \frac{L(n)+R(n)}{2} = \Delta x \left[\frac{f(x_0)}{2}+f(x_1)+f(x_2)+\cdots+f(x_{n-1})+\frac{f(x_n)}{2}\right]$$

The Midpoint Rule evaluates the function at the midpoint of each of the subintervals. Denoting the midpoint of the ith interval by $\bar{x}_i = \frac{x_{i-1}+x_i}{2}$ we obtain:

$$\text{Mid}(n) = \Delta x \left[f(\bar{x}_1)+f(\bar{x}_2)+\cdots+f(\bar{x}_n)\right]$$

Simpson's Rule is a weighted average of the Midpoint Rule and the Trapezoid Rule.

$$\text{Simp}(2n) = \frac{2\,\text{Mid}(n) + \text{Trap}(n)}{3}$$

when n is even, we may write:

$$\text{Simp}(n) = \frac{\Delta x}{3}\left(f(x_0)+4f(x_1)+2f(x_2)+4f(x_3)+\cdots+2f(x_{n-2})+4f(x_{n-1})+f(x_n)\right)$$

§4.8 Approximate Integration   CAFÉ   Page 367

WARMUP EXERCISES

1) Consider the integral $\int_0^1 \frac{1}{1+x} dx$ which illustrated our examples in this section.

   a) Divide the interval $[0,1]$ into four subintervals.
   State the coordinates of the equally spaced points x_0, x_1, x_2, x_3 and x_4 and evaluate the function at each of these points.

   b) Find the approximations $L(4)$, $R(4)$ and Trap(4).

   c) State the coordinates of the midpoints $\bar{x}_1, \bar{x}_2, \bar{x}_3$ and $\bar{x}_4$ and evaluate the function at each of these points.

   d) Find the approximations Mid(4) and Simp(8).

   e) Compare each approximation to the value for $\ln 2$ as displayed on a calculator. Which approximations are too large and which are too small?

   f) Draw a picture illustrating the Left, Right and Trapezoid approximations with $n = 4$.

2) Consider the integral $\int_0^1 \frac{1}{1+x^2} dx = \tan^{-1} x \Big]_0^1 = \frac{\pi}{4}$.

   By numerically approximating the integral, we can estimate the value of π.

   a) Divide the interval $[0,1]$ into four subintervals.
   State the coordinates of the equally spaced points x_0, x_1, x_2, x_3 and x_4 and evaluate the integrand at each of these points.

   b) Find the approximations $L(4)$, $R(4)$ and Trap(4).

   c) Find the approximations Mid(4) and Simp(8) using the points $\bar{x}_1, \bar{x}_2, \bar{x}_3$ and $\bar{x}_4$

   d) Multiply each of the five approximations by 4 to estimate π.
   Which gives the best answer?

3) It is impossible to find a nice closed-form antiderivative for the function $e^{-(x^2/2)}$.

   a) Estimate the integral $\int_0^1 e^{-(x^2/2)} dx$ using $L(4)$, $R(4)$, Trap(4), and Mid(4).

   b) Show that the function is both decreasing and concave downwards on this interval.

   c) State whether each approximation in (a) is an overestimate or an underestimate.

4) Consider the integral $\int_0^1 \cos x^2 \, dx$.

Find the approximations $L(10)$, $R(10)$, Trap(10), and Mid(10) for the integral.

---

5) A machine part has the dimensions shown.

  a) Find the exact area A of the part geometrically.

  b) Which of the five approximations $L(4)$, $R(4)$, Trap(4), Mid(4) or Simp(8) should give the exact answer for A?

  c) Find all five approximations and test your prediction.

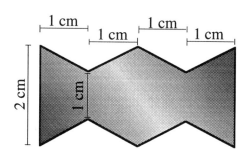

6) A machine part has the dimensions shown.

  a) Find the exact area A of the part geometrically.

  b) Predict which of the approximations $L(4)$, $R(4)$, Trap(4), Mid(4) or Simp(8) should give the exact answer? Which will be too large? Which too small?

  c) Find all five approximations and test your predictions.

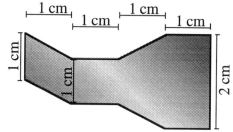

---

7) Draw a function such that any approximation using left-sided rectangles will be too small and a second function such that the approximation will always be too large.

8) Draw a function such that any approximation using right rectangles will be too small and a second function such that the approximation will always be too large.

9) Draw a function such that any approximation using the Trapezoid Rule will be too small and a second function such that the approximation will always be too large.

10) Draw a function such that any approximation using the Midpoint Rule will be too small and a second function such that the approximation will always be too large.

INTERMEDIATE EXERCISES

11) An automatic chlorination and purification system is to be designed for a large outdoor pool. The design will depend on the volume of water in the pool. The pool has a uniform depth of 2 meters and the width of the pool has been measured at 2 meter intervals as illustrated.

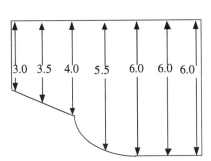

Width of outdoor pool at 2 meter intervals.

    a) Use Simpson's Rule to estimate the area of the pool.

    b) What is the approximate volume of water contained in the pool?

12) The pool illustrated in the last problem is proposed for the rooftop of a new luxury hotel. The architectural team has asked for an estimate of the weight of water in the pool when full.

    a) Use the Trapezoidal Rule to estimate the area of the pool.

    b) What is the volume of water contained in the pool using this estimate?

    c) Knowing that the density of water is 1000 kg/m^3, estimate the number of kilograms of water in the full pool.

    d) The architects have specifically asked that no estimation method be used which will underestimate the weight of the water. Which of the following methods will give an underestimate?

        i) Left Rectangles   **ii)** Right Rectangles
        iii) Trapezoidal Rule   **iv)** Midpoint rule

    e) Will the architects be pleased that you used the Trapezoidal Rule?

    f) In a situation like this, it is best to select a method that will give an overestimate so that a safety factor is built in. Which method would you use?

13) The speedometer reading on a car was measured at one minute intervals. The data is shown in the following table.

Time (min)	0	1	2	3	4	5	6
Speed (miles/hr)	30	35	40	45	50	50	50

    a) Use both Simpson's Rule and the Trapezoidal Rule to estimate the total distance traveled by the car.

    b) Assume the car has constant acceleration until it reaches the speed of 50 miles per hour and then cruises at this constant speed. What is the exact distance traveled during the above six minute interval?

    c) Identify one numerical rule that will underestimate the distance and one that will overestimate the distance.

14) A terrorist bomb has left a gigantic gaping hole in the concrete floor of one level of an underground parking lot. Spike, a volunteer fireman, has arrived just in time to help extinguish the resulting fire. To determine the power of the explosion, investigators need to know the area A of the hole blasted in the floor. The floor was constructed from concrete slabs measuring 10 meters square.

a) Estimate the area A of the hole by counting the number of slabs that have been destroyed. You may count certain partially destroyed slabs as one half a slab.

b) Approximate the area A of the hole using the Trapezoid rule by two different methods. First estimate the vertical width of the blasted hole at 10 meter intervals. Then estimate the horizontal width of the hole at 10 meter intervals.

c) In the previous part you found five vertical width measurements and six horizontal width measurements. Which set of measurements are appropriate for Simpson's method? Use Simpson's Rule and the appropriate width measurements to estimate the area.

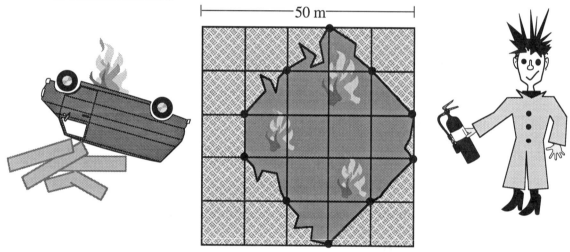

Hole blasted in floor by terrorist bomb

15) An architectural design proposes a row of granite columns to enhance the entrance to a proposed courthouse. Since the columns will be directly over an underground service area, there is considerable concern about their weight. The <u>diameter</u> of the columns is given in the illustration at 3 foot intervals.

  a) Estimate the volume V of each column using the Trapezoidal rule and Simpson's Rule.

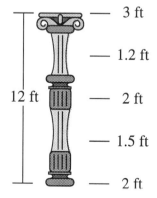

§4.8 Approximate Integration   CAFÉ   Page 371

Hint: The volume is the integral of the cross-sectional area.

$$V = \int_0^{12} A(x)\,dx$$

Since the data gives the diameter D of the column instead of A, you will have to first calculate the cross-sectional area using $A = \dfrac{\pi D^2}{4}$ at 3 foot intervals before applying either numerical rule.

b) Estimate the weight in pounds of each column if the weight density of granite is $170\,\dfrac{\text{lb}}{\text{ft}^3}$.

16) Every July 4th, Spike competes in a hot air balloon race. The lifting power of the balloon is determined by the volume V of hot air that it can contain. The diameter D of the balloon is indicated in the diagram at five foot intervals. Estimate the volume of the balloon using Simpson's Rule and the Trapezoid Rule. (Hint: The volume is the integral of the cross-sectional area.

$$V = \int_0^{20} A(x)\,dx$$

Since the data gives the diameter D of the balloon instead of the area, you will have to first calculate the area using $A = \dfrac{\pi D^2}{4}$.)

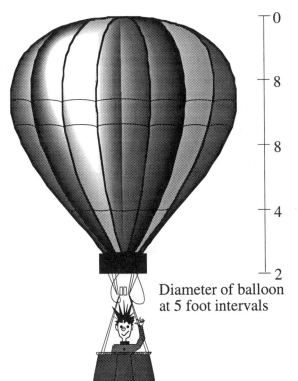

Diameter of balloon at 5 foot intervals

ADVANCED EXERCISES

17) The area under a function and over the interval $1 \leq x \leq 2$ has been estimated numerically using the left, right, trapezoid and midpoint rules. (Each method used the same number n of subintervals.) In an arbitrary order, the approximations found were 1, 1.2, 1.3 and 1.4. Note that the function decreases on the interval and is concave down.

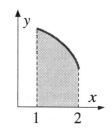

Function is concave down and decreases.

a) Which method generated each approximation?
b) What is the smallest interval in which we can conclude that the actual answer lies?

18) Suppose the numbers generated are 1, 1.1, 1.2, 1.4 and that the function is increasing and concave up,
a) Which method generated each approximation?
b) What is the smallest interval in which we can conclude that the actual answer lies?

Chapter 5
Geometric Applications of Integration

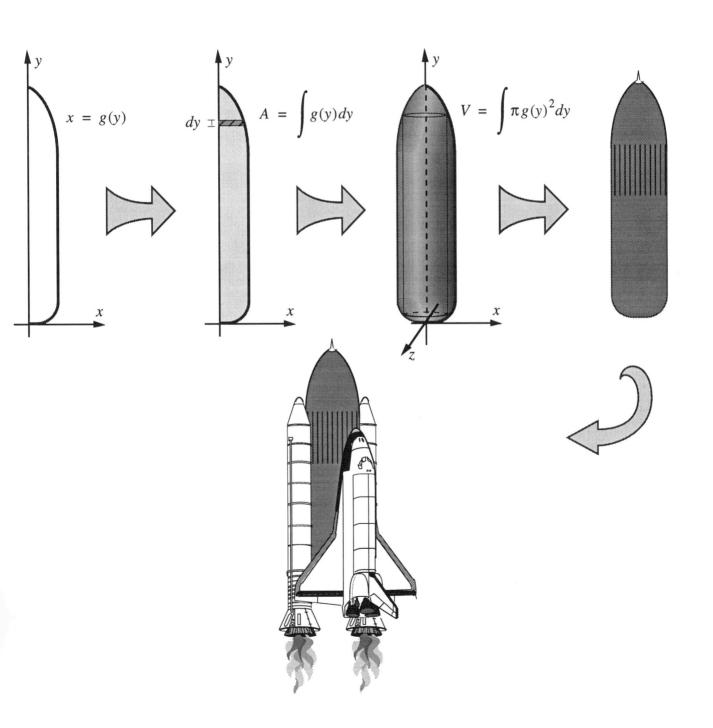

§ 5.1 Volume Part 1

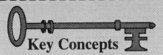
Key Concepts

- ❖ Generalized Cylinders
- ❖ Solids of Known Cross-Section
- ❖ Solids of Revolution
- ❖ Method of Discs and Washers

Without the aid of integral calculus, we could only know the volumes of a few simple geometric shapes. For example:

1) The volume of a <u>block</u> of length l, height h and width w is $V = lwh$.
2) The volume of a right circular <u>cylinder</u> of radius r and height h is $V = \pi r^2 h$.

Both the block and the cylinder are only a few of examples of a type of solid known as a <u>generalized cylinder</u>. All generalized cylinders have a volume V equal to the area A of their base times their height h:

$$V = Ah$$

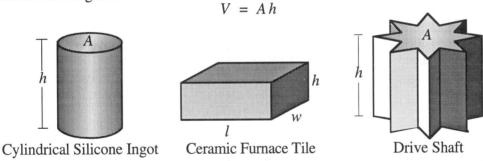

Cylindrical Silicone Ingot   Ceramic Furnace Tile   Drive Shaft

Generalized cylinders are constructed as follows. Any planar region of area A can serve as the bottom of the generalized cylinder. The top is formed by lifting the planar region upwards out of the plane to a height h. The solid object between the identical top and bottom copies is called a generalized cylinder and has volume:

> **Volume of a Generalized Cylinder**
> $$V = Ah$$

For example, the above block is a generalized cylinder formed by lifting its bottom rectangular surface vertically upwards to a height h. The bottom rectangle has area $A = lw$ so that the volume of the block is indeed: $V = (lw)h = Ah$. Similarly the right circular cylinder is a generalized cylinder generated by lifting the circular base to a height h. The base circle has radius r and area $A = \pi r^2$ so that the volume of the cylinder can be rewritten as $V = (\pi r^2)h = Ah$. This notion of a generalized cylinder will be our key to defining the volume of any object for which we know the area of its planar cross-sections.

Volumes of known cross-sectional area A

The previous formula $V = Ah$ for the volume of a generalized cylinder was so easy because the cross-sections of a generalized cylinder are identical and hence all have the same area. How can we find the volume of an object whose cross-sectional area is not constant, but varies? Using integral calculus and the formula for the volume of a generalized cylinder, we will now obtain a formula for finding the volume of any object!!

Divide and Conquer

Consider the magnified view of the brilliant diamond shown in the adjacent illustration. The number of carats in the diamond depends on its weight and hence its volume V. Notice however, that the cross-sectional area of the diamond is not constant but is a function of the height x! How can we calculate the volume of the diamond?

This specific question suggests the general problem of finding the volume of any object whose cross-sectional area $A(x)$ varies.

The cross-sectional area $A(x)$ of the diamond is not constant but depends on the height x.

Our fundamental approach in tackling this general volume problem could be called the "Divide and Conquer" strategy.
Imagine dividing the diamond up into infinitesimal slices of thickness dx.
The cross-sectional area of any particular thin slice can be treated as if it were <u>constant</u>. Thus each thin slice is approximately a generalized cylinder. Suppose the slice located at a given value of x, has thickness dx and cross-sectional area $A(x)$. Its volume will then be:

$$dV = A(x)\,dx$$

For a given solid we will have to actually find the cross-sectional area $A(x)$ from information about the geometry of the solid. Having 'divided' the solid into infinitesimal slices, we now 'conquer' the problem of calculating its total volume by recombining all the slices. Of course, this is achieved by integrating over the extent of the body.

> **Formula for the Volume of a Solid of Known Cross-Section.**
>
> $$V = \int dV = \int_a^b A(x)\,dx$$

Notice that for the diamond in the illustration, the integration variable x ranges from $x = a$ (the bottom tip) to $x = b$ (the top facet).

Example 1 - Aluminum Apex on the Washington Monument

The top of the Washington Monument is tipped with an aluminum apex in the shape of a pyramid measuring 5.6 inches on each of the four sides of its base and 8.9 inches high. Find the volume V of the aluminum pyramid.

Solution

Select an infinitesimal horizontal slice of thickness dx at a distance x from the tip. The infinitesimal slice can be viewed as a 'generalized cylinder' having thickness dx and a square cross-section. Using similar triangles, we see that each side of the square has length
$L(x) = \frac{5.6}{8.9} x = .6292\, x$, where x is the distance from the tip down to the infinitesimal slice. Thus the area of the square cross-section is:

$$A(x) = (0.6292\, x)^2 = 0.3959\, x^2$$

Aluminum Apex on the Washington Monument

Since the infinitesimal slice is a generalized cylinder with area $A(x)$ and height dx, its volume is $dV = A\, dx = 0.3959 x^2 dx$. The total volume V can now be found by integrating from the top ($x = 0$) to the bottom ($x = 8.9$) of the aluminum apex.

$$V = \int_0^{8.9} A(x)\, dx = \int_0^{8.9} 0.3959\, x^2\, dx = 0.3959 \left(\frac{(8.9)^3}{3}\right) = \boxed{93.0 \text{ in}^3}$$

Comment - The aluminum apex was cast in Philadelphia and because of its high conductivity, serves as an integral part of a system of lightening rods for the Washington Monument. Aluminum was also chosen because it would not tarnish or stain the marble of the monument.

Example 2 - Volume of a Right Circular Cone

Find the volume V of a cone of height h and having a circular base of radius r.

Solution

Note that each horizontal cross-section of the cone is a disc. Using similar triangles, we see that the radius y of this disc at a distance x from the apex is $y = \frac{r}{h} x$. Thus the area $A(x)$ of the cross-section at a distance x from the tip is

$$A(x) = \pi y^2 = \pi \left(\frac{r}{h} x\right)^2$$

The total volume V can now be found by integrating the cross-sectional area A from the tip of the cone ($x = 0$) to its bottom ($x = h$).

$$V = \int_0^h A(x)\,dx = \int_0^h \pi\left(\frac{r}{h}x\right)^2 dx = \pi\frac{r^2}{h^2}\int_0^h x^2\,dx = \pi\left(\frac{r^2}{h^2}\right)\left(\frac{h^3}{3}\right) = \boxed{\frac{1}{3}\pi r^2 h}$$

Thus the volume of the cone is $V = \frac{1}{3}Ah$ where A is the area of its circular base.

Example 3 - Conical Rain gauge

A conical rain gauge has height h inches and the radius of its circular opening is r inches. What is the maximum amount of rainfall in inches that the cone can measure before overflowing? Does the radius of the opening affect the answer?

Conical raingage

Solution   Let overflow occur when H inches of rain have fallen. Then the volume of rain that has entered the circular top of the gage is $V = AH$ where A is the area of the circular top. By the result of the last example the volume of water in the full rain gauge is $V = \frac{1}{3}Ah$.

Equating the two expressions for V we see that $V = \frac{1}{3}Ah = AH$ so that $\boxed{H = \frac{h}{3}}$
The maximum rainfall H before the gage overflows is only 1/3 the height of the gage! Thus a rain gauge of height 6 inches will begin to overflow when only 2 inches of rain have fallen.   The answer does not depend on the radius of the opening!

Example 4

Spike is removing a tree which is four feet in diameter at the base so that he can plant a healthy one. Using a chain saw, he has made two planar cuts into the trunk of the tree forming a wedge shape. One face of the wedge is horizontal, and the other is inclined to the horizontal at an angle of 45°. What is the volume V of the wedge removed from the tree?

Wedge sawed into trunk of tree

Solution

An interesting aspect of this wedge shape is that we can find its volume using either triangular or rectangular cross-sections.

Method 1 - Triangular Cross-sections
Introduce a coordinate system as shown in the figure, then partition the wedge into slices perpendicular to the x-axis. Note that each cross-section is an isosceles right triangle. Consider a typical cross-section such as triangle ABC shown in the illustration. The triangle is isosceles since $\angle CAB = 45°$ implies $AB = BC$.

§5.1 Volume Part I

Using the Pythagorean theorem, and noting that the tree has a radius of 2 feet we conclude that:

$$AB = BC = \sqrt{2^2 - x^2}$$

Therefore the area of the triangular cross-section is:

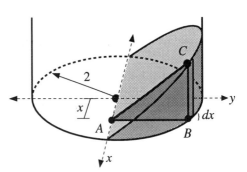

$$A = \tfrac{1}{2} \text{base} \times \text{height} = \tfrac{1}{2}(2^2 - x^2)$$

The volume V of the wedge is found by integrating the cross-sectional area A from $x = -2$ to $x = +2$.

$$V = \int_{-2}^{2} A(x)\,dx = \int_{-2}^{2} \tfrac{1}{2}(2^2 - x^2)\,dx$$

Area of triangular cross-section is
$$A = \tfrac{1}{2}(2^2 - x^2)$$

$$= \left(2x - \tfrac{1}{6}x^3\right)\bigg|_{-2}^{2} = \boxed{\tfrac{16}{3} \text{ ft}^3}$$

Method 2 - Rectangular Cross-sections

A second approach is to partition the wedge into slices that are perpendicular to the y-axis so that each slice is a rectangular slab as shown. A typical infinitesimal slice at y is illustrated. Notice that the value of y ranges from 0 to 2.

The height of the rectangle is y and the length of the rectangle is $2\sqrt{4 - y^2}$. (Use the Pythagorean theorem.)

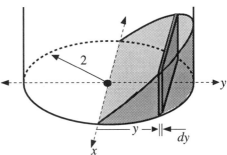

Area of rectangular cross-section is
$$A = 2y\sqrt{4 - y^2}$$

Thus the area A of the rectangle is:

$A = \text{length} \times \text{height} = 2y\sqrt{4 - y^2}$

The total volume V of the wedge is obtained by integrating the cross-sectional area A with respect to y over the interval from 0 to 2.

$$V = \int_0^2 A(y)\,dy = \int_0^2 2y\sqrt{4 - y^2}\,dy$$

This is easily integrated by making the substitution: $u = 4 - y^2$ so $du = -2y\,dy$.

$$V = \int_0^2 2y\sqrt{4 - y^2}\,dy = -\int_4^0 u^{1/2}\,du = \int_0^4 u^{1/2}\,du = \tfrac{2}{3}u^{3/2}\bigg|_0^4 = \left(\tfrac{2}{3}\right)8 = \boxed{\tfrac{16 \text{ ft}^3}{3}}$$

Of course, the result is the same as that obtained using triangular cross-sections.

Volumes of Solids of Revolution

The basic formula in volume calculations states that the volume V is the integral of the cross-sectional area function A. If the integration variable is x, this can be written:

$$V = \int_a^b A(x)\,dx$$

where the object is bounded by the planes $x = a$ and $x = b$.
We now consider some special cases of this formula.

Solids of Revolution

Consider the graph of a function $y = f(x)$ over the interval $a \leq x \leq b$. If the region between the curve and the x-axis is rotated about the x-axis, the solid object generated is called a <u>solid of revolution</u>.

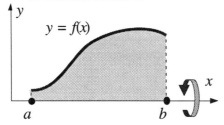

The area under the curve is rotated about the x-axis creating the adjacent solid of revolution.

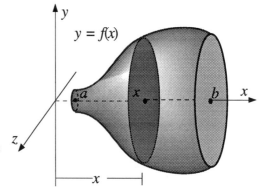

Each cross-section of a solid of revolution is a disc.

Notice that each cross-section perpendicular to the x-axis is a disc.
For a given value of x, the corresponding circular cross-section has radius $y = f(x)$ so that its area is: $A(x) = \pi y^2 = \pi f^2(x)$. Now that we know the cross-sectional area function $A(x)$ for the solid of revolution, we just use the general volume formula to compute:

Formula for the Volume V of a Solid of Revolution (Disc Method)

$$V = \int_a^b A(x)\,dx = \int_a^b \pi y^2\,dx = \int_a^b \pi f(x)^2\,dx$$

Since each cross-section is a disc, this formula is often called the Disc Method.

Example 5 - Volume of Sphere

A sphere is a typical solid of revolution. Test that the new formula introduced above by using it to confirm that the volume V of a sphere of radius R is $V = \frac{4}{3}\pi R^3$.

Solution

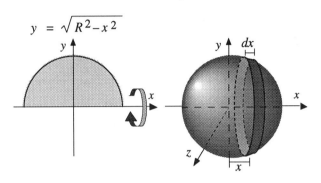

The entire sphere can be generated by rotating about the x-axis the region under the curve $y = \sqrt{R^2 - x^2}$ and over the interval $-R \leq x \leq R$. Thus the sphere is a solid of revolution. For a given value of x, the cross-section is a circle of radius $y = \sqrt{R^2 - x^2}$. Thus its area is:

$$A(x) = \pi y^2 = \pi \left(\sqrt{R^2 - x^2}\right)^2 = \pi (R^2 - x^2)$$

Integrating over the interval $-R \leq x \leq R$, we find the total volume V to be:

$$V = \int_{-R}^{R} \pi (R^2 - x^2)\, dx = \pi \left(R^2 x - \frac{x^3}{3}\right)\Big]_{-R}^{R} = \boxed{\frac{4}{3}\pi R^3}$$

Example 6

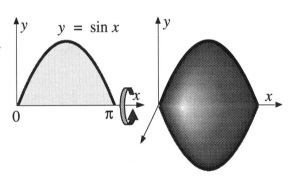

Consider the region bounded by one arch of the sine curve and the x-axis. Note that the sine curve forms a complete arch on the interval $[0, \pi]$. Find the volume V of the solid generated by rotating the region about the x-axis.

Solution

Using the disc method, the volume of this solid of revolution is:

$$V = \int_0^\pi \pi y^2\, dx = \pi \int_0^\pi \sin^2(x)\, dx$$

To integrate $\sin^2(x)$ recall the double angle identity $\sin^2(x) = \frac{1 - \cos(2x)}{2}$. Using this identity to rewrite the term $\sin^2(x)$ we find:

$$V = \pi \int_0^\pi \frac{1 - \cos(2x)}{2}\, dx = \pi \int_0^\pi \frac{1}{2}\, dx - \frac{\pi}{2} \int_0^\pi \cos(2x)\, dx = \boxed{\frac{\pi^2}{2}}$$

Above, the integral involving the cosine evaluates to zero.

Example 7 - Floating Mine (Volume of a Spherical Zone)

A spherical mine of radius R is designed to float with only a small portion sticking above the surface. A trigger is placed on the tip of this portion in the hope of destroying enemy ships. Find the volume of the portion of the mine floating above the surface. A portion of a sphere like this is called a spherical zone.

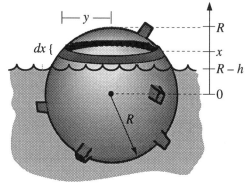

Solution

Assume the mine extends a distance h above the surface. The portion of the mine above the surface is a solid of revolution having circular cross-sections.

The horizontal infinitesimal slice with thickness dx at a distance x from the center of the mine is a circular disk of radius $y = \sqrt{R^2 - x^2}$. To find the volume of the portion above the surface, x must range from the water's surface (where $x = R - h$) to the top of the mine (where $x = R$). Using the disc method, the volume above the water surface is:

$$V = \int_{R-h}^{R} \pi y^2 \, dx = \int_{R-h}^{R} \pi (R^2 - x^2) \, dx = \pi R^2 h - \frac{\pi}{3}\left(R^3 - (R-h)^3\right)$$

Simplifying the algebra gives:

$$V = \frac{\pi}{3}\left(3R h^2 - h^3\right) = \boxed{\frac{\pi}{3} h^2 (3R - h)}$$

Check - As a check on this result, note that when $h = 2R$ the entire mine would be floating on the surface as if the mine were weightless. Substituting this value of h into our result gives $V = \frac{4}{3}\pi R^3$ which is the volume of a complete sphere of radius R as expected.

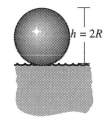

§5.1 Volume Part I   CAFÉ   Page 383

WARMUP EXERCISES

Cross-sectional areas (perpendicular to the x-axis) are given for each of the following solids.

Find the volume of each solid using the formula $V = \int_a^b A(x)\,dx$.

1) $A(x) = x^2$ for $0 \le x \le 1$
2) $A(x) = \pi(1 - x^2)$ for $-1 \le x \le 1$
3) $A(x) = \pi \sin^2 x$ for $0 \le x \le \pi$

The cross-sections $A(x)$ for each of the following solids are easy to calculate from their geometry. Find an expression for $A(x)$, then determine the volume of each solid. Note that the data does not uniquely determine the objects' shapes.

4) Each cross-section of this solid is a square with edges of length $L(x) = \sin x$. The solid is bounded by the planes $x = 0$ and $x = \pi$.

5) Each cross-section of this solid is a circle with radius $r(x) = x$. The solid is bounded by the planes $x = 0$ and $x = 1$.

6) Each cross-section of this solid is a equilateral triangle with edges of length $L(x) = x^2$. The solid is bounded by the planes $x = 1$ and $x = 3$.

7) Each cross-section of this solid is a right triangle with base $b(x) = x$ and height $h(x) = x^2$. The solid is bounded by the planes $x = 1$ and $x = 3$.

8) Each cross-section of this solid is a rectangle with base $b(x) = x$ and height $h(x) = \frac{1}{x}$. The solid is bounded by the planes $x = 1$ and $x = 3$.

Solids of Revolution and the Disc Method
Sketch the solid generated by revolving the region under the given curve about the x-axis. Then use the disk method to find the volume of each solid.

9) $y = \sqrt{16 - x^2}$ for $0 \le x \le 4$

10) $y = x^2$ for $1 \le x \le 2$

11) $y = \sec x$ for $-\pi/4 \le x \le \pi/4$

12) $y = e^x$ for $0 \le x \le 1$

13) $y = x + 1/x$ for $1/2 \le x \le 2$

14) $y = \dfrac{1}{x + 1}$ for $0 \le x \le 1$

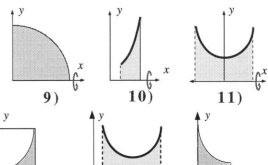

9)   10)   11)

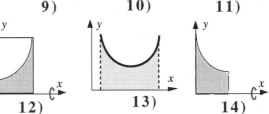

12)   13)   14)

15) A rather potent New Year's Punch is being enjoyed at a family reunion. The punch is contained in a hemispherical bowl of radius of 12 inches. If the beverage fills the bowl to a height of six inches, how much <u>punch</u> is in the bowl?

16) A cocktail glass is filled to the brim. What is the volume of <u>liquid</u> in the glass?

17) In a sample problem we showed that the volume of a spherical mine floating above the surface is $V = \frac{\pi}{3} h^2 (3R - h)$ where h is the distance the mine extends above the surface. Archimedes Principle implies that the weight W of the mine is equal to the weight of the volume of water it displaces. Find an equation which relates the height h to the weight W of the mine. Assume $R = \frac{1}{3}$ meter and note that the density of water is 1000 kg/m³.

18) A uniform wood ball of diameter 10 centimeters floats in water with 3 centimeters extending above the surface.
  a) What is the volume of the water displaced by the ball?
  b) What is the density of the wood?
  (The density of water is 1000 kg/m³.)

  (Hint: Archimedes principle implies that the weight of the ball is equal to the weight of the water it displaces.)

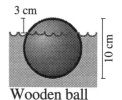

Wooden ball

Disc Method Revisited
A solid of revolution is generated by revolving the shaded region with respect to the indicated region. Use the disk method to find the volume of each solid.

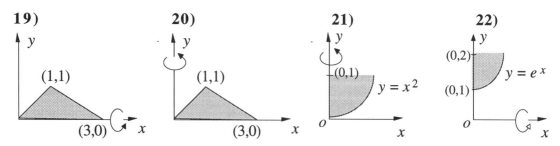

23) Squaring the Circle

The base of the solid is a circle of radius one in the xy-plane. The cross-sections perpendicular to the y-axis are squares whose bases run from the bottom to the top of the circle.
Find the volume of the solid.

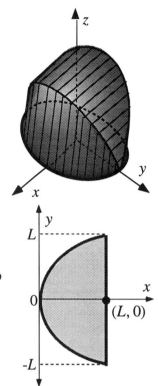

24)

The base of a solid is the region bounded by the parabola $y^2 = Lx$ and a line segment perpendicular to the axis of the parabola. The line segment has length $2L$ and is at a distance L from the vertex of the parabola. Find the volume of the solid if every cross-section perpendicular to the axis of the parabola is:

a) a semicircle.

b) an equilateral triangle.

Base of solid

25) Ellipsoid

The area of the planar ellipse

$$\frac{x^2}{a^2} + \frac{y^2}{b^2} = 1 \text{ is } A = \pi ab.$$

The three dimensional analog of the ellipse is the ellipsoid which has the standard equation

$$\frac{x^2}{a^2} + \frac{y^2}{b^2} + \frac{z^2}{c^2} = 1.$$

The volume of the ellipsoid can be found by integrating the cross-sections with respect to any of the three coordinate axes. The cross-sections are all ellipses.

Show that the volume of the ellipsoid is $V = \frac{4}{3}\pi abc$ by integrating the cross-sectional areas with respect to each of the three axes. All three integrations should give the same result.

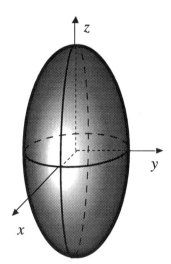

26) **Beer Keg**
A large keg of beer of height $2h$ is in the shape of an ellipsoid with the top and bottom ends cut off. If the radius of the keg at the top and bottom is r and the large central radius is R, find the volume of the keg.

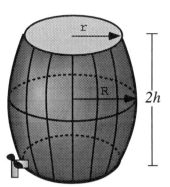

27) Molten metal is poured from a hemispherical crucible of radius 2 meters into ingots. Find the volume V of the molten metal remaining in the container when the crucible has been tilted by $45°$. Note the remaining molten metal in the crucible forms a spherical zone. Is the volume V remaining more or less than half the volume of a full crucible?

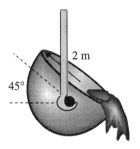

Molten metal pouring from hemispherical crucible

28) Furnace oil used for winter heating is stored in a spherical underground tank. Our problem is to determine the unknown radius R of the tank using only a dipstick. The dipstick passes through the center and displays the depth of the oil left in the tank.
Originally, the dipstick indicates that the fuel fills the tank to a depth of 76 centimeters. After one half cubic meter of fuel is added, the depth increases to 80 cm. Find the unknown radius R of the tank.

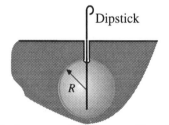

Underground furnace oil tank

29) A spherical mine of diameter one meter and average density $700 \text{ kg}/\text{m}^3$ is attached to the sea floor by a thin cable. Assume the density of sea water is $1040 \text{ kg}/\text{m}^3$.

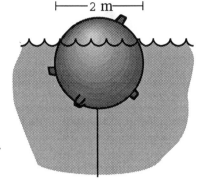

a) Determine how high the mine would float above the surface if the cable were not present. (Hint: use Newton's method to approximate the answer.)

b) What force F must the cable exert on the mine so that it floats just beneath the surface?

30) Weighing an Elephant

A famous Chinese proverb describes how one man solved his father's challenge to find the weight of an elephant. His solution involves coaxing the elephant onto a boat and measuring the draft of the boat. The draft of a boat is defined as the height of its submerged portion. Here is a problem based on the proverb.

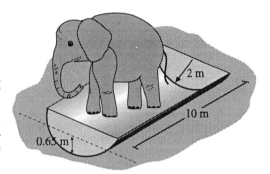

A large steel drum of length 10 meters and radius 2 meters and has been cut in half lengthwise to form a makeshift boat in the form of a semicylindrical shell with semicircular ends as shown. When empty, the initial draft is 0.5 meters. After the elephant has been coaxed onto the boat, the draft increases to 0.65 m. Calculate the weight of the boat and the weight of the elephant. The density of water is 1000 kg/m³.

Notes

§ 5.2 Volume Part 2

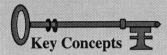

Key Concepts

- Area in Polar Coordinates
- Volume of Generalized Cones
- Morphs
- Method of Cylindrical Shells

Area in Polar Coordinates

In this section we will use polar coordinates to discover volume formulas for many interesting kinds of solids. Each of the solids we consider has a symmetry which makes their description easy in terms of polar coordinates. In each case, we will use polar coordinates to find an expression for the cross-sectional area of the solid and then simply integrate this area to find the volume. In this spirit, let us first see how to calculate areas in polar coordinates.

Area of a Circular Sector

A <u>circular</u> <u>sector</u> is just a pizza-shaped slice from a circle. If the sector has radius r and subtends an angle θ, then its area is:

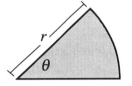

Area of a Circular Sector
$A = \frac{1}{2}\theta r^2$

Circular sector of radius r subtending an angle θ.

As a check, notice that the area of the sector goes to zero as θ does and that when $\theta = 2\pi$, the sector forms a complete circle and the area formula gives the correct result $A = \pi r^2$.

Don't Forget

Note that the angle θ must be measured in radians, not in degrees.

For example, consider a pizza of radius 6 inches which has been divided into 6 equal slices. Let's find the area A of each slice using the above formula. Using $\theta = \frac{\pi}{3}$ and $r = 6$ inches we find the area of each slice to be $A = \frac{1}{2}\theta r^2 = \frac{1}{2}\frac{\pi}{3}(6)^2 = 6\pi$ in^2. Of course, this is just $\frac{1}{6}$ the area of the whole pizza.

Area of a Planar Region in Polar Coordinates

Consider a planar region R whose boundary is described by the polar equation $r = f(\theta)$. If we choose the origin so that it lies inside the region, then the angle θ varies over the interval $0 \leq \theta \leq 2\pi$.

What is the area of this region in terms of the function $r = f(\theta)$?

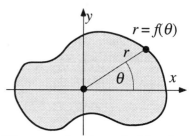

Planar region bounded by the curve $r = f(\theta)$.

Because the radius $r = f(\theta)$ is not constant, we will need to use calculus to solve this problem. Imagine slicing the region into infinitesimal sectors each subtending an angle $d\theta$. The infinitesimal sector centered at an angle θ has radius $r = f(\theta)$ and so its infinitesimal area is:

$$dA = \tfrac{1}{2} r^2 d\theta = \tfrac{1}{2} f^2(\theta) d\theta$$

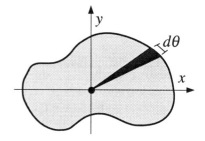

A typical infinitesimal sector

Integrating to add up all the infinitesimal contributions as θ varies over the interval $0 \leq \theta \leq 2\pi$ we discover that the area of the polar region can be written:

Area of a Planar Region in Polar Coordinates

$$A = \tfrac{1}{2} \int_0^{2\pi} r^2 d\theta = \tfrac{1}{2} \int_0^{2\pi} f^2(\theta) d\theta$$

The integration limits in the above formula are only valid if the region <u>entirely</u> surrounds the origin. A more general formula is shown in the exercises.

Example 1 - Cardioid

Find the area of the region bounded by the polar curve $r = 1 + \cos\theta$. Because it is shaped like a heart, this curve is called a cardioid.

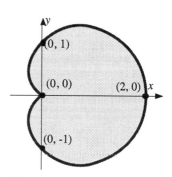

The cardioid $r = 1 + \cos\theta$

Solution

To find the area we only need evaluate the integral:

$$A = \frac{1}{2}\int_0^{2\pi} r^2\, d\theta = \frac{1}{2}\int_0^{2\pi}(1+\cos\theta)^2\, d\theta = \frac{1}{2}\int_0^{2\pi}\left(1+2\cos\theta+\cos^2\theta\right)d\theta$$

Integrating term by term and noticing that the integral of the cosine term is zero gives:

$$A = \pi + 0 + \frac{1}{2}\int_0^{2\pi}\cos^2\theta\, d\theta$$

Using the double angle identity $\cos^2\theta = \dfrac{1+\cos 2\theta}{2}$ we find that the remaining integral is

$\displaystyle\int_0^{2\pi}\cos^2\theta\, d\theta = \pi$. Thus the area of the cardioid is $A = \pi + \dfrac{\pi}{2} = \boxed{\dfrac{3\pi}{2}}$

In this section we will show that in addition to giving the area of planar regions, the formula $A = \dfrac{1}{2}\int_0^{2\pi} r^2\, d\theta$ can also be used to find the cross-sectional areas of many interesting solids. By integrating the resulting cross-sectional areas we can find the volumes of these solids. As a first example we consider those solids known as generalized cones.

Generalized Cones (Tapered Shapes)

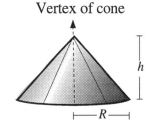

Vertex of cone

Directrix of a circular cone is a circle.

A circular cone is formed by connecting every point on the circular base to the vertex of the cone with a straight line. If the circle has area A and the vertex is a perpendicular distance h from the circle then the volume of the cone is:

$V = \dfrac{1}{3}Ah$

Of course, in general, there is no reason why in this construction the base region has to be a circle.

Definition of a Generalized Cone

Consider any planar region of area A and select a point P a perpendicular distance h from this region. The shape we obtain by connecting every point in the planar region to the point P is called a generalized cone. The point P is called the vertex of the cone and the curve defining the base region is called the directrix. The line segment through the vertex and perpendicular to the directrix is called the axis of the cone.

The figure illustrates a hexagonal cone formed by connecting every point in the base hexagon to the vertex point P. Such forms are common for the roofs of churches, spires and towers. Various spikes, daggers and electrodes also have this shape.

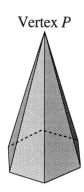

Vertex P

Generalized cone having a hexagonal directrix.

In the next example, we show that regardless of the shape of the base region, the volume of a generalized cone is:

Volume of a Generalized Cone

$$V = \frac{1}{3}Ah$$

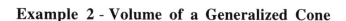

Example 2 - Volume of a Generalized Cone

Show that the volume of a generalized cone is $V = \frac{1}{3}Ah$ where A is the area of the directrix and h is the <u>perpendicular</u> distance from the vertex to the plane containing the base region.

Solution

First note that all the cross-sections of the generalized cone are similar in shape to the directrix (base region). As we progress towards the vertex, each cross-section is just a scaled down version of the base region. For example, in the case of the hexagonal cone discussed above, each cross-section is a perfect hexagon just like the base hexagon. We will obtain an expression for the cross-sectional shape at each height using polar coordinates.

Suppose the boundary of the base region is given by the polar equation $r = f(\theta)$. The cross-section at a perpendicular distance x from the vertex is scaled down from this base region by a factor of $\frac{x}{h}$. Indeed, using similar triangles one can see that the polar equation of this cross-section is $r = \frac{x}{h} f(\theta)$.

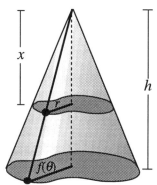

Using the polar area formula, we find that the area $A(x)$ of this cross-section is:

$$A(x) = \frac{1}{2}\int_0^{2\pi} \left(\frac{x}{h} f(\theta)\right)^2 d\theta = \left(\frac{x^2}{h^2}\right)\left(\frac{1}{2}\int_0^{2\pi} f^2(\theta)\, d\theta\right) = \frac{x^2}{h^2} A$$

Thus $A(x) = \dfrac{x^2}{h^2} A$ where A is the area of the directrix.

Integrating the cross-sectional area over the extent of the cone, we find that the volume is:

$$V = \int_0^h A(x)\, dx = A \int_0^h \dfrac{x^2}{h^2}\, dx = \boxed{\dfrac{1}{3} A h}$$

The Volume of Morphs

You are already familiar with the idea of 'morphing' from music videos and Hollywood movies with state of the art special effects as in Terminator II. In morphing, one shape is gradually transformed into another. Engineers have been using this concept to design structural components for years.

Linear Morphs

A linear morph is defined by its bottom and top surfaces. As we progress up through the morph, the cross-sections gradually change from the bottom shape to the top shape.

Let the bottom surface of the morph lie in the xy-plane and suppose it can be given in polar coordinates by the equation $r = f(\theta)$. Suppose the top surface of the morph lies in the plane $z = h$ and that it is described by the polar equation $r = g(\theta)$. Thus we have:

Bottom: $\quad r = f(\theta) \quad\quad 0 \le \theta \le 2\pi$
Top: $\quad\quad r = g(\theta) \quad\quad 0 \le \theta \le 2\pi$

A linear morph
The circular base is transformed into a five-pointed star at the top of the morph!

In a <u>linear</u> morph, for each angle $0 \le \theta \le 2\pi$, corresponding points on the bottom and top surfaces are connected by lines thus forming a solid. Our goal will be to find an expression for the volume V of such morphs. Note that the formulas for $f(\theta)$ and $g(\theta)$ will depend on the <u>orientation</u> of the bottom and top regions. For example, rotating a square will change its polar equation.

Equation for the cross-sections of a linear morph

However, we must first consider the equation describing the shape of each cross-section of the morph. The shape of the cross-section at height z is given by the equation:

> **General Equation of a Linear Morph**
> $$r_z(\theta) = \dfrac{(h-z)}{h} f(\theta) + \dfrac{z}{h} g(\theta)$$

For each fixed value of z, the above equation describes the cross-section at height z.

For example, when $z = 0$, we obtain the bottom curve and when $z = h$ we obtain the top curve. When $z = h/2$ the radius at a given angle θ is the average of the radius at the corresponding points on the top and bottom curves.

$$r_{(\frac{h}{2})}(\theta) = \frac{f(\theta) + g(\theta)}{2}$$

If we knew the cross-sectional area A as a function of the height z, the volume of the morph would be given by the fundamental volume formula:

$$V = \int_0^h A(z)\, dz$$

Thus to find the volume of the morph, we must first find the area $A(z)$ of each cross-section. For a given height z, the cross-section is described by the polar equation $r_z(\theta) = \frac{(h-z)}{h} f(\theta) + \frac{z}{h} g(\theta)$. Thus we must use the polar formula for calculating area.

$$A(z) = \frac{1}{2}\int_0^{2\pi} r_z^2(\theta)\, d\theta = \frac{1}{2}\int_0^{2\pi} \left(\frac{(h-z)}{h} f(\theta) + \frac{z}{h} g(\theta)\right)^2 d\theta$$

$$= \frac{(h-z)^2}{h^2} A(0) + \frac{z^2}{h^2} A(h) + \frac{z(h-z)}{h^2} \int_0^{2\pi} f(\theta)\cdot g(\theta)\, d\theta$$

where $A(0)$ is the area of the bottom shape while $A(h)$ is the area of the top shape. The above integral formula gives the cross-sectional area $A(z)$ at any height z within the morph. To find the total volume of the morph, we simply integrate $A(z)$ and obtain:

$$V = \frac{A(0)}{h^2}\int_0^h (h-z)^2\, dz + \frac{A(h)}{h^2}\int_0^h z^2\, dz + \frac{1}{h^2}\int_0^{2\pi} f(\theta)\cdot g(\theta)\, d\theta \cdot \int_0^h z(h-z)\, dz$$

Working out each of the integrations with respect to z we finally obtain:

Volume of a Linear Morph

$$V = \frac{h}{3}\left\{A(0) + A(h) + \frac{1}{2}\int_0^{2\pi} f(\theta)\cdot g(\theta)\, d\theta\right\}$$

Notice that in the case of a generalized cylinder, the top and bottom are the same so, $f(\theta) = g(\theta)$ and $A(0) = A(h) = A$. Thus the volume formula reduces to $V = hA$.

To gain confidence using the morph equation, let us apply it to derive the classical formula for the volume of a cone.

§5.2 Volume Part II — CAFÉ — Page 395

Example 3

A cone of base radius R and height h can be considered as a linear morph in which the base is a circle of radius R and the top is a degenerate circle of radius 0. Find the volume of the cone by considering it as a linear morph.

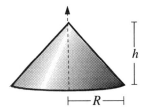

Solution

The polar equations for the bottom and top surfaces are:

Bottom circle: $r = f(\theta) = R$ Top point: $r = g(\theta) = 0$

The morph equation for the volume of the cone is:

$$V = \frac{h}{3}\left\{ A(0) + A(h) + \frac{1}{2}\int_0^{2\pi} f(\theta)\cdot g(\theta)\, d\theta \right\}$$

Now $A(0)$ is the area of the bottom circle so $A(0) = \pi R^2$.
Also $A(h)$ is the area of the degenerate circle (a point) so $A(h) = 0$.
Since the top point is described by the equation $r = g(\theta) = 0$ we can conclude that $\frac{1}{2}\int_0^{2\pi} f(\theta)\cdot g(\theta)\, d\theta = 0$. Thus the volume of the cone using the morph equation is:

$$V = \frac{h}{3}\left\{ A(0) + A(h) + \frac{1}{2}\int_0^{2\pi} f(\theta)\cdot g(\theta)\, d\theta \right\} = \frac{h}{3}\left\{ \pi R^2 + 0 + 0 \right\} = \boxed{\frac{\pi}{3} h R^2}$$

Many morph shapes are found in sheet metal ducting as the next example illustrates.

Example 4 - Coal Chute

Coal is fed into a furnace down a chute whose cross-section changes from square to circular through a transitional segment of length $h = 3$ feet. The diameter of the cylindrical portion and the length of each edge of the square portion are both equal to 2 feet. Find the volume V of the transitional segment joining the two portions of the chute. Assume it is a linear morph.

Solution

Since the transitional segment is assumed to be a linear morph, its volume is given by the formula

$$V = \frac{h}{3}\left\{ A(0) + A(h) + \frac{1}{2}\int_0^{2\pi} f(\theta)\cdot g(\theta)\, d\theta \right\}$$

where $h = 3$, $A(0) = \pi$ is the area of the circular bottom and $A(h) = A(3) = 4$ is the area of the square top.

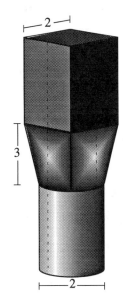

Coal Chute

Thus to find the volume of the morph we need to evaluate the integral: $\int_0^{2\pi} f(\theta)\cdot g(\theta)\, d\theta$

where $r = f(\theta) = 1$ is the polar equation of the circular base and $r = g(\theta)$ is the equation of the square top. Since the square has four sides, $g(\theta)$ must be defined piecewise for each side. To avoid this complication, note that the right edge of the square is a segment of the line $x = 1$ so that the polar equation of just this right edge of the square is $r\cos\theta = 1$ or $r = \sec\theta$. Evaluating the integral we obtain:

$$\int_0^{2\pi} f(\theta)\cdot g(\theta)\, d\theta = 8\int_0^{\pi/4} \sec\theta\, d\theta = 8\ln\left(\sec\theta + \tan\theta\right)\Big]_0^{\pi/4} = 8\ln(\sqrt{2}+1).$$

Notice that we used the symmetry of the square to write the integral as eight times the integral obtained by integrating over the interval from 0 to $\pi/4$. Inserting these values into the formula for the volume of a morph we find:

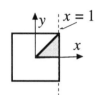

Square cross-section

$$V = \frac{h}{3}\left\{ A(0) + A(h) + \frac{1}{2}\int_0^{2\pi} f(\theta)\cdot g(\theta)\, d\theta \right\}$$

$$= \frac{3}{3}\left\{ \pi + 4 + \frac{8}{2}\ln(\sqrt{2}+1) \right\} = \boxed{\pi + 4 + 4\ln(\sqrt{2}+1)}$$

Using a calculator, we find that the volume is approximately $V \approx 10.667$ cubic feet.

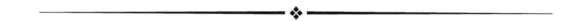

Method of Cylindrical Shells for Solids of Revolution

There is another method that uses polar coordinates to find the volume of solids. It applies only to solids of revolution. Although we have already seen how to find the volume of solids of revolution by slicing the solid into infinitesimal discs, the new method we are about to consider sometimes results in easier integrations.

In the method of cylindrical shells, instead of slicing the volume into discs at right angles to the axis of revolution, the volume is sliced into concentric cylindrical shells as illustrated in the adjacent figure. Instead of integrating with respect to the angular variable θ, as in the area formula $A = \frac{1}{2}\int_0^{2\pi} f^2(\theta)\, d\theta$ we will be integrating with respect to the radial variable r.

§5.2 Volume Part II

Consider a typical shell of thickness dr at a distance r from the axis of revolution. Its height h is a function of r and will be denoted $h(r)$. The area A of the cylindrical shell is $A = 2\pi rh$ since by cutting the shell as shown, we can form a rectangle of height h and base $2\pi r$.

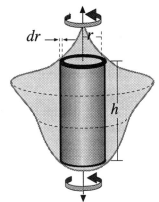

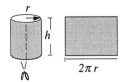

A solid of revolution can be cut up into concentric cylindrical shells.

Thus the volume of the typical infinitesimal shell is: $dV = A(r)\, dr = 2\pi r\, h(r)\, dr$.

Adding up all the small volumes leads to the volume for the entire solid:

Volume by Shell Method

$$V = \int_{r_1}^{r_2} A(r)\, dr = \int_{r_1}^{r_2} 2\pi r h(r)\, dr$$

The limits of integration correspond to the radii of the closest and the most distant shells from the axis of revolution. The method of cylindrical shells is illustrated in the following examples.

Example 5

The region between the curve $y = 2 + x^2$ and the x-axis over the interval $0 \le x \le 1$ is revolved about the y-axis. Use the method of cylindrical shells to find the volume V of the resulting solid of revolution.

Solution

The first step is always to sketch a typical cylindrical shell and then find expressions for its height and infinitesimal volume. Consider an infinitesimal shell at a distance r from the axis of revolution and having thickness dr. It has the following properties.

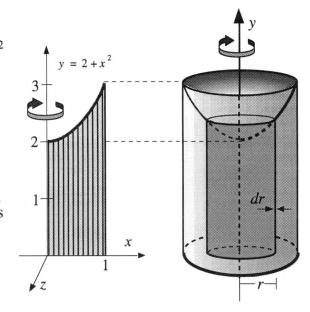

Radius: r
Circumference: $2\pi r$
Height: $y = 2 + r^2$
Infinitesimal Volume: $dV = (2\pi r)(2 + r^2)\, dr$

The total volume can now be found by integrating over the extent of the solid of revolution.

$$V = \int dV = \int_0^1 (2\pi r)(2+r^2)\, dr = 2\pi \int_0^1 (2r + r^3)\, dr = 2\pi\left(r^2 + \frac{r^4}{4}\right)\bigg]_0^1 = \boxed{\frac{5}{2}\pi}$$

Example 6 - Pearl Necklace

A pearl of radius 3 mm has a hole bored through its center of radius 1 mm so it can be worn on a necklace. What is the volume V of the beaded pearl?

Solution

Imagine slicing the beaded pearl into concentric cylindrical shells. The dimensions of a typical shell of radius r and thickness dr are:

Radius: r
Thickness: dr
Height: $h = 2\sqrt{9-r^2}$
Area: $A = 2\pi r h = 4\pi r \sqrt{9-r^2}$

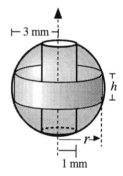

Beaded Pearl

The closest shell to the axis is at $r_1 = 1$, the farthest is at $r_2 = 3$. Thus the volume (in cubic millimeters) of the beaded pearl is:

$$V = \int_1^3 A(r)\, dr = \int_1^3 4\pi r\sqrt{9-r^2}\, dr = -2\pi \int_8^0 \sqrt{u}\, du = \frac{4\pi}{3} u^{3/2}\bigg]_0^8 = \boxed{\frac{64\sqrt{2}\,\pi}{3}}$$

where we made the obvious substitution $u = 9-r^2$, $du = -2r\, dr$.

§5.2 Volume Part II   CAFÉ   Page 399

Exercises on Generalized Cones

Recall that a generalized cone is generated by joining every point in a planar region of area A to a point a distance h from the planar region. The resulting generalized cone has area $V = \frac{1}{3}Ah$. Using this formula, find the volume V for each of the following examples of generalized cones.

1) **Pyramid** - The base of a pyramid is a square with edges of length 100 feet. The pyramid rises to a height of 100 feet above the base.

2) **Tetrahedron** - The solid is a tetrahedron with triangular base determined by the three points $(0,0,0)$, $(1,0,0)$ and $(0,1,0)$. All these points are connected to the vertex point $(0,0,h)$ on the z-axis.

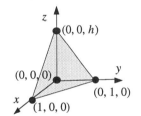

3) **Spike** - The directrix of a spike is an equilateral triangle with edges of length 1 cm. The vertex is located 10 cm to the left of the triangle's center.

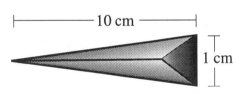

4) **Octagonal Spire** - The base of a church spire is an octagon with edges of length 2 meters. The vertex of the spire is located 10 meters above the center of the octagon.

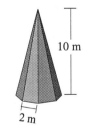

5) **Stalactite** - A stalactite is a mineral deposit, usually of calcite or aragonite, projecting downward from the roof of a cavern as a result of the dripping of mineral rich water. The base (where it attaches to the roof) is oddly shaped but has an area $A = 900$ cm². The length of the stalactite is 1 meter. Find its volume assuming the shape is reasonably modeled as a generalized cone.

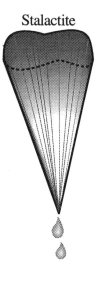
Stalactite

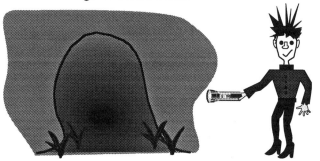

Consider the planar curve $r = f(\theta)$ defined over the interval $a \leq \theta \leq b$. The shaded region between this curve and the origin O has area

$$A = \frac{1}{2}\int_a^b r^2 d\theta.$$

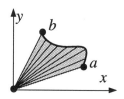

Find the area of the region between the origin and the curve $r = f(\theta)$ over the interval $a \leq \theta \leq b$ for each of the following exercises.

6) $\quad r = 1 \qquad 0 \leq \theta \leq \dfrac{\pi}{2}$

7) $\quad r = 2\cos(\theta) \quad -\dfrac{\pi}{2} \leq \theta \leq \dfrac{\pi}{2}$

8) $\quad r = \theta \qquad 0 \leq \theta \leq 3\pi/4$

9) $r = 1 - \sin\theta \quad 0 \leq \theta \leq 2\pi$

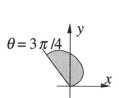
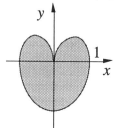

§5.2 Volume Part II — CAFÉ

Sketch and find the area of the region enclosed by each of the curves.

10) $r = 3 + 3\cos\theta$ Cardioid   11) $r = 3 + 2\cos\theta$ Limaçon

12) $r = 10$ Circle   13) $r = \cos 2\theta$ Four-Leaf Rose

14) $r = 5 - 4\cos(2\theta)$ Figure Eight   15) $r^2 = \cos 2\theta$ Figure Infinity

Drill Exercises on Morphs

Find the volume V of the following linear morphs.
In each case, the height h and polar equations for the bottom $r = f(\theta)$ and top $r = g(\theta)$ surfaces are given.

Use the morph equation $V = \dfrac{h}{3}\left\{A(0) + A(h) + \dfrac{1}{2}\displaystyle\int_0^{2\pi} f(\theta)\cdot g(\theta)\,d\theta\right\}$.

16) $f(\theta) = 4$ $g(\theta) = 3 + \cos\theta$ height = 4

17) $f(\theta) = R$ $g(\theta) = r$ height = h (A conical frustum).

18) $f(\theta) = 3 + \cos\theta$ $g(\theta) = 3 + \sin\theta$ height = 6

Shell Method

Find the volume of the solid of revolution generated when the region underneath the curve over the indicated interval is rotated about the y-axis.

Use the method of cylindrical shells: $V = \displaystyle\int_{r_1}^{r_2} 2\pi r h(r)\,dr$

19) $y = x$, $0 \le x \le 1$ 20) $y = |x|$, $-1 \le x \le 1$

21) $y = \dfrac{1}{1+x}$, $0 \le x \le 1$ 22) $y = |x^2 - 1|$, $0 \le x \le 2$

23) $y = e^{x^2}$, $0 \le x \le 1$ 24) $y = \ln x$, $1 \le x \le 2$

25) $y = \sin x$, $0 \le x \le \pi/2$ 26) $y = \dfrac{\sin x}{x}$, $0 \le x \le \pi$

27) $y = \sinh x$, $0 \le x \le \pi/2$ 28) $y = \lfloor x \rfloor$, $0 \le x \le 5/2$

Find the volume of the solid of revolution generated when the region underneath the curve over the indicated interval is rotated about the x-axis.

Use the method of cylindrical shells: $V = \int_{r_1}^{r_2} 2\pi r h(r)\, dr$

29) $y = 2x$, $\quad 0 \le x \le 1$

30) $y = x$, $\quad 1 \le x \le 2$

31) $y = 1 + 2x$, $\quad 0 \le x \le 1$

32) $y = x^2$, $\quad 1 \le x \le 2$

33) $y = e^x$, $\quad 0 \le x \le 1$

34) $y = \ln x$, $\quad 1 \le x \le 2$

35) $y = \lfloor x \rfloor$, $\quad 0 \le x \le 5/2$

36) $y = \dfrac{1}{1+x}$, $\quad 0 \le x \le 1$

In the following a region is given and the axis of rotation is indicated. Use the method of cylindrical shells to find the volume of the solid of revolution.

37)

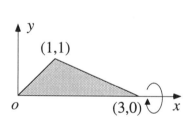

38)

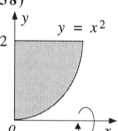

39)

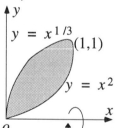

40)

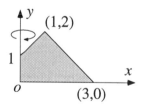

41) **Model Volcano** - A manufacturer of educational science products sells a demonstration volcano. The volcano begins production as a clay cone of height h and base radius r. Then a cylindrical hole of radius a is bored down the center of the cone to represent the volcano's vent. When filled with vinegar and baking soda, the volcano ejects copious amounts of "smoke" to the delight of small school children. Find the volume V of clay material in the model volcano.

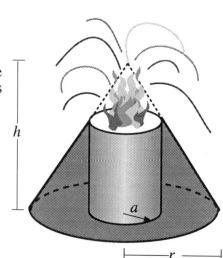

§5.2 Volume Part II

42) **Newton's Rings** - A planar convex lens is formed from a glass sphere of radius R by cutting off a cap of width h. In mathematics, such a shape is called a spherical zone. The planar face of the lens is a circle of radius a which can be expressed in terms of h and R using the relation:

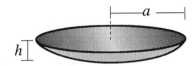

Planar convex lens is an example of a spherical zone.

$$(R-h)^2 + a^2 = R^2$$

Use the method of cylindrical shells to find the volume V of the lens.

Comment: An important use of such a lens is in quality control for manufacturing other optical lenses. Used in this fashion the lens is called the Control Lens. When the control lens is placed on a second lens whose quality is to be tested, circular fringes called Newton's Rings are visible due to interference between the light reflected from the curved surface of the control lens and that of the lens being tested. Only if the tested lens is ground almost flawlessly will the pattern of interference fringes appear perfectly circular.

43) Consider the linear morph of height 10 inches whose bottom surface is a circle of radius 4 inches and whose top surface is given by the polar equation:

$$r = g(\theta) = 3 + \cos(6\theta)$$

Find the volume V of the morph.

44) Consider the linear morph of height 1 meter whose bottom surface is a circle of radius 0.5 meter and whose top surface is an equilateral triangle with each side of length 0.3 meter. Find the volume V of the morph.

45) Consider the morph described by the equation $r_z(\theta) = \frac{(h-z)}{h} f(\theta) + \frac{z}{h} g(\theta)$. In this exercise, we obtain a second formula for the volume V of the morph in terms of its cross-sectional area at the midpoint $z = \frac{h}{2}$.

a) Show that the cross-sectional area at the midpoint $z = \frac{h}{2}$ of the morph is:

$$A(h/2) = \frac{1}{4}\left(A(0) + A(h) + \int_0^{2\pi} f(\theta) \cdot g(\theta)\, d\theta\right)$$

b) Show that the volume of a linear morph can be written:

$$V = \frac{h}{6}\left(A(0) + 4A\left(\frac{h}{2}\right) + A(h)\right)$$

Hint: Combine the expression for the cross-sectional area $A(h/2)$ with our original volume formula:

$$V = \frac{h}{3}\left\{A(0) + A(h) + \frac{1}{2}\int_0^{2\pi} f(\theta) \cdot g(\theta)\, d\theta\right\}$$

46) Lecture Hall

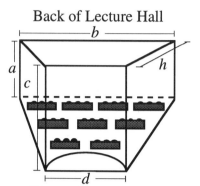

Back of Lecture Hall

Front of Lecture Hall

Many large lecture and concert halls have the following shape. The front of the room on which the board is located is a tall, narrow rectangle. In contrast, the back of the room is a short broad rectangle. Because of the need for each student to be able to see over the head of the students in front, the floor of the room rises towards the back. Find the volume of the lecture room whose dimensions are shown in the illustration. The perpendicular distance between the front and back walls is h. Is the shape of this lecture hall an example of a <u>morph</u>?

(Hint: Consider the formula $V = \frac{h}{6}\left(A(0) + 4A(\frac{h}{2}) + A(h)\right)$ discovered in the previous exercise.)

§ 5.3 Arclength

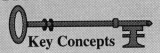

Key Concepts

- Length of Polygonal Paths
- Arclength of Curved Paths
- Arclength in Polar Coordinates
- Arclength in Parametric Coordinates

The length L of the line segment joining two points (x_1, y_1) and (x_2, y_2) in the plane is easy to find using the famous formula of Pythagoras:

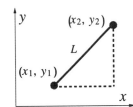

The Theorem of Pythagoras
$$L = \sqrt{(x_2-x_1)^2 + (y_2-y_1)^2}$$

For example, the length of the line segment joining the point $(1,1)$ to the point $(4,5)$ is:

$$L = \sqrt{(4-1)^2 + (5-1)^2} = \sqrt{9 + 16} = 5.$$

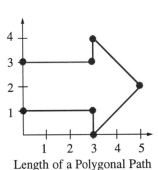

Length of a Polygonal Path

This elegant result from ancient Greece can be extended to find the length of any polygonal path. A <u>polygonal path</u> is a continuous curve composed out of line segments. The total length of a polygonal curve is just the sum of the lengths of each of its individual line segments. For example, the polygonal path shown to the right is composed of six line segments joining the points $(0, 1)$, $(3, 1)$, $(3, 0)$, $(5, 2)$, $(3, 4)$, $(3, 3)$, and $(0, 3)$. Applying the formula of Pythagoras to each of the six line segments in turn we find the total length L of the polygonal path to be:

$$L = 3 + 1 + 2\sqrt{2} + 2\sqrt{2} + 1 + 3 = 4\sqrt{2} + 8.$$

Why we need calculus for arclength.

What is the length of a smooth curve such as an ellipse or a figure eight which is not composed of line segments? What is the length of a coiled wire or the length of a support cable on a suspension bridge? Finding the arclength of curves like these requires the power of integral calculus. To see how calculus can be used to find arclengths for smooth curves, we first consider an example which clarifies what we mean by the length of a smooth curve.

Example 1 - Length of a Snake

A snake has been photographed slithering along the edge of a wire fence. Estimate the length L of the snake from the tip of his nose to the tip of his tail if the wire fencing forms a grid of one inch squares.

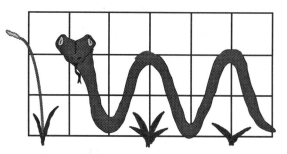

Solution

In this case, we have a clear idea as to what the length of the snake should be. The length of the snake can be measured if we gently and carefully hold the snake by its head and tail, straighten it out, and place it next to a ruler. It can be seen in the figure below that when the snake is straightened out, it has a total length of approximately 11.5 inches.

Numerical estimate of the snake's length.

The above notion of straightening out a curve into a line segment and then measuring the length of the resulting line segment clarifies our intuitive notion of the arclength of a curve. What we need is a way to calculate this arclength for any given curve. To hint at how calculus provides just such a formula we calculate the snake's length by another procedure. Notice how the shape of the snake's body can be <u>approximated</u> as a <u>polygonal path</u> of five line segments connecting the vertices (0,2), (1,0), (2,2), (3,0), (4,2) and (5,0).

Using the Pythagorean formula, $\Delta s = \sqrt{\Delta x^2 + \Delta y^2}$, each of the five line segments is found to have the same length $\Delta s = \sqrt{5}$. Thus the total length of the snake is approximately

$L = 5\Delta s = 5\sqrt{5} \approx 11.2$ inches.

Because of the curves in the snake's body, its true length will actually be a little longer than this estimate. However, the basic idea of obtaining the length of a curve by approximating the curve by a polygonal path will provide a formula for calculating the length of almost any curve.

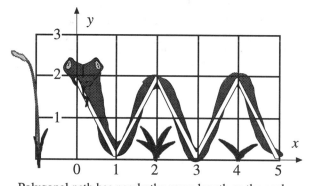

Polygonal path has nearly the same length as the snake.

§5.3 Arclength

Arclength Formula for a Planar Curve.

Consider the curve $y = f(x)$ over the interval $a \leq x \leq b$. Let $P = (x, y)$ and $Q = (x + dx, y + dy)$ denote two points on the curve separated by the infinitesimal displacement (dx, dy). Using the Theorem of Pythagoras, the infinitesimal distance dL between these points is:

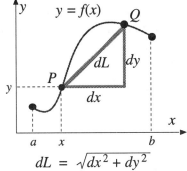

$$dL = \sqrt{dx^2 + dy^2}$$

Infinitesimal Arclength Element
$$dL = \sqrt{dx^2 + dy^2}$$

This expression for the <u>infinitesimal arclength</u> is the fundamental starting place for many arclength calculations. We will now see how it can be used to find the total arclength L of the curve $y = f(x)$ over the interval $[a, b]$. Through a simple <u>algebraic</u> manipulation, the infinitesimal arclength may be expressed in terms of $\dfrac{dy}{dx}$, the derivative of the function y with respect to x.

$$dL = \sqrt{(dx)^2 + (dy)^2} = \sqrt{\left(1 + \left(\frac{dy}{dx}\right)^2\right) dx^2} = \sqrt{1 + \left(\frac{dy}{dx}\right)^2}\, dx$$

Thus, the total arclength L of the curve $y = f(x)$ over the interval $[a, b]$ is:

Arclength Formula for a Planar Curve
$$L = \int dL = \int_a^b \sqrt{1 + \left(\frac{dy}{dx}\right)^2}\, dx$$

Example 2 - Test drive the new formula!

To gain confidence with the new formula, we will test it in a case where we already know the answer. Calculate the length of the line segment $y = 2x$ over the interval $[0, 1]$ using the arclength formula.

Solution

The curve is a line segment extending from the point $(0, 0)$ to the point $(1, 2)$. By the Pythagorean Formula the length of this line segment is

$$L = \sqrt{\Delta x^2 + \Delta y^2} = \sqrt{(1-0)^2 + (2-0)^2} = \boxed{\sqrt{5}}$$

Does our new arclength formula give the same result?

$$L = \int dL = \int_0^1 \sqrt{1 + \left(\frac{dy}{dx}\right)^2}\, dx = \int_0^1 \sqrt{1 + (2)^2}\, dx = \sqrt{5}\, x \Big|_0^1 = \boxed{\sqrt{5}}$$

Of course the answers are the same!

Now we will tryout the new arclength formula on a curve which is not composed of line segments. Without calculus, we would not be able to find its length.

Example 3

Find the arclength L of the curve $y = \frac{2}{3} x^{3/2}$ over the interval $1 \leq x \leq 3$.

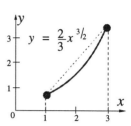

Solution

To use our arclength formula we will need the function's derivative which is $\frac{dy}{dx} = \sqrt{x}$. The arclength formula leads to the integral:

$$L = \int dL = \int_1^3 \sqrt{1 + \left(\frac{dy}{dx}\right)^2}\, dx = \int_1^3 \sqrt{1+x}\, dx$$

We can easily evaluate this integral by making the substitution $u = 1 + x$.

Changing the endpoints and noting that in this case $dx = du$ we find that the arclength is

$$L = \int_1^3 \sqrt{1+x}\, dx = \int_2^4 \sqrt{u}\, du = \left.\frac{2}{3} u^{3/2}\right]_2^4 = \boxed{\frac{16 - 4\sqrt{2}}{3} \approx 3.4477}$$

Checking the Result

Looking at the above graph of our curve, we see that it only slightly deviates from the straight line segment joining the endpoints of the curve. Thus, the calculated arclength should be just a little more than the length of the line segment joining the endpoints $\left(1, \frac{2}{3}\right)$ and $(3, 2\sqrt{3})$ of our curve. Using the formula of Pythagoras we find that the length of the line segment to four decimals is 3.4388 which is indeed just a little less than 3.4477

Example 4

Find the arclength L of the curve defined by $y = \frac{1}{6}x^3 + \frac{1}{2x}$ on the interval $\left[\frac{1}{2}, 2\right]$.

Solution

The curve is plotted to the right. To use the arclength formula we first find the derivative of y.

$$\frac{dy}{dx} = \frac{1}{2}x^2 - \frac{1}{2x^2} = \frac{1}{2}\left(x^2 - \frac{1}{x^2}\right)$$

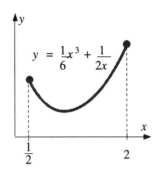

Inserting $\frac{dy}{dx}$ into the arclength formula yields:

$$L = \int_{1/2}^{2} \sqrt{1 + \left(\frac{dy}{dx}\right)^2}\, dx = \int_{1/2}^{2} \sqrt{1 + \frac{1}{4}\left(x^2 - \frac{1}{x^2}\right)^2}\, dx$$

Amazingly, the term under the square root sign can be written as a <u>perfect square</u>!

$$1 + \frac{1}{4}\left(x^2 - \frac{1}{x^2}\right)^2 = 1 + \frac{1}{4}\left(x^4 - 2 + \frac{1}{x^4}\right) = \frac{1}{4}\left(x^2 + \frac{1}{x^2}\right)^2$$

This enables us to remove the square root to obtain:

$$L = \int_{1/2}^{2} \sqrt{\frac{1}{4}\left(x^2 + \frac{1}{x^2}\right)^2}\, dx = \frac{1}{2}\int_{1/2}^{2} \left(x^2 + \frac{1}{x^2}\right) dx = \frac{1}{2}\left(\frac{1}{3}x^3 - \frac{1}{x}\right)\Bigg|_{1/2}^{2} = \boxed{\frac{33}{16}}$$

The curve is a little over two units long.

ALERT

The next problem requires the challenging integration:

$$\int \sqrt{1 + x^2}\, dx = \frac{x}{2}\sqrt{1 + x^2} + \frac{1}{2}\sinh^{-1}x + C$$

This can be found in tables or discovered by making the substitution $u = \sinh x$. Except for this tough integral, the problem is just a straight forward illustration of the arclength formula in an engineering context.

Example 5 - Bridge Repair

Many old bridges and arches were built with parabolic cross-sections, if for no other reason than that the form is so pleasing to the eye. Unfortunately, the great age of these beautiful structures naturally leads to significant crumbling and falling of debris. A steel net is to be placed along the underside of the arch illustrated in the picture to prevent serious injury due to falling debris. What length of net will be needed?

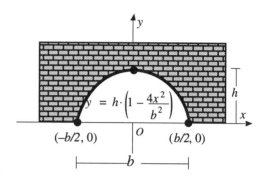

Solution

The parabolic arch is defined by the equation $y = h\left(1 - \frac{4x^2}{b^2}\right)$.

We will need the function's derivative which is $\frac{dy}{dx} = -\left(\frac{8h}{b^2}\right)x$.

Inserting $\frac{dy}{dx}$ into the arclength formula leads to the challenging integral:

$$L = \int dL = 2\int_0^{b/2} \sqrt{1 + \left(\frac{dy}{dx}\right)^2}\, dx = 2\int_0^{b/2} \sqrt{1 + \left(\frac{8h}{b^2}\right)^2 x^2}\, dx$$

Notice we used the symmetry of the arch to write its total length as twice the length of the right side only. We can reduce this integral to the form mentioned in the alert before this problem with the substitution $u = \left(\frac{8h}{b^2}\right)x$.

Changing the endpoints and noting that $du = \left(\frac{8h}{b^2}\right)dx$ we obtain:

$$L = \frac{b^2}{4h}\int_0^{4h/b} \sqrt{1 + u^2}\, du$$

From tables of integrals we know that $\int \sqrt{1 + x^2}\, dx = \frac{x}{2}\sqrt{1 + x^2} + \frac{1}{2}\sinh^{-1} x$.

Thus the length of the parabolic arch is:

$$L = \frac{b^2}{4h}\left(\frac{2h}{b}\sqrt{1 + \left(\frac{4h}{b}\right)^2} + \frac{1}{2}\sinh^{-1}\left(\frac{4h}{b}\right)\right) = \boxed{\frac{b}{2}\sqrt{1 + \left(\frac{4h}{b}\right)^2} + \frac{b^2}{8h}\sinh^{-1}\left(\frac{4h}{b}\right)}$$

The answer looks a little complicated, but in the case of an arch where the base is four times the height ($b = 4h$) it simplifies to $L = 2h\left(\sqrt{2} + \sinh^{-1}(1)\right)$.

§5.3 Arclength

Modified Arclength Formula $\{x = g(y)\}$

In certain situations, it is better to consider x as a function of y rather than the more usual convention where y is a function of x. If the curve has the equation $x = g(y)$ over the interval $c \leq y \leq d$, then by interchanging the roles of x and y in our previous arclength formula we obtain the following expression for its length.

$$L = \int_c^d \sqrt{1 + \left(\frac{dg}{dy}\right)^2}\, dy = \int_c^d \sqrt{1 + \left(\frac{dx}{dy}\right)^2}\, dy$$

Example 6

Find the length of the curve $x = \dfrac{y^3}{3} + \dfrac{1}{4y}$ over the interval $1 \leq y \leq 3$.

Solution

Notice that it would be difficult to express y as a function of x. It is easier to consider x as a function of y and use the arclength formula. Since $x = \dfrac{y^3}{3} + \dfrac{1}{4y}$, we see that $\dfrac{dx}{dy} = y^2 - \dfrac{1}{4y^2}$. Thus the arclength is

$$L = \int_1^3 \sqrt{1 + \left(\frac{dx}{dy}\right)^2}\, dy = \int_1^3 \sqrt{1 + \left(y^2 - \frac{1}{4y^2}\right)^2}\, dy = \int_1^3 \sqrt{\left(y^2 + \frac{1}{4y^2}\right)^2}\, dy$$

$$= \int_1^3 \left(y^2 + \frac{1}{4y^2}\right) dy = \left(\frac{y^3}{3} - \frac{1}{4y}\right)\Bigg|_1^3 = \left(9 - \frac{1}{12}\right) - \left(\frac{1}{3} - \frac{1}{4}\right) = \boxed{8\,\tfrac{5}{6}}$$

Arclength in Polar Coordinates

Although the arclength formula is very simple to visualize geometrically, it is often very difficult to calculate the resulting integral because of the square root term in the integrand. Occasionally, it may be easier to evaluate arclength using polar coordinates.

Consider a polar curve $r = r(\theta)$ over the interval $\alpha \le \theta \le \beta$. The arclength for this polar curve is easy to derive geometrically from the adjacent figure.

Consider two points $P = (r, \theta)$ and $Q = (r+dr, \theta+d\theta)$ on the polar curve which subtends an infinitesimal angle $d\theta$. The length dL between P and Q corresponds to the hypotenuse of the infinitesimal right-angled triangle shown. One edge of the triangle has length dr and the other has length $rd\theta$. The Pythagorean identity reveals

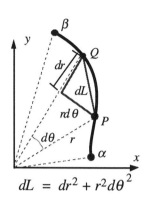

$$dL = \sqrt{dr^2 + r^2 d\theta^2} = \sqrt{\left(\frac{dr}{d\theta}\right)^2 + r^2}\; d\theta$$

$$dL = \sqrt{dr^2 + r^2 d\theta^2}$$

Since the angle θ varies over the interval $\alpha \le \theta \le \beta$, the total arclength L of the polar curve is:

Arclength Formula for a Polar Curve

$$L = \int dL = \int_\alpha^\beta \sqrt{\left(\frac{dr}{d\theta}\right)^2 + r^2}\; d\theta$$

To gain confidence with this new formula, let us apply it in a case where we know the answer.

Example 7 - Circumference of a Circle.

Show that the circumference of a circle of radius R is $C = 2\pi R$ using the arclength formula in polar coordinates.

Circle of radius R.

Solution

The polar equation for a circle of radius R is simply $r = R$.
Since the radius is constant, its derivative is zero.
According to the polar arclength formula, the circumference should be

$$C = \int_0^{2\pi} \sqrt{r^2 + \dot{r}^2}\; d\theta = \int_0^{2\pi} \sqrt{R^2 + 0^2}\; d\theta = \int_0^{2\pi} R\, d\theta = \boxed{2\pi R}$$

Thus $C = 2\pi R$ as expected!

§5.3 Arclength

The next example is a simple engineering application of the arclength formula in polar coordinates.

Example 8 - Arclength of a cardioid

Cardioid-shaped cams are used in exercise equipment to create a smooth balanced resistance during each exercise set. A thin plastic band is wrapped around the perimeter of the metal cam as protection against the sharp edges introduced during manufacture.

Find the length L of the safety band required if the cam is described by the cardioid $r = 1 + \cos\theta$.

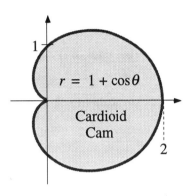

Solution

From geometry, the cardioid is symmetric with respect to the x-axis, hence we may write,

$$L = \int_0^{2\pi} \sqrt{r^2 + \left(\frac{dr}{d\theta}\right)^2}\, d\theta = 2\int_0^{\pi} \sqrt{r^2 + \left(\frac{dr}{d\theta}\right)^2}\, d\theta$$

Inserting $r = 1 + \cos\theta$ and $\frac{dr}{d\theta} = -\sin\theta$ we obtain

$$L = 2\int_0^{\pi} \sqrt{(1 + \cos\theta)^2 + (-\sin\theta)^2}\, d\theta = 2\int_0^{\pi} \sqrt{(1 + 2\cos\theta + \cos^2\theta) + \sin^2\theta}\, d\theta$$

$$= 2\int_0^{\pi} \sqrt{1 + (\cos^2\theta + \sin^2\theta) + 2\cos\theta}\, d\theta = 2\int_0^{\pi} \sqrt{2(1 + \cos\theta)}\, d\theta$$

Since $1 + \cos\theta = 2\cos^2\left(\frac{\theta}{2}\right)$ we have,

$$L = 2\int_0^{\pi} \sqrt{4\cos^2\left(\frac{\theta}{2}\right)}\, d\theta = 4\int_0^{\pi} \left|\cos\left(\frac{\theta}{2}\right)\right| d\theta \quad \text{since} \quad \left\{\sqrt{\cos^2\frac{\theta}{2}} = \left|\cos\frac{\theta}{2}\right|\right\}.$$

We can then obtain:

$$L = 4\int_0^{\pi} \cos\left(\frac{\theta}{2}\right) d\theta \quad \text{since} \quad \left\{\cos\left(\frac{\theta}{2}\right) \text{ is positive on } [0, \pi] \text{ so that} \left|\cos\left(\frac{\theta}{2}\right)\right| = \cos\left(\frac{\theta}{2}\right)\right\}$$

$$= \boxed{8\sin\left(\frac{\theta}{2}\right)\Big|_0^{\pi} = 8}$$

Thus, the protective band must be 8 units long.

Archimedean Spiral

A spiral curve that arises in many engineering contexts is the Archimedean Spiral. In polar coordinates, the equation of this spiral is $r = a + b\theta$. Each successive loop of the spiral is the same distance $d = \dfrac{b}{2\pi}$ from the adjacent loops. The spiral arises in many situations where material is rolled onto a cylinder. Such examples include rolls of sheet metal, newsprint and audio tape.

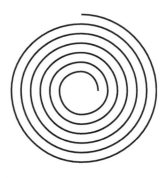

An Archimedean Spiral

Example 9 - Length of an Archimedean Spiral

Find the length of n complete loops of the Archimedean spiral $r = \theta$. Assume the first loop is the portion of the spiral for $0 \le \theta \le 2\pi$, the second loop is the portion for $2\pi \le \theta \le 4\pi$ and so on.

Solution

The angle θ increases by 2π radians for each complete loop. Thus the arclength of n complete loops is

$$L = \int dL = \int_0^{2\pi n} \sqrt{r^2 + \left(\frac{dr}{d\theta}\right)^2}\, d\theta = \int_0^{2\pi n} \sqrt{\theta^2 + 1}\, d\theta$$

Now this is the <u>same</u> nasty integral we encountered in the problem with the parabolic arch! Recalling the result $\int \sqrt{1+x^2}\, dx = \dfrac{x}{2}\sqrt{1+x^2} + \dfrac{1}{2}\sinh^{-1}x + C$, we find that the length of n complete loops is:

$$\boxed{L = n\pi\sqrt{1 + 4n^2\pi^2} + \frac{1}{2}\sinh^{-1}(2n\pi)}$$

When the number of loops n is large, it can be shown that the first term in the above example is much larger than the inverse sinh term so that the total arclength is about

$$L = n\pi\sqrt{1 + 4n^2\pi^2} \approx \boxed{2\pi^2 n^2}$$

Example 10 - Mosquito Coils

An interesting application of the Archimedean spiral is seen in the design of mosquito coils. These coils are lit at one end and slowly burn over a period of several hours releasing chemicals that repel and kill pesky mosquitoes.

One brand of mosquito coil is designed as an Archimedean spiral given by the curve $r = b\theta$ centimeters with $n = 4$ complete loops. (Assume $b = \frac{1}{2\pi}$ so that subsequent loops are 1 centimeter apart.)

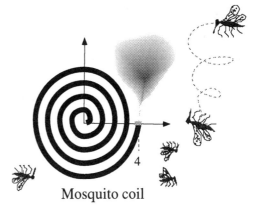

Mosquito coil

If this particular brand of mosquito coil burns at the rate of 5 centimeters per hour, how long before it is all burned and another has to be lit?

Solution

Using the arclength formula in polar coordinates we find:

$$L = \int_0^{2\pi n} \sqrt{r^2 + \left(\frac{dr}{d\theta}\right)^2}\, d\theta = \int_0^{2\pi n} \sqrt{(b\theta)^2 + b^2}\, d\theta = b\int_0^{2\pi n} \sqrt{\theta^2 + 1}\, d\theta$$

The integral is the same as in the previous example except it is multiplied by b! Using the previous result:

$$L = b\left(n\pi\sqrt{1 + 4n^2\pi^2} + \frac{1}{2}\sinh^{-1}(2n\pi)\right)$$

Substituting in the values $n = 4$ and $b = \frac{1}{2\pi}$ we find the length of the coil to be:

$$\boxed{L = 2\sqrt{1 + 64\pi^2} + \frac{1}{4\pi}\sinh^{-1}(8\pi)}$$

Evaluating this out to two decimal points with a calculator yields $L \sim 50.62$ centimeters. Since the coils burns about 5 centimeters per hour, it should last a little over ten hours.

Arclength of a Parametric Curve

Consider the curve represented by the parametric equations

$$x = x(t), \quad y = y(t), \quad a \leq t \leq b$$

How can we find the arclength L of this curve? Consider two points P and Q on the curve separated by the infinitesimal displacement (dx, dy). The infinitesimal length dL between the points P and Q is:

$$dL = \sqrt{dx^2 + dy^2}$$

Since x and y are functions of t, we have:

$$dx = \dot{x}(t)\,dt, \qquad dy = \dot{y}(t)\,dt$$

where $\dot{x}$ and $\dot{y}$ denote the derivatives of x and y with respect to t. Hence,

$$dL = \sqrt{dx^2 + dy^2} = \sqrt{(\dot{x}\,dt)^2 + (\dot{y}\,dt)^2} = \sqrt{\dot{x}^2 + \dot{y}^2}\,dt.$$

We obtain the total arclength L of the parametric curve over the interval $a \leq t \leq b$ by integrating this infinitesimal length.

Arclength Formula for a Parametric Curve

$$L = \int_a^b dL = \int_a^b \sqrt{\dot{x}^2 + \dot{y}^2}\,dt$$

Example 11

Find the arclength of the parametric curve

$$x = t^3, \quad y = t^2, \quad 0 \le t \le 2$$

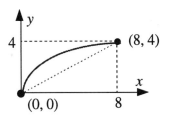

Solution

We will need the derivatives $\dot{x} = 3t^2$ and $\dot{y} = 2t$.
Inserting these into the arclength formula in parametric coordinates we find:

$$L = \int_0^2 \sqrt{\dot{x}^2 + \dot{y}^2}\, dt = \int_0^2 \sqrt{9t^4 + 4t^2}\, dt = \int_0^2 t\sqrt{9t^2 + 4}\, dt$$

To evaluate the above integral we use the substitution $u = 9t^2 + 4$ so that $du = 18t\,dt$.
Notice that when $t = 0$, $u = u(0) = 4$ and when $t = 2$, $u = u(2) = 40$.

$$L = \frac{1}{18}\int_0^2 \sqrt{9t^2 + 4}\,(18t\,dt) = \frac{1}{18}\int_4^{40} \sqrt{u}\,(du)$$

$$= \frac{1}{27} u^{3/2} \bigg]_4^{40} = \boxed{\frac{1}{27}\left(40^{3/2} - 4^{3/2}\right) \approx 9.0734}$$

As a check, we note that L should be a little larger than the distance D between the endpoints $(0, 0)$ and $(8, 4)$ of the curve.

$$D = \sqrt{8^2 + 4^2} = \boxed{\sqrt{80} \approx 8.9}$$

Indeed 8.9 is a little less that 9.0734.

WARMUP EXERCISES

1) Estimate the length L of each object by aligning a piece of string along the object and then straightening out the string. Assume the objects were drawn on a background grid of 1 inch squares before being reduced in size. Compare your estimate for the length of the circle with the exact answer of 4π. (When aligning the string be careful not to stretch it!)

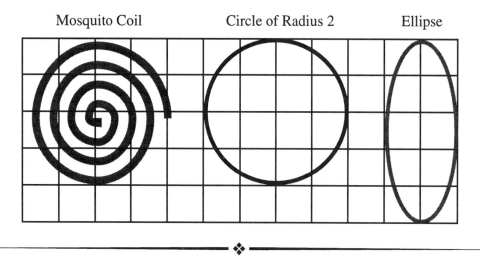

Mosquito Coil   Circle of Radius 2   Ellipse

In the following exercises, set up, but do <u>not</u> evaluate, integrals for the lengths of the given curve.

2) $y = \frac{x^3}{3}$, $0 \leq x \leq 2$ 3) $y = 1/x$, $1 \leq x \leq 10$

4) $y = \sin x$, $0 \leq x \leq \pi$ 5) $y = \tan x$, $0 \leq x \leq \frac{\pi}{4}$

6) $y = \frac{1}{x^2}$, $0 \leq x \leq 2$ 7) $y = \ln^2(x)$, $\frac{1}{2} \leq x \leq 1$

8) $y = e^x \sin x$, $0 \leq x \leq \pi$ 9) $y = e^x$, $0 \leq x \leq 1$

In the following exercises, set up, but do <u>not</u> evaluate, integrals for the lengths of the given curve. Use the formula for arclength in either polar or parametric form.

10) **Hyperbolic Spiral:** $r = \frac{1}{\theta}$ $\frac{1}{2} \leq \theta \leq 2$

11) **Four-Leaf Rose:** $r = \cos(2\theta)$ $0 \leq \theta \leq 2\pi$

12) **Limaçon:** $r = 1 - 2\cos(\theta)$ $0 \leq \theta \leq 2\pi$

13) **Three - Leaf Rose:** $r = \sin(3\theta)$ $0 \leq \theta \leq \pi$

§5.3 Arclength — CAFÉ — Page 419

14) Ellipse: $\begin{cases} x = 2\cos(\theta) \\ y = \sin(\theta) \end{cases}$ $\quad 0 \leq \theta \leq 2\pi$

15) Hyperbola: $\begin{cases} x = \cosh(t) \\ y = \sinh(t) \end{cases}$ $\quad 0 \leq t \leq 1$

16) Lissajous Curve: $\begin{cases} x = \sin(5t) \\ y = \cos(8t) \end{cases}$ $\quad 0 \leq t \leq 2\pi$

17) Lissajous Curve: $\begin{cases} x = \sin(t) \\ y = \sin(2t) \end{cases}$ $\quad 0 \leq t \leq 2\pi$

INTERMEDIATE EXERCISES

Use the appropriate arclength formula to find the length of following curves. Now you must evaluate the integrals.

18) $y = \frac{2}{3} x^{3/2}$, $\quad 3 \leq x \leq 24$

19) $y = \frac{4}{5} x^{5/4}$ $\quad 1 \leq x \leq 10$
(Hint: You will have to make the substitution $u = \sqrt{x}$.)

20) Figure 8 $\quad r = \cos^2\theta \quad 0 \leq \theta \leq 2\pi$

21) Spiral: $\quad r = \dfrac{\theta^2}{2} \quad 0 \leq \theta \leq 2$

22) Cardioid: $\quad r = 1 + \sin\theta \quad 0 \leq \theta \leq 2\pi$

23) Logarithmic Spiral: $\quad r = e^{a\theta}, a \neq 0 \quad 0 \leq \theta \leq 2\pi$

24) Circle: $\quad r = \cos\theta \quad -\dfrac{\pi}{2} \leq \theta \leq \dfrac{\pi}{2}$

25) Archimedean Spiral: $\quad r = a + b\theta$, $b \neq 0 \quad 0 \leq \theta \leq 2\pi$

26) Find the arclength of one arch of the cycloid. Here r is a constant.

$\begin{cases} x = r(\theta - \sin(\theta)) \\ y = r(1 - \cos(\theta)) \end{cases}$ $\quad 0 \leq \theta \leq 2\pi$

27) Find the total arclength of the astroid.

$\begin{cases} x = a\cos^3(\theta) \\ y = a\sin^3(\theta) \end{cases}$ $\quad 0 \leq \theta \leq 2\pi$

28) Find the arclength of the curve.

$\begin{cases} x = e^t\cos(t) \\ y = e^t\sin(t) \end{cases}$ $\quad 0 \leq t \leq \pi$

29) Find the arclength of the Cornu's Spiral.

$\begin{cases} x = \int_0^t \cos(u^2)\, du \\ y = \int_0^t \sin(u^2)\, du \end{cases}$ $\quad 0 \leq t \leq 30$

30) Show that the curves $y = f(x)$ and $y = f(x) + C$ have the same arclength over any interval $[a, b]$.

31) Is the arclength over the interval $[a, b]$ of the curve $y = 2f(x)$ double the arclength of the curve $y = f(x)$ over the same interval?

32) **Infinitesimal Arclength in Polar Coordinates**

In deriving the formula for arclength in polar coordinates we arrived at the expression $dL = \sqrt{(dr)^2 + r^2(d\theta)^2}$ by geometric arguments. The same result can be obtained by the following algebraic approach.

a) Use the identities $x = r\cos\theta$ and $y = r\sin\theta$ to show that

$$dx = dr\cos\theta - r\sin\theta \, d\theta \quad \text{and} \quad dy = dr\sin\theta + r\cos\theta \, d\theta$$

b) Starting with the identity $dL = \sqrt{dx^2 + dy^2}$, show that
$$dL = \sqrt{(dr)^2 + r^2(d\theta)^2}.$$

Perfect Squares

The arclength formula usually leads to challenging integrations because of the troublesome square root in the integrand. In the following special cases however, the integration can be performed because the term $1 + \left(\dfrac{dy}{dx}\right)^2$ is a perfect square, allowing the square root to be canceled. Find the arclength of the given curve.

33) $y = \dfrac{x^2}{2} - \dfrac{\ln x}{4}$ $1 \le x \le 2$ 34) $y = \dfrac{x^3}{3} + \dfrac{1}{4x}$ $1 \le x \le 2$

35) $y = \dfrac{x^4}{4} + \dfrac{1}{8x^2}$ $2 \le x \le 4$ 36) $y = \dfrac{x^5}{5} + \dfrac{1}{12x^3}$ $2 \le x \le 4$

37) Verify that if y is the average of two functions, $y = \dfrac{f(x) + g(x)}{2}$, satisfying $\dfrac{df}{dx}\dfrac{dg}{dx} = -1$, then the term $1 + \left(\dfrac{dy}{dx}\right)^2$ is a perfect square. The arclength over the interval $[a, b]$ becomes in this case $L = \displaystyle\int_a^b \left|\dfrac{f'(x) - g'(x)}{2}\right| dx$. Check that this is how the last four exercises were generated and then try to find your own examples of curves such that the term $1 + \left(\dfrac{dy}{dx}\right)^2$ is a perfect square.

38) $y = \dfrac{1}{2}\left(\ln x - \dfrac{x^2}{2}\right)$ $\dfrac{1}{2} \le x \le 2$ 39) $y = \dfrac{1}{2x^2} + \dfrac{x^4}{16}$ $\dfrac{1}{2} \le x \le 2$

§5.3 Arclength   CAFÉ   Page 421

40) Another example of the perfect square scenario described above results from the problem of hanging cables. It can be shown that a uniform cable assumes a shape described by the equation $y = \dfrac{e^x + e^{-x}}{2}$. Find the length of such a cable over the interval $-1 \leq x \leq 1$.

Using Simpson's rule with $n = 20$, calculate the arc length of each curve over the given interval.

41) $y = \tan x$ $0 \leq x \leq \dfrac{\pi}{4}$ **42)** $y = \sinh x$ $0 \leq x \leq 1$

43) $y = x^4$ $0 \leq x \leq 1$ **44)** $y = \dfrac{\sin x}{x}$ $0 \leq x \leq \pi$, $y(0) = 1$

45) A sine arch $y = \sin x$ $0 \leq x \leq \pi$

46) An ellipse with major axis 2 and minor axis 1.

47) Find the arclength of one leaf in a 3-leaf rose $r = 2\cos(3\theta)$.
To generate one leaf, we may use $\dfrac{-\pi}{6} \leq \theta \leq \dfrac{\pi}{6}$.

48) $\begin{cases} x = 2\sin(4t) \\ y = 3\cos(5t) \end{cases}$ $0 \leq t \leq 2\pi$

49) $\begin{cases} x = t^3 \\ y = t^4 \end{cases}$ $0 \leq t \leq 1$

50) An electric heater is constructed from nichrome wire of resistance 4 ohms per meter. Thus, if L denotes the length of nichrome used, the total resistance R of the heating element is $R = 4L$ ohms. The coil is bent in the shape of the Archimedean spiral
$r = a + b\theta$ where a and b are constants.
  a) Find a and b using the dimensions given in the illustration.

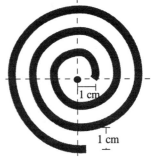

  b) The heater is designed to deliver 1500 watts of power P when connected across a voltage $V = 110$ volts. Using $P = \dfrac{V^2}{R}$, find the required length L of nichrome wire.
  c) Estimate to the nearest integer the number of loops in the heating element.

51) A high voltage transmission line is supported between two towers which are 1000 meters apart on level terrain. The transmission line sags by 20 meters at its midpoint. The shape of the transmission line is a <u>catenary</u> which is defined by the equation $y = \frac{1}{k}\cosh(kx) + C$, where k and C are constants. Calculate the length L of the transmission line.
Use Newton's method to calculate k to an accuracy of 4 decimal places.
[Hint: Choose the lowest point of the transmission line as the origin.]

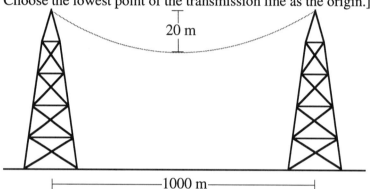

52) The Golden Gate Bridge in San Francisco has a main span of 4200 feet. The support cable is approximately <u>parabolic</u> in shape because of the heavy uniform static loading of the bridge. Calculate the length L of the support cable. [Hint: Choose the lowest point of the cable as the origin and assume the shape of the cable to be $y = ax^2$. Use the given geometrical configuration to determine the parameter a.]

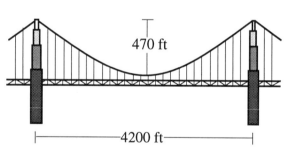

§ 5.4 Area of Surfaces of Revolution

Key Concepts

❖ Area of Conical Frustums   ❖ Area of Surfaces of Revolution

The calculation of the surface area of an object is important in many engineering applications. The surface area reveals a lot of important information to the engineer, such as how much material is needed for a coating or for packaging, and the efficiency or performance of the object. For example, in the case of a high speed aircraft, the surface area of the wing is made as small as possible to reduce drag, thus allowing the wing to "slice" through the air.

Surface Area without Calculus
To illustrate the concept of surface area, we begin with a few simple examples which do not require calculus.

Example 1 - Cylinder

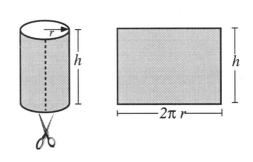

Consider first a hollow, open ended cylinder of height h and base radius r. The lateral surface area A of the cylinder is the same as the area of the rectangle formed by cutting the cylinder as shown. Since the rectangle has height h and base $2\pi r$ (the circumference of the circle), we see that the area of the cylinder is:

When cut, the cylinder can be flattened into a rectangle having the same area.

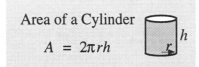

Area of a Cylinder
$$A = 2\pi r h$$

Example 2 - Cone

Consider a cone of base radius r and height h. The cone is open at the base so that is does not have a bottom surface. By the theorem of Pythagoras, the length of its slanted edge is $L = \sqrt{r^2 + h^2}$.

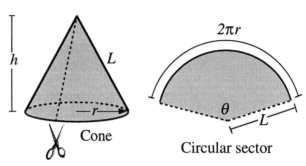

Cone Circular sector

Cutting the cone along a line as shown, we can form a circular sector of radius L subtending an angle of $\theta = \dfrac{2\pi r}{L}$. The surface area A of the cone is the same as the planar area of the sector.

Since the area of the circular sector is $A = \dfrac{\theta L^2}{2}$ we see that the surface area of the cone is:

$$A = \frac{\theta L^2}{2} = \left(\frac{2\pi r}{L}\right)\frac{L^2}{2} = \boxed{\pi r L}$$

> **Surface Area of a Cone**
>
> $A = \pi r L$

Example 3 - Teepee

How much buffalo hide is used in the construction of a teepee having height $h = 12$ feet and base radius $r = 5$ feet? Assume no hide is used for the floor of the teepee.

Solution

The teepee can be idealized as a cone.
The length of the slanted edge is

$$L = \sqrt{r^2 + h^2} = \sqrt{5^2 + 12^2} = 13 \text{ ft}$$

Thus the total area A of buffalo hide required is: $\boxed{A = \pi r L = 65\pi \text{ ft}^2}$

Lateral Surface Area of a Conical Frustum

Definition of a Conical Frustum
A conical frustum is a truncated cone or a piece of a cone.
(The word 'frustum' is a Latin word denoting a piece or portion.)

(In this section we only consider right circular cones and frustums.)

Consider the conical <u>frustum</u> shown in the labeled figure. Its top defines a small circle of radius r and its bottom defines a larger circle of radius R. The length of the slanted edge is L. The figure illustrates how the frustum can be extended upwards so as to form a complete cone. Note that the extension is itself a cone with radius r. We will denote this smaller cone's slant length by l. The illustration shows how the surface area A of the frustum can be found by subtracting the areas of these two cones giving:

$$A = \pi R(L+l) - \pi r l$$

$$A \quad = \quad \pi R(L+l) \quad - \quad \pi r l$$

We would like to eliminate the slant length l from this equation because it refers to the small cone instead of the frustum. To eliminate the length l, we note by similar triangles that $\frac{l}{r} = \frac{l+L}{R}$. Thus $l = \frac{rL}{R-r}$. Substituting this value for l into the formula for A we find:

$$A = \pi R(L+l) - \pi r l = \pi R L + \pi (R-r)l = \pi R L + \pi r L = 2\pi \frac{R+r}{2} L$$

If we let $\bar{r} = \frac{R+r}{2}$ denote the <u>average radius</u> of the frustum, then its area is: $A = 2\pi \bar{r} L$
Thus the area of a conical frustum is 2π times the product of its average radius and its slant length.

Lateral Surface Area of a Conical Frustum

$$A = 2\pi \bar{r} L$$

Don't Forget

Remember. <u>Lateral</u> surface area means only the area of the sides and does not include the area of any horizontal top or bottom surfaces.

We will soon see that this formula for the area of a conical frustum is the key to discovering an integral formula to find the surface area of many solids of revolution using calculus. Since it is so important, let us stop to do a quick exercise to be sure we understand how it works.

Exercise 4 - Conical Frustum

A cone of base radius 10 cm and height 24 cm is cut at the halfway point forming both a smaller cone (on top) and a conical frustum (on bottom). How many times larger is the lateral surface area of the frustum than that of the conical cap?

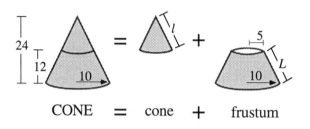

CONE = cone + frustum

Solution

The length of the slanted edge of the original large cone is $\sqrt{10^2 + 24^2} = 26$.
The circular cross-section at the halfway point has radius $r = 5$.
Since the cone is cut at the halfway point; $L = l = 13$, where L and l refer to the slanted edges of the frustum and the smaller cone respectively.

Since the smaller cone has radius $r = 5$ and slanted edge of length $l = 13$, its surface area is $A_{cone} = \pi r l = 65\pi$.

Since the frustum has average radius $\bar{r} = \dfrac{5 + 10}{2} = 7.5$ and slanted edge of length $L = 13$, its surface area is $A_{frustum} = 2\pi \bar{r} L = 195\pi$.

Thus the lateral surface area of the frustum is <u>three</u> times that of the conical cap!
$$195\pi = 3 \cdot (65\pi)$$

Can you prove this for a right circular cone of base radius R and slanted edge of length L that has been cut at the halfway point?

Finding Surface Areas using Calculus

Although we have been able to find the surface areas A of simple shapes like cylinders, cones and frustums without using calculus, there are many situations where one must resort to the power of integral calculus. A food processing engineer must know the surface area of a beverage bottle since this determines the amount of plastic or glass to be used in its production. An aerospace engineer may need to know the surface area of the space shuttle's external tank to calculate the weight of an applied paint. (Originally, this tank used to be painted white. Since the paint added considerable weight to the shuttle, this practice was soon abandoned.) Using calculus, we will be able to calculate the surface area of many different shapes. The formula from elementary geometry that we need along the way is none other than our formula for the area of a conical frustum!

Surfaces of Revolution

Consider a nonnegative function $y = f(x)$ defined over an interval $a \leq x \leq b$. If this curve is revolved about the x-axis, it traces out a surface in space called a <u>surface of revolution</u>. Many engineering objects can be considered as surfaces of revolution. Examples include ellipsoidal storage tanks and objects produced on a lathe. The figure below shows a surface of revolution obtained by this process of rotating a curve about the x-axis.

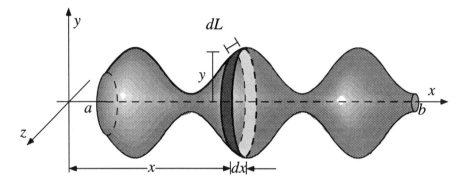

Surface of Revolution

To find the area A of this surface we consider an infinitesimal band of the surface located between x and $x + dx$. This band can be viewed as an <u>infinitesimal conical frustum</u>. Recall that the surface area of a conical frustum is 2π times the product of its average radius and its slant length. The infinitesimal band highlighted in the figure has average radius y and a slanted edge of length $dL = \sqrt{dx^2 + dy^2}$. Thus its infinitesimal area is:

$$dA = 2\pi y \, dL$$

You may recall that the infinitesimal length dL was encountered before in our treatment of arclength.

There we saw that this quantity could be written $\qquad dL = \sqrt{dx^2 + dy^2}$.

Factoring dx out of the radical we get: $\qquad dL = \sqrt{1 + \left(\dfrac{dy}{dx}\right)^2} \, dx$

Since the integration variable x ranges from a to b (see above illustration) the surface area is:

Area of a Surface of Revolution

$$A = \int_a^b 2\pi y \sqrt{1 + \left(\frac{dy}{dx}\right)^2} \, dx$$

Don't Forget

Note the above formula only applies to a surface generated by rotation around the x-axis.

Test Driving the New Formula

To gain confidence with this new formula, let's use it to find the surface area of an object whose area we already know. A cone provides a good test since we found the surface area of a cone without using calculus in the beginning of this section.

Example 5 - Surface Area of Cone

Let A denote the surface area of a cone of length H and base radius R. By slicing the cone into a circular sector, we showed at the beginning of this section that:

$$A = \pi R L$$

where $L = \sqrt{H^2 + R^2}$ is the length of the slanted edge of the cone. Rederive this result using the integral surface area formula.

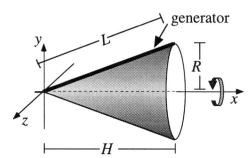

Solution

The cone can be generated by rotating the line $y = \dfrac{R}{H} x$ about the x-axis over the interval $0 \leq x \leq H$. Using the surface area formula we find that:

$$A = \int_0^H 2\pi y \sqrt{1 + \left(\dfrac{dy}{dx}\right)^2}\, dx = \int_0^H 2\pi \left(\dfrac{R}{H} x\right) \sqrt{1 + \dfrac{R^2}{H^2}}\, dx$$

$$= 2\pi \dfrac{R \sqrt{H^2 + R^2}}{H^2} \int_0^H x\, dx = \pi R \sqrt{H^2 + R^2}$$

Thus the surface area of the cone is indeed $A = \pi R \sqrt{H^2 + R^2} = \boxed{\pi R L}$

Example 6 - Surface Area of Sphere

Using the surface area formula, verify that the surface area A of a sphere of radius R is $A = 4\pi R^2$.

Solution

A sphere of radius R can be considered as the surface of revolution generated by rotating the curve $y = \sqrt{R^2 - x^2}$ about the x-axis over the interval $-R \leq x \leq R$.

To calculate the surface area we will need the derivative $\dfrac{dy}{dx} = \dfrac{-x}{\sqrt{R^2 - x^2}}$.

Inserting this into the integral for surface area we find:

$$A = \int_{-R}^{R} 2\pi y \sqrt{1+\left(\frac{dy}{dx}\right)^2}\, dx = \int_{-R}^{R} 2\pi \sqrt{R^2 - x^2} \sqrt{1 + \frac{x^2}{R^2 - x^2}}\, dx$$

$$= \int_{-R}^{R} 2\pi R\, dx = \boxed{4\pi R^2}$$

Notice the integrand reduces to a constant in this case.

Example 7 - The Golden Sphere and the Pirates.

In the last example we saw that the surface area integrand for a sphere reduced to a constant since $2\pi y \sqrt{1+\left(\frac{dy}{dx}\right)^2} = 2\pi R$.
This constancy has a wonderful application to the problem of the golden sphere and the pirates. A <u>hollow</u> sphere of diameter $D = 2R$ is made of pure gold. The N pirates who 'found' it wish to divide it evenly amongst themselves. Show that if the sphere is cut into bands having the same width $\Delta x = \frac{D}{N}$, then each pirate will receive the same amount of gold. (Assume that the sphere is very thin.)

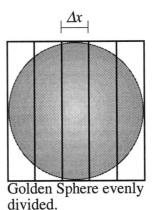

Golden Sphere evenly divided.

Solution

We will show that all bands of a fixed width Δx have the same surface area (and hence contain the same amount of gold.) If the left edge of the band is located at the point $x = a$, then the right edge is at the point $x = a + \Delta x$ since its width is Δx.

Using the previous result that $2\pi y \sqrt{1+\left(\frac{dy}{dx}\right)^2} = 2\pi R$, the surface area of the band over the interval $[a, a+\Delta x]$ is:

$$A = \int_{a}^{a+\Delta x} 2\pi R\, dx = 2\pi R \Delta x$$

Since $\Delta x = \frac{D}{N}$ and $2R = D$, we may write,

$$\boxed{A = \frac{\pi D^2}{N}}$$

Notice that this area does not depend on the initial point a of the interval. We conclude that if the sphere is cut into N bands of the same width, then each pirate will receive the same amount of gold.

Page 430

Example 8

Rotation of the curve $y = x^3$, $0 \leq x \leq 1$ about the x-axis generates a beautiful bell-shaped surface.
Find the surface area A of the bell.

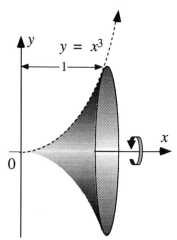

Solution

We will need the derivative $\dfrac{dy}{dx} = \dfrac{d}{dx}(x^3) = 3x^2$.

Using the integration formula for surface area we find:

$$A = \int_0^1 2\pi y \sqrt{1 + \left(\dfrac{dy}{dx}\right)^2}\, dx = 2\pi \int_0^1 x^3 \sqrt{1+9x^4}\, dx$$

Now make the substitution $u = 1 + 9x^4$. This implies $du = 36x^3 dx$.
Since $u = 1 + 9x^4$, we see that $u(0) = 1$ and $u(1) = 10$. Thus the integral becomes:

$$A = \dfrac{\pi}{18} \int_1^{10} \sqrt{u}\, du = \dfrac{\pi}{27} u^{(3/2)} \Big]_1^{10} = \boxed{\dfrac{\pi}{27}(10\sqrt{10} - 1)} \approx 3.56$$

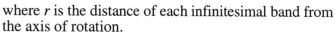

General Surface Area Formula

Our formula for surface area is more general than the previous discussion may imply.
If any curve is rotated about an axis O we may write

$$A = \int dA = \int 2\pi r\, dL$$

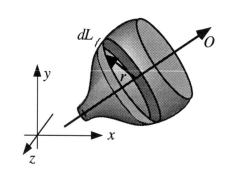

where r is the distance of each infinitesimal band from the axis of rotation.
We will now discuss several special cases.

For rotation about the x-axis, $r = y$ so the general surface area formula reduces to:

General Surface Area Formula for Rotation about the x-axis

$$A = \int 2\pi y\, dL$$

§5.4 Area of Surfaces ...

For rotation about the y-axis, $r = x$ so the general surface area formula reduces to:

> **General Surface Area Formula for Rotation about the y-axis**
> $$A = \int 2\pi x \, dL$$

For both cases, one may use either

$$dL = \sqrt{1+\left(\frac{dy}{dx}\right)^2}\,dx \quad \text{or} \quad dL = \sqrt{1+\left(\frac{dx}{dy}\right)^2}\,dy$$

A wise choice may make the integration easier.

Example 9 - Solar Furnace

A solar furnace is a <u>parabolic</u> reflector that is used to focus the energy of the sun at a point achieving temperatures as high as 4000°C. Knowing the total surface area of the collector is important both due to the high cost of the reflective material and for the analysis of the efficiency and power of the collector. Calculate the surface area A of the solar furnace illustrated in the diagram.

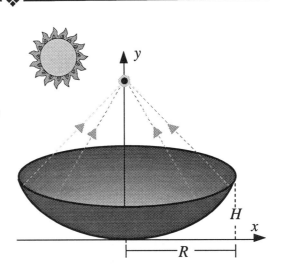

Solution

The depth of the illustrated collector is H and the maximum radius of the collector is R. Notice that the solar collection surface is not obtained by revolving a curve about the x-axis but instead is obtained by revolving about the y-axis.

Thus we must use the formula $A = \int 2\pi x \, dL$.

Method 1 - Integrate with respect to y

In this first solution, we will express everything in terms of the variable y. The parabolic surface is generated by rotating the parabolic arc $y = H\left(\frac{x}{R}\right)^2$ about the y-axis.

Expressing x as a function of y gives $x = \frac{R}{\sqrt{H}}\sqrt{y}$. Differentiating we find: $\frac{dx}{dy} = \frac{R}{2\sqrt{Hy}}$. Using the surface area formula (for rotation about the y-axis) and choosing y as the integration variable so that $dL = \sqrt{1+\left(\frac{dx}{dy}\right)^2}\,dy$, we find:

$$A = \int_0^H 2\pi x \sqrt{1+\left(\frac{dx}{dy}\right)^2}\,dy = \int_0^H 2\pi \left(\frac{R}{\sqrt{H}}\sqrt{y}\right)\sqrt{1+\frac{R^2}{4Hy}}\,dy$$

$$= \quad 2\pi \frac{R}{\sqrt{H}} \int_0^H \sqrt{y + \frac{R^2}{4H}}\, dy \quad = \quad \frac{4\pi R}{3\sqrt{H}} \left(y + \frac{R^2}{4H}\right)^{3/2} \Big|_0^H$$

$$= \quad \frac{\pi R}{6H^2} \left[(4H^2 + R^2)^{3/2} - R^3\right]$$

Thus the surface area of the solar furnace is $\boxed{A = \dfrac{\pi R^4}{6H^2}\left[\left(1 + \dfrac{4H^2}{R^2}\right)^{3/2} - 1\right]}$

Method 2 - Integrate with respect to x

In this second approach, we will express everything in terms of the variable x.

Thus we choose the form $dL = \sqrt{1 + \left(\dfrac{dy}{dx}\right)^2}\, dx$ for the infinitesimal arclength dL.

Recalling that $y = H\left(\dfrac{x}{R}\right)^2$, we find the derivative $\dfrac{dy}{dx} = \dfrac{2xH}{R^2}$. Thus,

$$A \quad = \quad \int_0^R 2\pi x \sqrt{1 + \left(\frac{dy}{dx}\right)^2}\, dx \quad = \quad \int_0^R 2\pi x \sqrt{1 + \left(\frac{2xH}{R^2}\right)^2}\, dx$$

Making the substitution $u = 1 + \left(\dfrac{2xH}{R^2}\right)^2$ and noting that $du = \dfrac{8H^2}{R^4}\, x\, dx$ we obtain:

$$A \quad = \quad 2\pi \frac{R^4}{8H^2} \int_1^{1 + 4H^2/R^2} \sqrt{u}\, du \quad = \quad \boxed{\dfrac{\pi R^4}{6H^2}\left[\left(1 + \dfrac{4H^2}{R^2}\right)^{3/2} - 1\right]}$$

Of course, the two approaches give the same answer.

Developing good checking skills. (Extreme or limiting values)

How can we be sure of a result like the expression for the area A of the solar collector in the last problem. (Could we have made a mistake?) Of course we did the calculation two different ways and that is always a very good way to assure that your answers are correct. Another approach is to consider the behavior of the answer as the parameters take on <u>extreme</u> or <u>limiting</u> values. For example, consider what happens as the depth H of the solar furnace approaches zero. In this extreme, the collector approximates a flat disc of radius R and hence its area A should approach $A_0 = \pi R^2$. Checking that our answer has the correct limit in this extreme is another good way to assure that one has not made any error.

Show that $\quad \lim\limits_{H \to 0} \dfrac{\pi R^4}{6H^2}\left[\left(1 + \dfrac{4H^2}{R^2}\right)^{3/2} - 1\right] = \pi R^2.$

Solution

Applying the linear approximation $(1+x)^n \sim 1 + nx$ to the term $\left(1 + \dfrac{4H^2}{R^2}\right)^{3/2}$ we find that:

$$\left(1 + \frac{4H^2}{R^2}\right)^{3/2} - 1 \sim \frac{6H^2}{R^2}$$

Thus the limit of the surface area as the height H of the collector approaches zero is:

$$\lim_{H \to 0} \frac{\pi R^4}{6H^2}\left[\left(1 + \frac{4H^2}{R^2}\right)^{3/2} - 1\right] = \lim_{H \to 0} \frac{\pi R^4}{6H^2}\left[\frac{6H^2}{R^2}\right] = \pi R^2$$

With the answer derived by two different methods and with the added check of the correct limiting value we can be very confident of our answer.

WARMUP EXERCISES

1) For which of the following objects could the external surface reasonably be described as a surface of revolution?

    a) Basketball   **b)** Wine Bottle   **c)** A 2×4 length of lumber

    d) Ice cube   **e)** Glass Portion of Light Bulb   **f)** Football

2) **Painting a Church Spire**
A church spire in the shape of a cone, has height $H = 10$ meters and the radius at the base is $R = 1$ meter. What is the surface area A of the spire? What quantity of paint Q is required to cover the spire if the paint is applied with a uniform thickness of 1/2 millimeter?

Find the lateral surface area A for each of the following conical frustums using the formula:
$$A = 2\pi \bar{r} L$$

3)

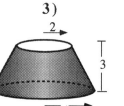

4)

5)

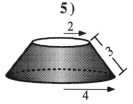

6)

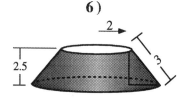

§5.4 Area of Surfaces ...

7) Each frustum shown in the previous exercise can be extended into a complete cone by adding a smaller conical cap. Find the lateral surface area of the conical cap and of the completed cone in each case. Check that the difference of the two conical areas gives the area of the frustum.

Each of the following surfaces of revolution are cones or frustums of cones. Calculate the lateral surface area for each exercise using an appropriate integration formula, and then verify your result by using the geometric formula $A = \pi R L$ or $A = 2\pi \bar{r} L$.

8) The line $y = x + 1$ revolved about the x-axis over the interval $0 \le x \le 1$.

9) The line $y = 2x$ revolved about the y-axis over the interval $0 \le y \le 1$.

10) The line $y = 4 - x$ revolved about the x-axis over the interval $1 \le x \le 2$.

11) The line $y = 2x + 1$ revolved about the y-axis over the interval $2 \le y \le 3$.

Each surface of revolution can be generated by revolving a curve $y = f(x)$ about the x-axis over an interval $a \le x \le b$. Give an explicit formula for the function $f(x)$ and identify the appropriate interval.

12) Rocket shape.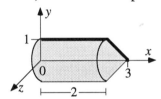

13) Cone of radius 1.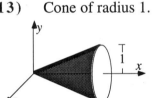

14) Conical frustum.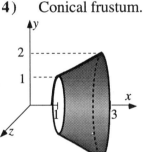

In the following exercises, a surface is generated by rotating the given curve about the x-axis. Set up the integral representing the surface area. Do <u>not</u> evaluate the integral.

15) A sine arch $y = \sin(x)$ on $[0, \pi]$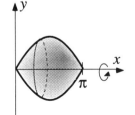

16) A parabolic arch $y = 1 - x^2$ on $[-1, 1]$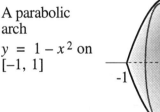

17) $y = \sec(x)$ on $\left[-\frac{\pi}{4}, \frac{\pi}{4}\right]$

18) $y = \ln(x)$ on $[1, 2]$

19) $y = xe^{-x}$ on $[0, 2]$

20) $y = \frac{1}{x}$ on $[1, 2]$

INTERMEDIATE EXERCISES

In the following exercises, a surface is generated by rotating the given curve about the *x*-axis. Sketch each surface of revolution and then find its surface area using integration.

21) $y = \sqrt{x}$ $1 \le x \le 4$ 22) $y = x^3$ $2 \le x \le 3$

23) $y = \sin x$ $0 \le x \le \pi$ 24) $y = \cos x$ $0 \le x \le \pi/2$

25) $y = \cosh x$ $0 \le x \le a$ 26) $y = e^x$ $0 \le x \le 1$

In the following exercises, a surface is generated by rotating the given curve about the *y*-axis. Sketch each surface of revolution and then find its surface area using integration.

27) $y = \sqrt[3]{x}$ $1 \le y \le 4$ 28) $y = 4 - x^2$ $0 \le x \le 2$

29) $y = x^{3/2}$ $1 \le y \le 27$ 30) $x = \cosh y$ $-1 \le y \le 1$

Surface Area for Perfect Squares

Just as in the case of arclength, the surface area formula usually leads to challenging integrations because of the troublesome square root in the integrand. In the following special cases however, the integration can be performed because the term $1 + \left(\frac{dy}{dx}\right)^2$ is a perfect square, allowing the square root to be canceled.

Find the area of the surface which results from rotating the curve about the *x*-axis over the given interval.

31) $y = \frac{x^3}{3} + \frac{1}{4x}$ $1 \le x \le 2$ 32) $y = \frac{x^4}{4} + \frac{1}{8x^2}$ $2 \le x \le 4$

33) $y = \frac{x^5}{5} + \frac{1}{12x^3}$ $2 \le x \le 4$ 34) $y = \frac{x^2}{2} - \frac{\ln x}{4}$ $1 \le x \le 2$

35) Let *y* be the average of two functions, $y = \frac{f(x) + g(x)}{2}$, satisfying $f'(x) \cdot g'(x) = -1$ and $f(x) + g(x) \ge 0$.

a) Show that the area of the surface formed by rotating the graph of *y* about the *x*-axis over the interval [*a*, *b*] is given by the integral

$$A = \frac{\pi}{2} \int_a^b (f(x) + g(x)) \left| f'(x) - g'(x) \right| dx.$$

b) Simplify this integral assuming that $f'(x) - g'(x) \ge 0$.

§5.4 Area of Surfaces ...   CAFÉ   Page 437

36) Find the area of the surface obtained by rotating the curve $y = \dfrac{e^x + e^{-x}}{2}$, $-1 \le x \le 1$ about the *x*-axis.

37) Consider the ellipse $\dfrac{x^2}{a^2} + \dfrac{y^2}{b^2} = 1$. Assume $a > b$.
   a) Find the surface area of the ellipsoid generated by rotating this ellipse about the *x*-axis.
   b) Find the surface area of the ellipsoid generated by rotating this ellipse about the *y*-axis.
   c) Compare the two answers.

38) An old water tower is to be spruced up with a fresh coat of paint. It has been constructed in the shape of an ellipsoid formed by rotating the ellipse $\dfrac{x^2}{100} + \dfrac{y^2}{25} = 1$ about the *y*-axis. (The above units are in meters.) What volume of paint is needed if the paint is applied with a uniform thickness of 1/2 millimeter?

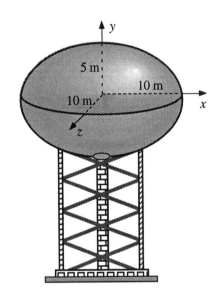

39) If a cone is cut at the midway point, the surface area of the top portion is one third the area of the bottom portion. Describe in general how a cone should be cut so that the two areas are equal.

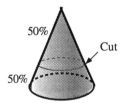

NOTES

Answers to Odd Numbered Problems

Chapter One: Elementary Engineering Functions

§ 1.1 - Linear Functions

1)   **a)** $y = 2x$ **b)** $y = 1$ **c)** $y = \left(\dfrac{d-b}{c-a}\right)x + \dfrac{bc-ad}{c-a}$ **d)** $y = x$

3)   **a)** $y = 3x + 7$ **b)** $y = 3x + b$ **c)** $y = 3x - 3c$

5) $\begin{cases} P_A = 20t + 10 \\ P_B = 20t + 20 \\ P_C = 20t + 30 \end{cases}$ $\dfrac{1}{20}$ m/s in directions opposite that of cars.

7)   10 turns   9)   **a)** $L = \dfrac{x}{3}$ **b)** $6\cot\theta$

11) $x(t) = \dfrac{0.1}{\pi R^2} t$ *(Graph: Height of Water H vs t, reaching H at $10\pi R^2 H$)*

13) $k = \dfrac{3}{2}$ lb/in $L = \left(10 + \dfrac{2}{3}F\right)$ inches   15)   **a)** $L = 8 + 2F$ **b)** $F = 1$ lb

17) $k = 5$ lb/in   19) $k = \dfrac{5\pi}{16}$ lb/in   21) $\dfrac{P_o}{\rho g}$

23) $0.2041x$ 25) $G = 66 \times 10^{-3} - 33 \times 10^{-5} T$

§ 1.2 - Intervals and Tolerance Analysis

1   **a)** $|l - 7.5| \le 1.5$, $|w - 100| \le 40$, $|v - 37.5| \le 2.5$
     b) $|l - 17.5| \le 2.5$, $|w - 1,100| \le 100$
     c) $|l - 47| \le 3$, $|w - 8,000| \le 2000$ 2000 lbs = 1 ton

3) $|x - 1.5| \le 1.5$ 5) $|x - 3| \le 1$ 7) $|x - 9| < 2$ 9) $|x - 50| \le 50$

11) $|x| < 1$ 13) $\begin{cases} 2\text{ cm} \pm 50\% \\ |L - 2| \le 1 \\ [1, 3] \end{cases}$ 15) $\begin{cases} 200 \pm 0.5\% \\ |L - 200| \le 1 \\ [199, 201] \end{cases}$

17) $\begin{cases} 100 \pm 10\% \\ |R - 100| \le 10 \\ [90, 110] \end{cases}$ 19) $\begin{cases} 50 \pm 20\% \\ |v - 50| \le 10 \\ [40, 60] \end{cases}$ 21)   B   23)   A-yes, B-no, C-no, D-no

25) $y = \dfrac{x + |x|}{2}$ 27)   *(Graph showing $y = x$ and $y = 2x$ on axes from -5 to 5)*

Page 440   CAFÉ   Answers to Odd Numbered Problems

29)   31)   33)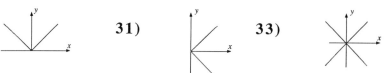

39) $y = \begin{cases} 0, & x < -1 \\ x+1, & -1 \le x \le 1 \\ 2, & 1 < x \end{cases}$, $y = \frac{1}{2}(|x+1| - |x-1|) + 1$

41) $y = \begin{cases} 0, & x < -1 \\ 2x+2, & -1 \le x < 0 \\ -2x+2, & 0 \le x \le 1 \\ 0, & x > 1 \end{cases}$, $y = ||x|-1| - |x| + 1$ 43) $x \in (0, 6)$ or $|x-3| < 3$

45) $x \in [2, \infty) \cup [-2, 1)$ 47) $x \in (1, \infty) \cup (-1, 0)$

49) $\min\{a, b\} = \dfrac{a+b-|a-b|}{2}$ 51) $|F - 86| \le 9$

53) $\left[\dfrac{55}{3}, \dfrac{190}{9}\right]$ 55) $[2.87, 2.89]$

§ 1.3 - Functions, Parabolas and Tangent Lines

1) a, b, c   3) b, d

5) This is not a function of x.   7) This is not a function of x

9) This is a function of x.   9) This is not a function of x

13) No   15) $f: [0, 14] \to [0, 7]$ $f(x) = \dfrac{x}{2}$ 17) A, C

19) $\max y = \dfrac{1}{4}$ @ $x = \dfrac{1}{2}$ 21) $\min y = -1$ @ $x = 0$

23) $x = 2$ or $x = -3$ 25) $x = \dfrac{1}{2}$ or $x = 1$ 27) $f(x)$ is never 0.

29) $(-1, 0)$ and $(1, 0)$ 31) slope $= 2$

37) **a)** No real roots for $|b| < 2$ **b)** Exactly one root for $|b| = 2$ **c)** Two roots for $|b| > 2$

39) $y = \left(\dfrac{D}{d}\right)^2 x$ 41) A is maximized at $x = 50, y = 50$. The size of the pool doesn't matter.

43) **a)** $A(x) = \dfrac{\sqrt{3}}{2} x^2 + 30x$ **b)** $\dfrac{350\sqrt{3}}{3}$ ft²

45) **a)** # divisors of $2^n = n+1$ **b)** # divisors of $6^n = (n+1)^2$
    c) # divisors of $30^n = (n+1)^3$ **d)** # divisors of $12^n = (2n+1)(n+1)$

47) $t = \dfrac{1}{4}, \dfrac{5}{2}, \dfrac{5\sqrt{10}}{2}$ sec

§ 1.4 - The Algebra of Functions

1) $x > 0$ **3)** $x > 0$ **5)** $|x| \geq 2$ **7)** $x \neq 4$

9) $f(x)$ defined only at $x = \pm 1$

13) **a)** $\begin{cases} (f+g)(x) = (m_1 + m_2)x + (b_1 + b_2) \\ (f-g)(x) = (m_1 - m_2)x + (b_1 - b_2) \\ (f \cdot g)(x) = m_1 m_2 x^2 + (b_1 m_2 + b_2 m_1)x + b_1 b_2 \\ \left(\dfrac{f}{g}\right)(x) = \dfrac{m_1 x + b_1}{m_2 x + b_2} \end{cases}$ **b)** $f+g, f-g$

c) $m_1 + m_2,\; m_1 - m_2$ **d)** $\begin{cases} \text{Domain }(f+g) = \text{Domain }(f-g) = \text{Domain }(f \cdot g) = \mathbb{R} \\ \text{Domain }\left(\dfrac{f}{g}\right) = \mathbb{R} \setminus \left\{\dfrac{-b_2}{m_2}\right\} \end{cases}$

15) [block diagrams showing $f(x)+g(x)$, $f(x)-g(x)$, and combined $f(x)+g(x)$ / $f(x)-g(x)$]

17) $\begin{cases} (f \circ g \circ h)(x) = \dfrac{2}{x^2} & (g \circ h \circ f)(x) = \dfrac{1}{4x^2} \\ (f \circ h \circ g)(x) = \dfrac{2}{x^2} & (h \circ f \circ g)(x) = \dfrac{4}{x^2} \\ (g \circ f \circ h)(x) = \dfrac{1}{2x^2} & (h \circ g \circ f)(x) = \dfrac{1}{4x^2} \end{cases}$

19) $\text{Pay}(t) = 15t + \dfrac{15(t-35)\,u(t-35)}{2}$

21) $f(x) = \begin{cases} 0 & x < 1 \\ x & 1 \leq x < 2 \\ x^2 + x & 2 \leq x \end{cases}$

23) $f(x) = \begin{cases} f_1(x) & x < a \\ f_1(x) + f_2(x) & x \geq a \end{cases}$

25) $A(t) = 200 + \dfrac{8T}{10^4} + \dfrac{8T^2}{10^{10}}$

27) **a)** $C(t) = \dfrac{16}{8+t}$, $0 \leq t \leq 4(20\pi - 2)$ **b)** $C(t) = \dfrac{16+t}{8+t}$, $0 \leq t \leq 4(20\pi - 2)$

29) **a)** $H = \dfrac{80}{(D-1)^2} + 1$ **b)** $V = \pi D^2 \left(\dfrac{20}{(D-1)^2} + \dfrac{1}{4}\right)$ **31)** $A = 2x\sqrt{R^2 - x^2}$

33) $H = \dfrac{\left(\dfrac{2t}{343} + \dfrac{2}{g}\right) - \sqrt{\left(\dfrac{2t}{343} + \dfrac{2}{g}\right)^2 - \dfrac{4t^2}{343^2}}}{\dfrac{2}{343^2}}$

35) **a)** $V = (3 - 2x)(2 - 2x)x$ **b)** $D = 0 \leq x \leq 1$ [graph of V vs x]   **c)** $V = 0$ @ $x = 0, 1$

37) $C(x) = 3.5 \times 10^6 \sqrt{(10-x)^2 + 25} + 7 \times 10^5 x$ **39)** $T = \dfrac{50000 - 80x}{2000 + x}$

41) $T = \dfrac{50000 + 4000t}{2000 + 50t}$

§ 1.5 - Trigonometric Functions

3)   a) $\cos x = 0$ iff $x = \dfrac{(2n+1)\pi}{2}$ b) $\tan x = 0$ iff $x = n\pi$

     c) $\sec x$ is never 0   d) $\csc x$ is never 0   e) $\cot x = 0$ iff $x = \dfrac{(2n+1)\pi}{2}$

5)   a) $\dfrac{\pi}{2}$ b) $\dfrac{\pi}{6}$ c) $\dfrac{\pi}{4}$ d) $\left(\dfrac{d^\circ}{180} \cdot \pi\right)$ radians

9)   a) $x = 1$, slope undefined, does not have y-intercept

     b) $y = \dfrac{1}{2}x - \dfrac{1}{2}$, slope $= \dfrac{1}{2}$, y-intercept $= -\dfrac{1}{2}$

11) $300 + 150 \sin\left(2\pi t + \dfrac{\pi}{8}\right)\cos\left(\dfrac{\pi}{8}\right)$

13)   a) $h = D\tan\theta$ b) $h = \dfrac{400}{\sqrt{3}}$ ft   c) $\theta = \tan^{-1}\left(\dfrac{555}{400}\right) \approx 0.946$ radians

15) $x = 0, \dfrac{\pi}{3}, \dfrac{2\pi}{3}, \pi, \dfrac{4\pi}{3}, \dfrac{5\pi}{3}, 2\pi, \dfrac{\pi}{10}, \dfrac{3\pi}{10}, \dfrac{5\pi}{10} \cdots \dfrac{19\pi}{10}$ 17) $x = \dfrac{\pi}{3}, \pi, \dfrac{5\pi}{3}$

19) $x = \cos^{-1}\dfrac{24}{25}, 2\pi - \cos^{-1}\dfrac{24}{25}, \pi$

21)   It is true whenever $\sin x \neq 0$ i.e. $x \in [0, 2\pi] \setminus \{0, \pi, 2\pi\}$

23)   [graph]   25)   a) $\beta = 0.415$ rad   b) $\beta = 0.500$ rad   c) $\beta = 0.4636$ rad

27)   beats/sec = 5   29)   a) $s = 0.917$ b) $s = 0.504$ c) $s = 0.307$

31)   a) $\theta = 32.027°$ north of east   b) 0.1088 hr   33)   178.59 ft

35)   a) $h = \dfrac{d\sin\theta_1 \sin\theta_2}{\sin(\theta_2 - \theta_1)}$ b) $D = \dfrac{d\cos\theta_1 \sin\theta_2}{\sin(\theta_2 - \theta_1)}$

§ 1.6 - Exponential and Logarithms

1) 2^7 3) 2^4 5) -5 7) x^2 9) 0 11) $t = \dfrac{1}{2}$

13) $t = \dfrac{\log b}{\log a}$ 15) $x = \dfrac{\log 2 + 2\log 5}{3\log 5 - \log 2}$ 17) $x = 999$

19) 6.31×10^{-5} moles per liter

21) a) 1000 times stronger   b) pH = 10 is 100,000 times less acidic than pH = 5

23) $x = \dfrac{1}{\log 2}\log\left(\dfrac{5 \pm \sqrt{21}}{2}\right)$ or $x = \pm 2.26$ 25) $x = 0, 1$

27) $x = \log\sqrt{3}$ 29) $x = \dfrac{1}{10}, x = 10$ 31) $x = 10^{100} - 3$

33) a) $t = 1$ year   b) $t = \dfrac{\log 2}{\log(1.1)} \approx 7.27$ years   c) $t = \dfrac{\log 2}{\log(1.01)} \approx 69.66$ years

35) a) $P = 10^6 \cdot 3^{24}$ b) $P = 10^6 \cdot 3^t$ c) $t = \dfrac{6}{\log 3} = 12.575$ hours

37) a) 0.871 grams   b) $2^{-t/5000}$ grams   c) 11610 years

39) a) 9   b) 369,693,100 digits   c) about 1.172 years

Answers to Odd Numbered Problems   CAFÉ   Page 443

Chapter Two: Limits and Derivatives

§ 2.1 - Tangent Lines and Derivatives

1)   a) $6.92\ °F/hr$ b) $2.8\ °F/min$ 3) $\ 2$ 5) $\ 3$ 7) $\ \frac{1}{2}$

9) $y'(x) = 2$ 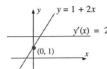   11) $\ y'(x) = 3x^2$

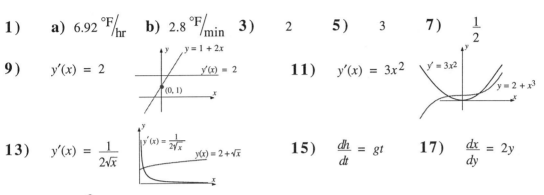

13) $\ y'(x) = \dfrac{1}{2\sqrt{x}}$ 15) $\ \dfrac{dh}{dt} = gt$ 17) $\ \dfrac{dx}{dy} = 2y$

19) $\ y = \dfrac{-2x}{9} + \dfrac{5}{9}$ 21) $\ y = -12x - 3$ 23)   A - 3, B - 1, C - 3, D - 2, 1

25) $\ f'(0)$ 27)   a) broke down at 7pm, repaired by 8pm.   b) $10^6 calls/hr$ c) 10^6 calls

29)
   a) Derivative not defined at 1, 2, 3
   b) The first hour
   c) The second and fourth hour
   d) The third hour

31) 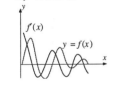   33) $f'(x)$ does not exist for $x = 1, 2, 3, 3.5$

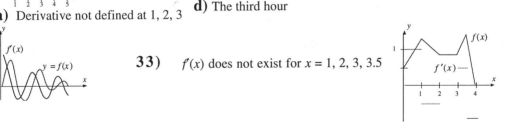

§ 2.2 - Limits

1)   a) $\lim\limits_{x \to 0^+} f(x) = 0$ $\lim\limits_{x \to 1} f(x) = 1$

        $\lim\limits_{x \to 2^-} f(x) = 2$ $\lim\limits_{x \to 2^+} f(x) = 1/2$ $\lim\limits_{x \to 3^-} f(x) = 0$

   b) $\lim\limits_{x \to 0^+} g(x) = 0$ $\lim\limits_{x \to 1} g(x) = 2$

        $\lim\limits_{x \to 2^-} g(x) = 0$ $\lim\limits_{x \to 2^+} g(x) = 2$ $\lim\limits_{x \to 3^-} g(x) = 1$

3)   a) Note, we're taking neighborhoods to be open   b) No such neighborhood can be drawn

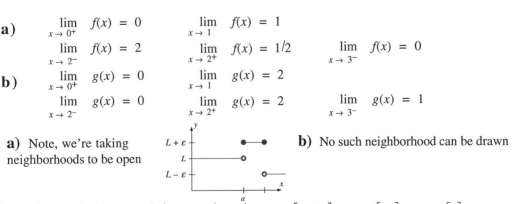

5) $\lfloor 2.718 \rfloor = 2,\ \lfloor -\pi \rfloor = -4,\ \lfloor 0 \rfloor = 0,\ \lfloor -1.1 \rfloor = -2,\ \lceil 2.718 \rceil = 3,\ \lceil -\pi \rceil = -3,\ \lceil 0 \rceil = 0$
   $\lceil -1.1 \rceil = -1$

7)   Let $f(x) = \dfrac{\lfloor x \rfloor + \lfloor -x \rfloor}{2}$, $g(x) = \dfrac{\lfloor x \rfloor - \lfloor -x \rfloor}{2}$, $h(x) = \dfrac{\lceil x \rceil + \lceil -x \rceil}{2}$, $J(x) = \dfrac{\lceil x \rceil - \lceil -x \rceil}{2}$

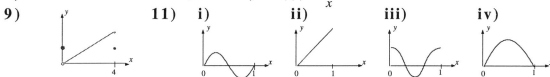

9) **a)** 12 **b)** 8 **c)** 20 **d)** 5 **e)** 1/5 **11)** 3 **13)** $\dfrac{-1}{2\pi + 1}$

17) $F(x) = 1.8 + 0.3\lfloor 6x \rfloor$ **19)** $C(t) = 100 + 20\lfloor t \rfloor$ **21)** -1 **23)** 0

25) -1 **27)** 0 **29)** $\dfrac{a}{b}$ **31)** $\dfrac{1}{2}$ **33)** $\dfrac{5}{2}$ **35)** 0

37) 0 **39)** -1 **41)** 1 **43)** 2 **45)** 0 **47)** 0

51) **a)** $\dfrac{2000}{3t + 100}$ **b)** 0 **53)** 0 **55)** 1 **57)** 1 **59)** 1

61) $\dfrac{1 + \sqrt{13}}{2}$ **63)** **a)** 0 **b)** 1

65) **a)** $A_2 = \dfrac{1}{2}(\pi - 2)$, $A_3 = \dfrac{1}{4}(\pi - 2)$ **b)** $2(\pi - 2)$

67) $\left(1 - \dfrac{\pi}{k}\cot\dfrac{\pi}{k}\right)\dfrac{A_p}{\sin^2\dfrac{\pi}{k}}$, where A_p is the area of the external regular k-gon.

§ 2.3 - Continuity of Functions

1) **a)** continuous at 0 and 3 **b)** continuous at 0 and 1 **3)** none

5) discontinuities at $x = -2, -1$ **7)** $f(x) = \dfrac{1}{x}$ is not continuous over $\mathbb{R}$

9) **11)** **i)** **ii)** **iii)** **iv)**

13) $k = 1$ **15)** $k = 1$

17) **a)** $x = \log\left(\dfrac{-1 + \sqrt{41}}{2}\right) \approx 0.4316$ **b)** $x = \log\left(\dfrac{-1 + \sqrt{201}}{2}\right) \approx 0.8188$

 c) $x = \log\left(\dfrac{-1 + \sqrt{401}}{2}\right) \approx 0.9783$

21) $m = -2, M = 4$ **23)** $m = -1, M = 3$ **25)** $m = -\dfrac{1}{4}, M = 2$

27) Removable discontinuity at $x = 0$.   Define $\tilde{f}(x) = \begin{cases} \dfrac{x^3}{|x|} & x \neq 0 \\ 0 & x = 0 \end{cases}$

29) Removable discontinuity at $x = -1$. $\tilde{f}(x) = \begin{cases} \dfrac{x^2 + 3x + 2}{x + 1} & x \neq -1 \\ 1 & x = -1 \end{cases}$

31) Discontinuity at $x = -2, x = -1$. Removable discontinuity at $x = -2$.

$\tilde{f}(x) = \begin{cases} \dfrac{x^2 - 4}{x^2 + 3x + 2} & x \neq -2, -1 \\ 4 & x = -2 \end{cases}$

Answers to Odd Numbered Problems   CAFÉ   Page 445

33) b) $T(x) = \frac{18}{19}x + \frac{122}{19}$ c) $122\,°F$ **41)** b) 12:00 noon and 12:00 midnight

§ 2.4 - Differentiation Rules

1) $f'(x) = 35x^4 + 3x^2$ **3)** $f'(x) = \frac{-8}{x^2} - \frac{14}{x^3}$ **5)** $f'(x) = -4x^3$

7) $f'(x) = \frac{4x}{(1-x^2)^2}$ **9)** $f'(x) = \frac{3x^2}{(1+x^3)^2}$

13) $h'(t) = 3t^2(2+t^2)(3+t) + 2t(1+t^3)(3+t) + (1+t^3)(2+t^2)$

15) $g'(x) = f'(x)(1+x^2)(2+x^2) + 2x(3+2x^2)f(x)$

17) $C'(t) = \frac{2}{(100-t)^2}$ **19)** a) $p = \frac{nRT}{V-b} - \frac{a}{V^2}$ b) $p'(V) = \frac{2a}{V^3} - \frac{nRT}{(V-b)^2}$

21) $y''(x) = 2 + 6x$ **23)** $x''(t) = \frac{2}{t^3}$ **25)** $x''(t) = g$

27) $x''(t) = a$ **29)** $\vec{V} = (2t, 3t^2)$, $V = t\sqrt{4+9t^2}$, $\vec{a} = (2, 6t)$

31) $\vec{V} = (V_x, V_y + at)$ $v = \sqrt{V_x^2 + (V_y + at)^2}$ $\vec{a} = (0, a)$

33) $f'(x) = 2x - 4xu(x+2) + 4xu(x-2)$, $x \ne \pm 2$

35) $f'(x) = (2x - 3x^2)(1 - 2u(x-1))$, $x \ne 1$

37) b) $f(x) = x$, $g(x) = \frac{1}{x-1}$ **39)** a) 0   b) $\frac{\sqrt{1+f'(0)^2}}{f'(0)}$ c) 1

§ 2.5 - Derivatives of Trigonometric Functions

5) $y' = \frac{-1}{1+\sin x}$ **7)** $\frac{dy}{dx} = 2\sin x \cos x$ **9)** $\frac{dy}{dx} = \sec^2 x - \csc^2 x$

11) $\frac{dy}{dx} = 0$, $x \ne \frac{n\pi}{2}$ **13)** $\frac{dy}{dx} = 0$, $x \ne \frac{n\pi}{2}$ **15)** Tangent line $y = x$

17) Tangent line $y = 2x + 1 - \frac{\pi}{2}$ **19)** Tangent line $y = \sqrt{2}x + \sqrt{2}\left(1 - \frac{\pi}{4}\right)$

21) $v = \dot{x} = 2\cos 2t$, $a = \dot{v} = -4\sin 2t$

23) $v = \dot{x} = gt - \omega\sin\omega t$, $a = \dot{v} = g - \omega^2\cos\omega t$

25) $\frac{dr}{d\theta} = 2\sec\theta\tan\theta$ **27)** $\frac{dr}{d\theta} = \frac{-2\cos\theta}{(3+\sin\theta)^2}$ **31)** tan, cot

33)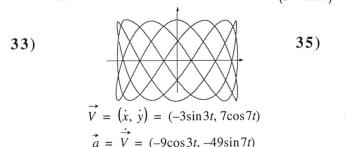

$\vec{V} = (\dot{x}, \dot{y}) = (-3\sin 3t, 7\cos 7t)$

$\vec{a} = \dot{\vec{V}} = (-9\cos 3t, -49\sin 7t)$

35)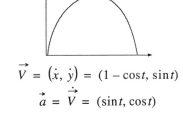

$\vec{V} = (\dot{x}, \dot{y}) = (1 - \cos t, \sin t)$

$\vec{a} = \dot{\vec{V}} = (\sin t, \cos t)$

37)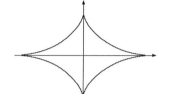

39) $\cos 3x = \cos x - 4\sin^2 x \cos x$ or
$\cos 3x = -3\cos x + 4\cos^3 x$

$\vec{V} = (\dot{x}, \dot{y}) = (-3\cos^2 t \sin t, +3\sin^2 t \cos t)$
$\vec{a} = (6\sin^2 t \cos t - 3\cos^3 t, 6\cos^2 t \sin t - 3\sin^3 t)$

41) a) $y^{(4)} = \sin x$ b) $y^{(4)} = \cos x$

43) $\dfrac{dP}{dt} = 172{,}800\,\pi\,\sin(120\pi t)\cos(120\pi t)$

45) $\dfrac{dH}{d\theta} = \dfrac{3kI}{r}\sin^2\theta\cos\theta$ **47)** 0 **49)** $\dfrac{7}{3}$ **51)** $\dfrac{5}{3}$ **53)** $\dfrac{1}{2}$

55) $\begin{cases} \text{max speed} = 0.04\,\pi\ \text{m}/\text{s} \\ \text{max acceleration} = 0.08\,\pi^2\ \text{m}/\text{s}^2 \end{cases}$

57) a) 1 b) 1 c) 1 d) 1 e) 1 f) 1

§ 2.6 - Chain Rule and Implicit Differentiation

1) $y' = 6(1 + 3x)$ **3)** $y' = 18x^2(1 + 3x^3)$ **5)** $f'(x) = 2(1 + 3x + 7x^2)(3 + 14x)$

7) $\dfrac{dy}{dx} = \dfrac{x}{\sqrt{1 + x^2}}$ **9)** $\dfrac{dy}{dx} = \dfrac{1}{2\sqrt{x + \sqrt{x + \sqrt{x}}}}\left(1 + \dfrac{1}{2\sqrt{x + \sqrt{x}}}\left(1 + \dfrac{1}{2\sqrt{x}}\right)\right)$

11) $y' = 2\cos 2x$ **13)** $y' = 2\sin x \cos x = \sin(2x)$

15) $y' = 2\sin(\sin x)\cdot\cos(\sin x)\cdot\cos x$ **17)** $y' = 1000\left(\dfrac{x-1}{x+1}\right)^{999}\left(\dfrac{2}{(x+1)^2}\right)$

19) $y' = 2x\tan(\cos x) - x^2\sec^2(\cos x)\sin x$ **21)** $\dfrac{dy}{dx} = \dfrac{3x^2}{1 + x^6}$ **23)** $y' = \dfrac{1}{\sqrt{-x - x^2}}$

25) $y' = -1$

27) a) $\begin{cases} f\circ g\circ h\,(x) = \dfrac{1}{(1 + 2x)^2} \quad g\circ f\circ h\,(x) = \dfrac{1}{(1 + 2x)^2} \\ h\circ f\circ g\,(x) = 1 + \dfrac{2}{x^2} \quad f\circ h\circ g\,(x) = \left(1 + \dfrac{2}{x}\right)^2 \\ g\circ h\circ f\,(x) = \dfrac{1}{1 + 2x^2} \quad h\circ g\circ f\,(x) = 1 + \dfrac{2}{x^2} \end{cases}$ b) $f\circ g\circ h = g\circ f\circ h$ and $h\circ f\circ g = h\circ g\circ f$

c) Compositions of rational functions leads to other rational functions

d) $\begin{cases} (f\circ g\circ h)' = \dfrac{-4}{(1 + 2x)^3} = (g\circ f\circ h)' \quad (g\circ h\circ f)' = \dfrac{-4x}{(1 + 2x^2)^2} \\ (h\circ f\circ g)' = \dfrac{-4}{x^3} = (h\circ g\circ f)' \quad (f\circ h\circ g)' = \dfrac{-4}{x^2}\left(1 + \dfrac{2}{x}\right) \end{cases}$

29) $\dfrac{dE}{d\theta} = \dfrac{60I}{r}\left[\dfrac{(\sin\theta)(\sin(2\pi N\cos\theta))(2\pi N\sin\theta) - (\cos\theta)\cos(2\pi N\cos\theta)}{\sin^2\theta}\right]$

31) a) $D = D_o + L_o - L$ b) $\begin{cases} \dfrac{dD}{dT} = -m \\ \dfrac{dD}{dL} = -1 \\ \dfrac{dL}{dT} = m \end{cases}$ **33)** $\dfrac{d(KE)}{dt} = 0$

Answers to Odd Numbered Problems   CAFÉ

35) $\dfrac{d(KE)}{dt} = m\omega^3 A^2 \sin\omega t \cos\omega t$ 39) $\dfrac{dy}{dx} = \dfrac{2a}{y}$ 41) $\dfrac{dy}{dx} = \dfrac{y - x^{n-1}}{y^{n-1} - x}$

43) $\dfrac{dy}{dx} = -\dfrac{x^3}{y^3}$ Derivative not defined at $(\pm 1, 0)$ 45) $y = -2x + 6$

47) a) [graph]   b) Horizontal tangent at $(\pm 3\sqrt{2}, 6)$ c) Vertical tangent at $(6, \pm 3\sqrt{2})$

49) a) $x = 3$ b) $y = 2$ and $y - 2 = -\dfrac{40}{9}(x - 2)$

51) a) $y = 2L\cos\theta$ b) $y = 2L\cos(\theta_0 \cos\omega t)$, stroke $= 2L(1 - \cos\theta_0)$
    c) $v = 2L\theta_0 \omega \sin(\theta_0 \cos\omega t)\sin\omega t$
    $a = 2L\theta_0 \omega\left[-\theta_0 \omega \cos(\theta_0\cos\omega t)\sin^2\omega t + \omega \sin(\theta_0\cos\omega t)\cos\omega t\right]$

Chapter Three: Applications of the Derivative

§ 3.1 - The Natural Exponential and Logarithm

1) 2^x 3) $\dfrac{x}{\ln a}$ 5) $\ln A + kt$ 7) $3^{x\log_3 2}$ 9) $\dfrac{\log_b x}{\log_b a}$

11) 434,295 digits, the leading digit of this number is 3   13) $a, b = \dfrac{1}{d}\cosh^{-1}\left(\dfrac{h}{a}\right)$

15) $\dfrac{dy}{dx} = (\ln 2)2^x + 2x$ $\dfrac{d^2y}{dx^2} = (\ln 2)^2 2^x + 2$ 17) $\dfrac{dy}{dx} = 2e^{2x+3}$ $\dfrac{d^2y}{dx^2} = 4e^{2x+3}$

19) $\dfrac{dy}{dx} = ax^{a-1} + (\ln a)a^x$ $\dfrac{d^2y}{dx^2} = a(a-1)x^{a-2} + (\ln a)^2 a^x$

21) $\dfrac{dy}{dx} = \dfrac{1}{x \ln x}$ $\dfrac{d^2y}{dx^2} = \dfrac{-1}{x^2 \ln x}\left(\dfrac{1}{\ln x} + 1\right)$ 23) $\dfrac{dy}{dx} = \dfrac{1}{\ln 2}$ $\dfrac{d^2y}{dx^2} = 0$

25) $y(x)$ is equal to $x^{\ln 3/\ln 2}$ $\dfrac{dy}{dx} = \dfrac{\ln 3}{\ln 2} x^{(\ln 3/\ln 2) - 1}$ $\dfrac{d^2y}{dx^2} = \dfrac{\ln 3}{\ln 2}\left(\dfrac{\ln 3}{\ln 2} - 1\right) x^{(\ln 3/\ln 2) - 2}$

27) $\dfrac{dy}{dx} = \dfrac{1 - \ln x}{x^2}$ $\dfrac{d^2y}{dx^2} = \dfrac{2\ln x - 3}{x^3}$ 29) $\dfrac{dy}{dx} = \text{sech}^2 x$ 31) $\dfrac{dy}{dx} = -\text{csch}^2 x$

33) $\dfrac{dy}{dx} = -\coth x \, \text{csch} x$ 35) $\dfrac{dy}{dx} = e^x \cosh x + e^x \sinh x$ $\dfrac{d^2y}{dx^2} = 2e^x(\sinh x + \cosh x)$

37) $\dfrac{dy}{dx} = k\cosh^2(kx) + k\sinh^2(kx)$ $\dfrac{d^2y}{dx^2} = 4k^2 \sinh(kx)\cosh(kx)$

39) $\dfrac{dy}{dx} = \dfrac{1}{x + \ln x}\left(1 + \dfrac{1}{x}\right)$ 41) $\dfrac{dy}{dx} = \dfrac{2e^x}{(1 - e^x)^2}$ 43) $\dfrac{dy}{dx} = \dfrac{e^{4t}\left(4\ln 4t - \dfrac{1}{t}\right)}{(\ln 4t)^2}$

45) $\dfrac{dy}{dx} = x^x(1 + \ln x)$ 47) $\dfrac{dy}{dx} = x^{(x^x)}\left(x^{(x-1)} + x^x(1 + \ln x)\ln x\right)$

49) $\dfrac{dy}{dx} = e^x(ex^{e-1} + x^e)$ 53) $\dfrac{dy}{dx} = \dfrac{e^x \sin y - 2}{1 - e^x \cos y}$

55) $\dfrac{dy}{dx} = y\left(\dfrac{12x^3}{x^4 - 1} + \dfrac{3x + 3/2}{x^2 + x + 1} - \dfrac{x}{x^2 + 1}\right)$ 57) $\dfrac{dy}{dx} = \dfrac{2x - y - [(x + y)\ln(x + y)]^{-1}}{x + [(x + y)\ln(x + y)]^{-1}}$

59) $\cosh 2x = \sinh^2 x + \cosh^2 x$ 61) $\cosh(x + y) = \cosh x \cosh y + \sinh x \sinh y$

63) a) $y' = e^x + xe^x$, $y'' = 2e^x + xe^x$, $y''' = 3e^x + xe^x$ b) $10^{100}e^x + xe^x$
    c) 100   d) $10^{(10^{100})}e^x + xe^x$

65) $\ln x$ 67) 41.94 years   69) a) 82.2 minutes   b) 110 minutes

71) 290 days **73)** the patient one's coffee is cooler

§ 3.2 - Linear Approximations and Newton's Method

1) 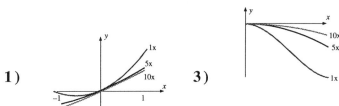 **3)**

5) a) no b) no c) yes d) no
7) a) yes b) no c) yes d) no
9) $\sqrt{x} \approx 4 + \frac{1}{8}(x-16)$ if x is close to 16 , $\sqrt{17} \approx 4.125$ from linearization at $x = 16$
11) $x^{1/3} \approx 4 + \frac{1}{48}(x-64)$ if x is close to 64 , $\sqrt[3]{65} \approx 4.02083$ **13)** cot and csc
17) a) $x = \frac{-b}{m}$ b) $x - \frac{mx+b}{m}$

19) $Z(x) = x - \frac{x^3 - x - 1}{3x^2 - 1}$ $Z(Z(Z(1))) = 1.325200399$

21) $f(x) = x^3 - 3$, $Z(x) = x - \frac{x^3 - 3}{3x^2}$, $\sqrt[3]{3} \approx 1.4422$ **23)** $x^4 - 10x^2 + 1 = 0$
25) $4x^6 - 36x^4 - 4x^3 + 108x^2 - 36x - 107 = 0$

27) $Z(x) = x - \frac{2\sin x - x}{2\cos x - 1}$ roots at $\approx \pm 1.8955$

29) a) method fails for r with $0 < r < \frac{1}{2}$ b) $r = 1$
31) There are two solutions: $r_1 = 21.6893$ $r_2 = 9.4315$ **33)** $d = 0.4623$ m

§ 3.3 - Related Rates

1) $200\pi \frac{\text{mm}}{\text{hr}}$ **3)** $-1 \frac{\text{cm}}{\text{sec}}$ **5)** $\frac{1}{288\pi} \frac{\text{cm}}{\text{sec}}$ **7)** $\frac{16}{25} \frac{\text{in}}{\text{min}}$
9) a) $0.045 \frac{\text{m}}{\text{s}}$ b) $0.0225 \frac{\text{m}}{\text{s}}$ **11)** Cross sectional area deceasing at $\frac{\pi}{10} \frac{\text{in}^2}{\text{yr}}$
13) a) $4 \frac{\text{in}}{\text{sec}}$ b) 0 **15)** $\frac{20}{3} \frac{\text{cm}}{\text{min}}$ **19)** $\frac{10^5}{6\pi} \frac{\text{N}}{\text{m}^2 \text{min}}$
21) a) $\frac{2}{\pi} \frac{\text{cm}}{\text{sec}}$ b) $\frac{dh}{dt} = 0.59029 \frac{\text{cm}}{\text{sec}}$ c) $\frac{25}{18\pi} \frac{\text{cm}}{\text{sec}}$

23) a) $-0.037712 \frac{\text{m}^3}{\text{sec}}$ b) $\frac{dL}{dt} = \begin{cases} \frac{\sqrt{2}}{20L}\sin\left(\frac{3\pi}{4} - \theta\right)\cos^2\theta , & 0 \leq \theta \leq \frac{\pi}{4} \\ \frac{\sqrt{2}}{20L}\sin\left(\frac{3\pi}{4} - \theta\right)\sin^2\theta , & \frac{\pi}{4} < \theta \leq \frac{\pi}{2} \end{cases}$

where $L = \left[\left(3 - 2\sqrt{2}\cos\left(\theta - \frac{3\pi}{4}\right)\right)\right]^{1/2}$

Answers to Odd Numbered Problems CAFÉ

25) **a)** $\dfrac{dh}{dt} = 0.05261 \dfrac{m}{min}$ @ $h = 0.25$ m **b)** $\dfrac{dh}{dt} = \dfrac{1}{20\pi} = 0.01592 \dfrac{m}{min}$ @ $h = 1.25$ m

27) **a)** $r'(20) \approx -0.18 \dfrac{km}{s}$ **b)** $v_x = 0.3487 \dfrac{km}{s}$

$\theta'(20) \approx -\dfrac{\pi}{180} \dfrac{rad}{s}$ $v_y = -0.2603 \dfrac{km}{s}$

29) **a)** $x = (2a+b+c)\left(\dfrac{a\sin\theta}{a+b-a\cos\theta}\right)$ **b)** $\dfrac{dx}{dt} = (2a+b+c)(a\omega)\left(\dfrac{(a+b)\cos\theta - a}{(a+b-a\cos\theta)^2}\right)$

c)   **31)** additional 11.4045 hours

§ 3.4 - Optimazation

1) local (and absolute) min @ $x = 1$ absolute max @ $x = 5$

3) local max at $x = -1$ local min at $x = 1/5$

5) local min at $x = \pm 1$

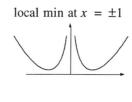

7) local max at $x = 0$ local min at $x = 4/7$

9) local max at $x = 1$

11) local min at $x = 1/e$

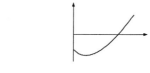

13) f has no extreme values on $\mathbb{R}$

15) local min at $x = 1/e$

17) local min at $x = \dfrac{1}{e^2}$

19) local min at $x = \dfrac{\pi}{2}, \dfrac{3\pi}{2}$ local max at $x = 0.2527 , 2.8889$

21) **a)** just one value of x so that so that $a^x = x^a$ **b)** just one value i.e., $x = e$

23) triangle is equilateral so that the length of the sides is $L/3$ **25)** $y = p/6$, $x = p/3$

27) the rectangle has length $\sqrt{2}\, r$ and height $\sqrt{2}\, r / 2$ **31)** $\dfrac{10\sqrt{10}}{\sqrt{3}}$ ft

33) $x = 25\sqrt{2}$ m , $y = 50\sqrt{2}$ m   **35)** best location $(1/2, 1/4)$, minimal length of rope $\sqrt{2}/4$

37) the cheetah escapes !   **39)** $x = 10\sqrt{2}$ cm , $y = 30/\sqrt{2}$ cm   **41)** $x \approx 1.6052$ mi

43) $r \approx 13.3650$ m, $h \approx 8.91$ m   **45)** Rent = $555/mo, # vacancies = 182

47) 41 trees should be planted per acre

49)   **51)** $x = 50 + 10\sqrt[3]{45}$ mi   **53)** No, minimal length is 25.40 ft

optimal speed

55) maximum volume is $\frac{100}{27}$ ft^3 obtained by taking $x = \frac{2}{3}$ ft

57) radius of the bottom circle is 0.89206 ft, height is 1.04512 ft and maximum volume is 3.844 ft^3

Chapter Four: Introduction to Integration

§ 4.1 - The Area Problem

1) $\frac{\sqrt{3}}{4} L^2$ **3)** $\frac{5L^2}{4} \tan 54°$ **5)** $\frac{nL^2}{4} \cot\left(\frac{\pi}{n}\right)$ **7)** $\frac{dA}{dx} = e^x \sqrt{x^2 + 1}$

9) $\frac{dA}{dx} = \frac{\sqrt{3x^2 + 1}}{\sqrt{x + 5} + e^x}$ **11)** $\frac{dA}{dx} = -f(x)$ **13)** $x + x^2 + x^3 + x^4 + k$

15) $-\frac{3}{x} + \ln|x| + k$ **17)** $-3\cos x + 5e^x + 2\ln|x| + k$ **19)** $\frac{1}{3}(x^3 - 1)$

21) $x + \frac{x^2}{2} + \frac{x^3}{3} - \frac{11}{6}$ **23)** 6π **25)** 4   **27)** $\frac{1}{6}$ **29)** 4

31) $\frac{9}{8}$ **33)** 4   **35)** $2e^2 - 4$ **37)** $2\sqrt{3} - 2 - \frac{\pi}{6}$ **39)** 3π

41) $n \approx 1.6138$ **43)** a) $\frac{2}{n+1}$ b) $\frac{2n}{n+1}$

§ 4.2 - Area and Definite Integrals

1) 102.874   **3)** 2.9897   **5)** $\sum_{i=1}^{n} \sin\left(\frac{\pi(i-1)}{n}\right)\left(\frac{\pi}{n}\right)$ **7)** 0, zero rule

9) $\frac{-1}{3}$ **11)** 0   **13)** 1   **15)** 1   **17)** $2 \ln 5$ **19)** $\frac{145}{6}$

21) $\frac{2}{3}(5^{3/2} - 2^{3/2})$ **23)** $2 \sin 1$ **25)** 1   **27)** $\frac{73}{3}$ **29)** 4

35) $2\sqrt{2}$ **37)** $\frac{32}{3}$ **39)** 10   **41)** 3

43) $24 - (3 + \sqrt{3})\sqrt{2} - \sqrt{3} - \sqrt{5} - \sqrt{7}$ **47)** $\frac{1}{6}$ **49)** $e - 1$

51) $\frac{2}{3}(3^{3/2} - 1)$ **53)** $2(\sqrt{3} - 1)$ **55)** $\frac{1}{e}$

§ 4.3 - Fundamental Theorems

Answers to Odd Numbered Problems   CAFÉ

1) $\dfrac{\sin x}{x^2+1}$ 3) $\dfrac{-\sin x}{\cos^2 x+1}$ 5) $\dfrac{-e^x}{x^4+5}$ 7) $\dfrac{9}{2}$ 9) $3e-5$

11) $\dfrac{28}{15}$ 13) $\dfrac{17}{6}$ 15) $\dfrac{2}{3}$ 17) 0 19) $\dfrac{1}{2}$ 21) $\dfrac{8}{3}$

23) $2(2\sinh^{-1}(\sqrt{3})-\sqrt{3})$ 25) $\dfrac{dy}{dx}=\dfrac{2\sin x^2}{x}$ 27) $\dfrac{dy}{dx}=\dfrac{\sin x}{2\sqrt{x}}$

29) $\dfrac{dy}{dx}=2(x\exp(x^4)-\exp((2x+1)^2))$ 31) $\dfrac{dy}{dx}=\exp(g^2(x))\cdot g'(x)-\exp(h^2(x))\,h'(x)$

33) $f(1)=\cos(1)-\sin(1)$ 35) $2\sqrt{2}$ 37) $f(g(x))\cdot g'(x)-f(h(x))\cdot h'(x)$

39) $\displaystyle\int_0^x f(t)\,dt = \begin{cases} 0, & 0\le x\le 1 \\ \dfrac{1}{2}(x^2-1), & x>1 \end{cases}$ No, $\dfrac{d}{dx}\displaystyle\int_0^x f(t)\,dt$ does not exist at $x=1$

41) 2 43) a) $\begin{cases} a=2 \\ v_0=2 \\ k=2 \end{cases}$ b) $x(t)=2t+8t^2-\dfrac{32}{3}t^3$ c) 0.8324

§ 4.4 - Substitution

1) a) $\dfrac{1}{2}\sin(1+x^2)+k$ b) $\sqrt{1+x^2}+k$ c) $\dfrac{1}{2}\tan(1+x^2)+k$

3) $\dfrac{1}{4002}(5^{2001}-3^{2001})$ 5) $\sin(\sin(x))+k$ 7) $\dfrac{1}{2}e^{x^2}+k$

9) $\dfrac{1}{3}(x^2+1)^{3/2}+k$ 11) $2\sqrt{\ln x}+k$ 13) $\dfrac{116}{15}$ 15) $\dfrac{-\pi}{30}\cos\left(\dfrac{60t}{2\pi}\right)+k$

17) $\dfrac{1}{\omega}\sin(\omega t+\phi)+k$ 19) $-\dfrac{1}{2}\ln|\cos(1+x^2)|+k$ 21) $2e^{\sqrt{x}}+k$

23) $\dfrac{1}{2}\ln|2x+1|+k$ 25) $\dfrac{1}{3}\ln|1+x^3|+k$ 27) $\ln|x+x^2|+k$

29) $\dfrac{2}{9}(x^3+1)^{3/2}+k$ 31) $-2\cos\sqrt{x}+k$ 33) $\dfrac{1}{3}(x^2-4)^{3/2}+k$

35) $\dfrac{2}{7}(1+x)^{7/2}-\dfrac{4}{5}(1+x)^{5/2}+\dfrac{2}{3}(1+x)^{3/2}+k$ 37) $\dfrac{2}{3}\sqrt{x^3+1}+k$

39) $\dfrac{1}{4}\tan(4x)+k$ 41) $\dfrac{1}{4}(2x-\sin(2x))+k$ 43) $\dfrac{\tan^2 x}{2}-\ln|\cos x|+k$

45) $\dfrac{1}{8}(4x+\sin 4x)+k$ 47) $\dfrac{\sin 2x}{4}+\dfrac{\sin 12x}{24}+k$ 49) 0 51) 0

53) 0 55) $\dfrac{(3+x^2)^4}{4}+k$ 57) $\dfrac{1}{12}(1+x^3)^4+k$ 59) $\dfrac{1}{2}F(2x)+k$

61) $F(\sin x)+k$ 63) $\dfrac{1}{3}F(1+3x)+k$ 65) $1000-500\ln 2$

§ 4.5 - Mean Value Theorem Part I

1) The Mean Value Theorem can not be applied.   3) $c=1$ 5) $c=2$

7) $c=0$ 9) $c=25$ 11) a) $c=2$ b) $c=2-\dfrac{\sqrt{2}}{2}$ 13) $c=50$

15) $c=4$ 17) $c=4$ 19) $c=1$ 21) $c=\ln\left(\dfrac{e^4-1}{4e}\right)$ 23) $c=\cos^{-1}\left(\dfrac{2}{\pi}\right)$

25) $c=\dfrac{5}{3\sqrt{3}}$ 27) b) For $[0,1]$, $c=\dfrac{1}{\sqrt{2}}$, For $[-1,1]$, $c=0$

29) This does not violate the MVT, since f is not differentiable at 0 in $(-1, 5)$

35) a) t_1 any time b) $t_2 = 1$ c) $t_3 = \frac{1}{2}\left(\sqrt{\frac{31}{3}} - 1\right)$

§ 4.6 - Mean Value Theorem Part II

1) a) 3 b) 3 c) 3 **3)** 0 **5)** $\frac{2}{\pi}$ **7)** a

9) $k + \frac{m}{2}(b + a)$ **11)** 55.9 **13)** $\frac{1 - \cos 2}{2}$ **15)** $\frac{\pi}{4}$ **17)** $\frac{3}{2}$

19) $\frac{\sqrt{3}}{4}$ **21)** a) $\frac{71}{30}$ b) $2 + \frac{2n+1}{6n}$ c) $\frac{7}{3}$ d) $\frac{7}{3}$ **23)** $\frac{1}{b-a}\int_a^b f(x)\,dx$

25) $c = \ln\left(\frac{e^2 - e^{-2}}{4}\right)$ **27)** $c = \frac{\pi}{2}$ **29)** $c = \frac{1}{\sqrt{3}}$

31) a) $70 + \frac{20}{\pi}$ b) 70

§ 4.7 - Integration by Parts

1) $\frac{x\,e^{4x}}{4} - \frac{e^{4x}}{16} + C$ **3)** $\frac{x \sin 3x}{3} + \frac{\cos 3x}{9} + C$ **5)** $\frac{x \sin 5x}{5} + \frac{\cos 5x}{25} + C$

7) $\frac{x^2 \ln(2x)}{2} - \frac{x^2}{4} + C$ **9)** $\frac{1}{4}(e^2 + 1)$ **11)** $\frac{\pi}{2} - 1$ **13)** $\tan^{-1}(2) - \frac{1}{4}\ln(5)$

15) $2x \sinh x - 2\cosh x + 3\sinh x + C$ **17)** $\frac{\pi}{2} - 1$

19) $\left(\frac{x^3}{3} + \frac{3x^2}{2} + x\right)\ln x - \frac{x^3}{9} - \frac{3x^2}{4} - x + C$ **21)** $-x^2 \cos x + 2x \sin x + 2\cos x + C$

23) $x^2 \cosh x - 2x \sinh x + 2\cosh x + C$ **25)** $\frac{x^2 e^{x^2}}{2} - \frac{e^{x^2}}{2} + C$

27) $\frac{e^{x^3}}{3}(x^3 - 1) + C$ **29)** $\frac{e^{-x}}{5}(2 \sin 2x - \cos 2x) + C$ **31)** $\frac{250}{3}\ln 5 - \frac{248}{9}$

33) $\frac{16}{e} - \frac{38}{e^2}$ **35)** $7 \sinh(1) - 9 \cosh(1) + 6$ **37)** -4

39) $-2\sqrt{x}\cos(1 + \sqrt{x}) + 2 \sin(1 + \sqrt{x}) + C$ **41)** $\frac{e}{2}\cos 1 + \frac{e}{2}\sin 1 - \frac{1}{2}$

43) a) $\int_0^1 (\ln x)^n \, dx = -n \int_0^1 (\ln x)^{n-1} \, dx$ b) -6

45) a) $\int_0^1 x^n(1-x)^m \, dx = \frac{n}{m+1}\int_0^1 x^{n-1}(1-x)^{m+1}\, dx$ b) $\frac{1}{4004}$

47) $4\pi\rho_o \frac{e^{-kR}}{k}\left[\frac{2e^{kR}}{k^2} - \frac{2}{k^2} - \frac{2R}{k} - R^2\right]$

§ 4.8 - Approximate Integration

1) a) $x_0 = 0$, $x_1 = \frac{1}{4}$, $x_2 = \frac{1}{2}$, $x_3 = \frac{3}{4}$, $x_4 = 1$

$f(x_0) = 1$, $f(x_3) = \frac{4}{7}$

$f(x_1) = \frac{4}{5}$, $f(x_4) = \frac{1}{2}$

$f(x_2) = \frac{2}{3}$

b) $L(4) = 0.7595$, $R(4) = 0.6345$, $\text{Trap}(4) = 0.6970$

Answers to Odd Numbered Problems   CAFÉ

c) $\bar{x}_1 = \frac{1}{8}, \bar{x}_2 = \frac{3}{8}, \bar{x}_3 = \frac{5}{8}, \bar{x}_4 = \frac{7}{8}$ $\quad f(\bar{x}_1) = \frac{8}{9}, f(\bar{x}_3) = \frac{8}{13}$
$f(\bar{x}_2) = \frac{8}{11}, f(\bar{x}_4) = \frac{8}{15}$

d) Mid(4) = 0.6912, Simp(8) = 0.6932

e) L(4), Trap(4) and Simp(8) are too large. R(4) and Mid(4) are too small.

3) a) L(4) = 0.9016, R(4) = 0.8033, Mid(4) = 0.8572, Trap(4) = 0.8525

c) $\begin{cases} \text{Overestimate: } L(4), \text{Mid}(4) \\ \text{Underestimate: } R(4), \text{Trap}(4) \end{cases}$

5) a) 6   b) Trap(4), Mid(4), Simp(8)   c) $\begin{cases} L(4) = 6 & \text{Mid}(4) = 6 \\ R(4) = 6 & \text{Simp}(8) = 6 \\ \text{Trap}(4) = 6 \end{cases}$

7) 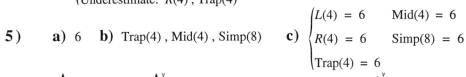   9)

11) a) $\frac{178}{3}$ m²   b) $\frac{356}{3}$ m³   13) a) $D_{simp} = \frac{13}{3}$ mi   b) $D_{trap} = \frac{13}{3}$ mi

c) Left hand rule will underestimate. Right hand rule will overestimate

15) a) $V_{trap} = 33.4344$ ft³ $V_{simp} = 28.085$ ft³
b) $W_{trap} = 5683.8480$ lbs $W_{simp} = 4774.5860$ lbs

17) a) $L(n) = 1.4 \quad \text{Trap}(n) = 1.2$
$R(n) = 1 \quad \text{Mid}(n) = 1.3$ b) $1.2 \leq$ actual area ≤ 1.3

§ 5.1 - Volume Part I

1) $\frac{1}{3}$ 3) $\frac{\pi^2}{2}$ 5) $\frac{\pi}{3}$ 7) 10   9) $\frac{128\pi}{3}$ 11) 2π

13) $\frac{57\pi}{8}$ 15) 360π 17) $W = 1000g\left[\frac{4\pi}{81} - \frac{\pi}{3}h^2(1-h)\right]$ 19) π

21) $\frac{\pi}{2}$ 23) $\frac{16}{3}$

27) $\frac{\pi}{3}(2-\sqrt{2})^2(4+\sqrt{2}) = 1.9456$, less that half the molten metal remains

29) a) $h = 0.764$ m   b) 13957.04 N

§ 5.2 - Volume Part II

1) $\frac{10^6}{3}$ ft³   3) $\frac{5\sqrt{3}}{6}$ cm³   5) 0.03 m³   7) π 9) $\frac{3\pi}{2}$

11) 11π 13) $\frac{\pi}{2}$ 15) 1

17) $\frac{\pi h}{3}[R^2 + r^2 + Rr]$ 19) $\frac{2\pi}{3}$ 21) $2\pi(1-\ln 2)$ 23) $\pi(e-1)$

25) 2π 27) $2\pi\left(\frac{\pi}{2}\cosh\left(\frac{\pi}{2}\right) - \sinh\left(\frac{\pi}{2}\right)\right)$ 29) $\frac{4\pi}{3}$ 31) $\frac{13\pi}{3}$

33) $\dfrac{\pi}{2}(e^2 - 1)$ 35) 3π 37) π 39) $\dfrac{2\pi}{5}$

41) $\dfrac{\pi}{3}(r^2 + ra - 2a^2)\left(\dfrac{h}{r}\right)(r - a)$ 43) 125π

§ 5.3 - Arclength

3) $\displaystyle\int_1^{10} \dfrac{\sqrt{x^4 + 1}}{x^2}\, dx$ 5) $\displaystyle\int_0^{\pi/4} \sqrt{1 + \sec^4 x}\, dx$ 7) $\displaystyle\int_{1/2}^1 \dfrac{\sqrt{x^2 + 4\ln^2 x}}{x}\, dx$

9) $\displaystyle\int_0^1 \sqrt{1 + e^{2x}}\, dx$ 11) $\displaystyle\int_0^{2\pi} \sqrt{\cos^2 2\theta + 4\sin^2 2\theta}\, d\theta$

13) $\displaystyle\int_0^\pi \sqrt{\sin^2(3\theta) + 9\cos^2(3\theta)}\, d\theta$ 15) $\displaystyle\int_0^1 \sqrt{\cosh(2t)}\, dt$

17) $\displaystyle\int_0^{2\pi} \sqrt{\cos^2(t) + 4\cos^2(2t)}\, dt$ 19) 16.1994   21) $\dfrac{4}{3}(2^{3/2} - 1)$

23) $\dfrac{\sqrt{1 + a^2}}{a}(e^{2\pi a} - 1)$

25) $\dfrac{|b|}{2}\left(\sinh^{-1}\left(\dfrac{a}{b} + 2\pi\right) + \left(\dfrac{a}{b} + 2\pi\right)\sqrt{1 + \left(\dfrac{a}{b} + 2\pi\right)^2} - \sinh^{-1}\left(\dfrac{a}{b}\right) - \left(\dfrac{a}{b}\right)\sqrt{1 + \left(\dfrac{a}{b}\right)^2}\right)$

27) $6a$ 29) 30   33) $\dfrac{3}{2} + \dfrac{1}{4}\ln 2$ 35) $\dfrac{7683}{128}$ 39) $\dfrac{735}{256}$

41) 1.27798   43) 1.60023   45) 3.820198   47) 4.45705

49) 1.42776   51) $L = 1001.0658$ m , $k = 0.0001599$

§ 5.4 - Area of Surfaces of Revolution

1) a, b, e, f   3) $6\pi\sqrt{13}$ 5) 18π

7) For problem (3): $\begin{cases} S_{\text{whole}} = 8\pi\sqrt{13} \\ S_{\text{cap}} = 2\pi\sqrt{13} \end{cases}$ For problem (4): $\begin{cases} S_{\text{whole}} = 2\pi\sqrt{13} \\ S_{\text{cap}} = \dfrac{\pi\sqrt{13}}{2} \end{cases}$

For problem (5): $\begin{cases} S_{\text{whole}} = 24\pi \\ S_{\text{cap}} = 6\pi \end{cases}$ For problem (6): $\begin{cases} S_{\text{whole}} = \pi\left(2 + \dfrac{\sqrt{11}}{2}\right)\left(3 + \dfrac{12\sqrt{11}}{11}\right) \\ S_{\text{cap}} = \pi(2)\left(\dfrac{12\sqrt{11}}{11}\right) \end{cases}$

9) $\dfrac{\pi\sqrt{5}}{4}$ 11) $\dfrac{3\sqrt{5}\,\pi}{4}$ 13) $y = \dfrac{1}{2}x$ on $[0, 2]$

15) $A = 2\pi \displaystyle\int_0^\pi \sin x \sqrt{1 + \cos^2 x}\, dx$ 17) $A = 2\pi \displaystyle\int_{-\pi/4}^{\pi/4} (\sec x)\sqrt{1 + \tan^2 x \sec^2 x}\, dx$

19) $A = 2\pi \displaystyle\int_0^2 x\, e^{-2x} \sqrt{e^{2x} + 1 - 2x + x^2}\, dx$ 21) $\dfrac{\pi}{6}(17^{3/2} - 5^{3/2})$

23) $2\pi\left(\sinh^{-1}(1) + \sqrt{2}\right)$ 25) $\pi\left(\dfrac{\sinh(2a)}{2} + a\right)$ 27) $\dfrac{\pi}{27}(2305^{3/2} - 10^{3/2})$

29) 947.75   31) $\dfrac{515\pi}{64}$ 33) $(41{,}903.7)\,\pi$

Answers to Odd Numbered Problems   CAFÉ

37) **a)** $2\pi b^2 + \dfrac{2\pi a^2 b}{\sqrt{a^2 - b^2}} \sin^{-1}\left(\dfrac{\sqrt{a^2 - b^2}}{a}\right)$ **b)** $2\pi a^2 + \dfrac{2\pi a b^2}{\sqrt{a^2 - b^2}} \sinh^{-1}\left(\dfrac{\sqrt{a^2 - b^2}}{b}\right)$

39) Cut at the height of $\dfrac{1}{\sqrt{2}} h$ above the base, where h is the height of the cone.